SOLID STATE PHYSICS

VOLUME 34

Contributors to This Volume

D. de Fontaine

S. E. Schnatterly

Glen A. Slack

SOLID STATE PHYSICS

Advances in

Research and Applications

Editors

HENRY EHRENREICH

Division of Applied Sciences
Harvard University, Cambridge, Massachusetts

FREDERICK SEITZ

The Rockefeller University, New York, New York

DAVID TURNBULL

Division of Applied Sciences
Harvard University, Cambridge, Massachusetts

VOLUME 34

1979

ACADEMIC PRESS ● NEW YORK SAN FRANCISCO LONDON

A Subsidiary of Harcourt Brace Jovanovich, Publishers

COPYRIGHT © 1979, BY ACADEMIC PRESS, INC.
ALL RIGHTS RESERVED.
NO PART OF THIS PUBLICATION MAY BE REPRODUCED OR
TRANSMITTED IN ANY FORM OR BY ANY MEANS, ELECTRONIC
OR MECHANICAL, INCLUDING PHOTOCOPY, RECORDING, OR ANY
INFORMATION STORAGE AND RETRIEVAL SYSTEM, WITHOUT
PERMISSION IN WRITING FROM THE PUBLISHER.

ACADEMIC PRESS, INC.
111 Fifth Avenue, New York, New York 10003

United Kingdom Edition published by
ACADEMIC PRESS, INC. (LONDON) LTD.
24/28 Oval Road, London NW1 7DX

LIBRARY OF CONGRESS CATALOG CARD NUMBER: 55–12200

ISBN 0–12–607734–7

PRINTED IN THE UNITED STATES OF AMERICA

79 80 81 82   9 8 7 6 5 4 3 2 1

Contents

CONTRIBUTORS TO VOLUME 34	vii
PREFACE	ix
SUPPLEMENTS	xi

The Thermal Conductivity of Nonmetallic Crystals

GLEN A. SLACK

I.	Introduction	1
II.	Absolute Value of the Thermal Conductivity	2
III.	Volume Dependence of the Thermal Conductivity	34
IV.	Temperature Dependence of the Thermal Conductivity	37
V.	Thermal Conductivity at the Melting Point	44
VI.	Minimum Thermal Conductivity	57
VII.	Further Problems	70

Configurational Thermodynamics of Solid Solutions

D. DE FONTAINE

I.	Introduction	74
II.	State of Order	75
III.	Internal Energy	88
IV.	Elastic Interactions	116
V.	Free Energy Models	146
VI.	Stable and Unstable States	179
VII.	Fluctuations and Kinetics	231
VIII.	Conclusion	272

Inelastic Electron Scattering Spectroscopy

S. E. SCHNATTERLY

I.	Introduction	275
II.	Elastic Scattering	278
III.	Inelastic Electron Scattering	282
IV.	Basic Spectrometer Description	296
V.	Sample Preparation	305
VI.	Experimental Examples	309

AUTHOR INDEX	359
SUBJECT INDEX	371

Contributors to Volume 34

Numbers in parentheses indicate the pages on which the authors' contributions begin.

D. DE FONTAINE,* *School of Engineering and Applied Science, University of California, Los Angeles, California 90024* (73)

S. E. SCHNATTERLY,** *Joseph Henry Research Laboratories, Princeton University, Princeton, New Jersey 08540* (275)

GLEN A. SLACK, *General Electric Research and Development Center, Schenectady, New York 12301* (1)

---

* Present Address: Department of Materials Science and Mineral Engineering, University of California, Berkeley, California 94720.
** Present Address: Physics Department, University of Virginia, Charlottesville, Virginia 22901.

Preface

The characteristics and theory of the thermal conductivity of solids was reviewed comprehensively in Volume 7 of this serial publication by Klemens. In a later article, appearing in Volume 12, Mendelssohn and Rosenberg surveyed the theory of and experience with low-temperature thermal conductivity of metals. The review by Slack in this volume updates the developments of the knowledge and understanding of the intrinsic thermal conductivity of pure nonmetallic solids and liquids at relatively high temperatures. Extensive comparisons of the more recent experimental results with theory are presented.

The second article in this volume, by de Fontaine, presents a comprehensive and thorough development of the theory for the atomic configurations of solid solutions. Applications of the theory to interpretations of the thermodynamics and kinetics of compositional ordering or clustering in alloys are shown. In the final article of this volume Schnatterly describes the technique of and underlying theory for inelastic electron scattering spectroscopy and surveys some of its recent applications.

HENRY EHRENREICH
FREDERICK SEITZ
DAVID TURNBULL

Supplements

Supplement 1: T. P. DAS AND E. L. HAHN
Nuclear Quadrupole Resonance Spectroscopy, 1958

Supplement 2: WILLIAM LOW
Paramagnetic Resonance in Solids, 1960

Supplement 3: A. A. MARADUDIN, E. W. MONTROLL, G. H. WEISS, AND I. P. IPATOVA, Theory of Lattice Dynamics in the Harmonic Approximation, 1971 (Second Edition)

Supplement 4: ALBERT C. BEER
Galvanomagnetic Effects in Semiconductors, 1963

Supplement 5: R. S. KNOX
Theory of Excitons, 1963

Supplement 6: S. AMELINCKX
The Direct Observation of Dislocations, 1964

Supplement 7: J. W. CORBETT
Electron Radiation Damage in Semiconductors and Metals, 1966

Supplement 8: JORDAN J. MARKHAM
F-Centers in Alkali Halides, 1966

Supplement 9: ESTHER M. CONWELL
High Field Transport in Semiconductors, 1967

Supplement 10: C. B. DUKE
Tunneling in Solids, 1969

Supplement 11: MANUEL CARDONA
Optical Modulation Spectroscopy of Solids, 1969

Supplement 12: A. A. ABRIKOSOV
An Introduction to the Theory of Normal Metals, 1971

Supplement 13: P. M. PLATZMAN AND P. A. WOLFF
Waves and Interactions in Solid State Plasmas, 1973

Supplement 14: L. LIEBERT, Guest Editor
Liquid Crystals, 1978

Supplement 15: ROBERT M. WHITE AND THEODORE H. GEBALLE
Long Range Order in Solids, 1979

SOLID STATE PHYSICS

VOLUME 34

The Thermal Conductivity of Nonmetallic Crystals

GLEN A. SLACK

General Electric Research and Development Center, Schenectady, New York

I. Introduction 1
II. Absolute Value of the Thermal Conductivity 2
    1. Previous Calculations, Acoustic Phonons 2
    2. Rare-Gas Crystals ($n = 1$) 5
    3. Adamantine Crystals ($n = 2$) 9
    4. Rocksalt Structure Crystals ($n = 2$) 17
    5. Heat Transport by Optic Phonons 21
    6. Other Crystals ($n \geq 3$) 23
    7. Generalizations 31
    8. Longitudinal and Transverse Phonons 33
III. Volume Dependence of the Thermal Conductivity 34
    9. Hydrostatic Stress 34
    10. Uniaxial Stress 36
IV. Temperature Dependence of the Thermal Conductivity 37
    11. Thermal Expansion Effects 37
    12. Scattering by Optic Phonons 39
V. Thermal Conductivity at the Melting Point 44
    13. Theory for Liquids 44
    14. Volume Dependence for Liquids 47
    15. Volume Dependence for Disordered Solids 49
    16. Grüneisen Parameter for Liquids 51
    17. Solid–Liquid Transition 52
    18. Summary 57
VI. Minimum Thermal Conductivity 57
    19. Rare-Gas Crystals ($n = 1$) 57
    20. Adamantine and Rocksalt Structure Crystals ($n = 2$) 59
    21. Fluorite Structure Crystals ($n = 3$) 63
    22. Other Crystals ($n \geq 4$) 66
    23. Liquids 69
VII. Further Problems 70

I. Introduction

The present work is a study of and a review of the thermal conductivity of nonmetallic crystals at temperatures comparable to or higher than the Debye temperature. Two previous reviews[1] of thermal conductivity by Klemens and by Mendelssohn and Rosenberg have been published in this series. The review by Klemens treats many of the problems found in nonmetallic crystals. The present review deals more extensively with intrinsic behavior of such pure crystals at high temperatures. In such crystals the dominant carriers of thermal energy are phonons and the dominant scattering mechanism to be considered is the intrinsic phonon–phonon scattering. This is a small section of the much larger problem of the thermal conductivity of nonmetallic solids and clearly it neglects possible heat transport by photons, charge carriers, polarons, magnons, etc. It also neglects other possible phonon scattering mechanisms such as isotopes, impurities, vacancies, charge carriers, dislocations, grain boundaries, crystal boundaries, etc. The limited scope of this review covers the absolute value of the thermal conductivity, K, as determined by phonon–phonon scattering, the temperature dependence of K, the volume dependence of K, the change in K upon melting, and the minimum value of K. The concept of this review is to collect what is known about these topics and to present them in a unified fashion using a simple type of analysis.

In Fig. 1 is shown a composite curve for the thermal conductivity versus temperature of pure KCl measured at a constant pressure of, say, one atmosphere. The data have been collected from various references which are listed later. The K results shown between 1°K and 172°K will not be considered. The concern here is with the absolute value of K at the Debye temperature, $\bar{\theta}_\infty$ or θ_∞, and the manner in which K changes up to and through the melting point both for KCl and for a number of other crystals.

II. Absolute Value of the Thermal Conductivity

1. Previous Calculations, Acoustic Phonons

The problem of calculating the absolute value of the thermal conductivity, K, of a nonmetallic crystal was studied by Debye[2] in 1914. He was concerned with the high temperature or classical region where the crystal temperature is above the Debye temperature. This region is the one we shall concentrate on here. The next serious study of the problem

---

[1] P. G. Klemens, *Solid State Phys.* **7**, 1 (1958); and K. Mendelssohn and H. M. Rosenberg, *Solid State Phys.* **12**, 223 (1963).
[2] P. Debye, *in* "Vorträge über die Kinetische Theorie der Materie und der Electrizität," p. 43. B. G. Teubner, Berlin, 1914.

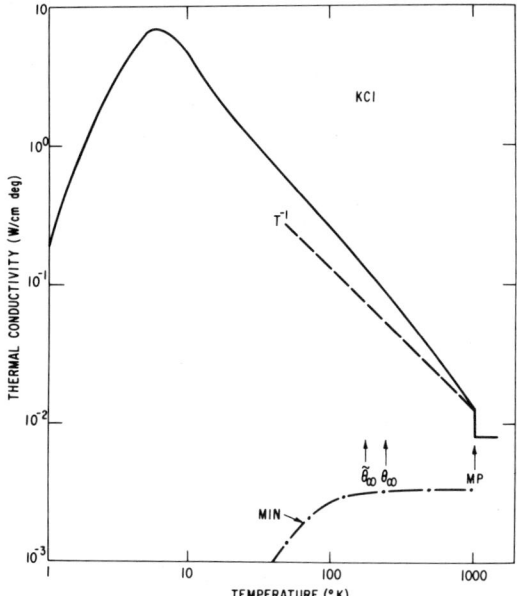

FIG. 1. A composite curve of the thermal conductivity versus temperature for a pure crystal of KCl about 0.3 cm in diameter. The two Debye temperatures and the melting point are indicated by arrows. The curve at temperatures near the melting point and in the liquid phase assumes that the internal photon heat transport is negligible. The dashed line has a slope of −1. The dot-dash line is the minimum value of K.

was made by Akhiezer[3] in 1940, although Papaetru[4] considered the problem earlier. Later work was carried out by Pomeranchuk,[5] Mizushima,[6] Leibfried and Schlömann,[7] Dugdale and MacDonald,[8] and many others.[9–19] Of all of these calculations the most useful in providing a

[3] A. Akhiezer, *Zh. Eksp. Teor. Fiz.* **10**, 1354 (1940).
[4] A. Papapetru, *Phys. Zeit.* **35**, 527 (1934).
[5] I. Pomeranchuk, *Phys. Rev.* **60**, 820 (1941); *J. Phys. USSR* **4**, 259 (1941); and *J. Phys. USSR* **7**, 197 (1943).
[6] S. Mizushima, *J. Phys. Soc. Jap.* **9**, 546 (1954).
[7] G. Leibfried and E. Schlömann, *Nach. Akad. Wiss. Göttingen, Math. Phys. Klasse* **4**, 71 (1954).
[8] J. S. Dugdale and D. K. C. MacDonald, *Phys. Rev.* **98**, 1751 (1955).
[9] T. A. Kontorova, *Zh. Tekh. Fiz. USSR* **26**, 2021 (1956) [*Sov. Phys.—Tech. Phys.* **1**, 1959 (1957)].
[10] A. W. Lawson, *J. Phys. Chem. Solids* **3**, 155 (1957).
[11] R. W. Keyes, *Phys. Rev.* **115**, 564 (1959).
[12] C. L. Julian, *Phys. Rev.* **137**, A128 (1965).
[13] L. G. Nagornykh, *Fiz. Tverd. Tela* **8**, 587 (1966) [*Sov. Phys.—Solid State* **8**, 467 (1966)].

convenient formula for calculating K from other known parameters of the crystal are those of Leibfried and Schlömann,[7] and Julian.[12] They are essentially similar calculations and use Ziman's[20] variational principle. Julian[12] considered the rare-gas crystals, as did Gluck,[14] Benin,[16] and Bennett.[19] This same variational principle has been applied[15-17] to calculating K for crystals of KBr,[15] LiF and NaF,[16] and Ge, Si, GaAs, MgO, NaCl, and KCl.[17] The agreement with experiment is generally to within ±20% for the rare-gas crystals, and is poorer for the others. In all of these calculations it has been assumed that the heat is carried by the acoustic phonons and they are scattered by interacting with other acoustic phonons via three-phonon processes. This model is acceptable for the rare-gas crystals that have only one atom per primitive crystallographic cell. It is open to some question for the adamantine (diamondlike) and rocksalt structures that have two atoms per primitive cell, and hence possess optical phonon branches. We shall return to this problem later.

Some of the simple formulae for computing K have been reviewed by White and Woods[21] and Drabble and Goldsmid.[22] The results of Leibfried and Schlömann[7] and Julian[12] for a rare-gas crystal can be written as

$$K'(\theta) = LM\delta\theta^2\gamma^{-2}, \qquad (1.1)$$

where M is the atomic mass in grams, δ^3 is equal to $(a_0^3/4)$, a_0 is the lattice constant in Ångstroms of the fcc unit cell containing four atoms, θ is the Debye temperature in degrees Kelvin, γ is the dimensionless Grüneisen constant, L is a constant, and $K'(\theta)$ is the theoretical value in W/cm deg of the thermal conductivity at a temperature equal to θ. The results of Leibfried and Schlömann[7] give $L = 5.720 \times 10^{-8}$. Julian[12] shows that their value of L is too large by a factor of 2 because of a

---

[14] P. Gluck, *Phys. Lett. A* **24**, 292 (1967).
[15] R. A. Cowley, *Proc. Phys. Soc. (London)* **90**, 1127 (1967).
[16] D. B. Benin, *Phys. Rev. Lett.* **20**, 1352 (1968); *Phys. Rev.* **B5**, 2344 (1972).
[17] R. A. Hamilton and J. E. Parrott, *Phys. Rev.* **178**, 1284 (1969); also *Phys. Lett. A* **29**, 556 (1969).
[18] P. G. Klemens *in* "Thermal Conductivity" (R. P. Tye, ed.), Vol. 1, p. 50. Academic Press, New York, 1969.
[19] B. I. Bennett, *Solid State Commun.* **8**, 65 (1970).
[20] J. M. Ziman, *Can. J. Phys.* **34**, 1256 (1956).
[21] G. K. White and S. B. Woods, *Phil. Mag.* **3**, 785 (1958).
[22] J. R. Drabble and H. J. Goldsmid, "Thermal Conduction in Semiconductors," ch. 5. Pergamon Press, London, 1961.

mistake in counting. His equation can be written as

$$K'(\theta_\infty) = \frac{M\delta\theta_\infty^2}{\gamma^2}\left[\frac{L\,0.849}{2(1 - 0.514\gamma^{-1} + 0.228\gamma^{-2})}\right]. \tag{1.2}$$

We have made use of the relationship that Λ as defined by Julian is given by

$$\Lambda = -(1/6\gamma).$$

If we make use of the fact that $\gamma \cong 2$ for many solids, the square bracket expression in Eq. (1.2) reduces to a new constant $B = 3.04 \times 10^{-8}$. Furthermore, the value of θ used by Julian is θ_∞, as explained by Domb and Salter.[23] This means

$$\theta_\infty^2 = \frac{5h^2}{3k^2}\frac{\int_0^\infty \nu^2 g(\nu)\,d\nu}{\int_0^\infty g(\nu)\,d\nu} \tag{1.3}$$

where h = Planck's constant, k = Boltzmann's constant, ν = phonon frequency, and $g(\nu)$ = phonon density-of-states function.

2. Rare-Gas Crystals ($n = 1$)

For the rare-gas crystals θ_∞ is clearly the high-temperature limit of θ for the acoustic modes, while θ_0 is the acoustic mode Debye temperature at absolute zero as calculated from the elastic constants. In Table I these values are given. The two values, θ_0 and θ_∞, are nearly equal, but we shall use θ_∞ in calculating $K'(\theta)$. The value of θ_∞ has been calculated from the $g(\nu)$ spectrum as determined by neutron scattering measurements. Thus we have

$$K'(\theta_\infty) = BM\delta\theta_\infty^2\gamma^{-2}. \tag{2.1}$$

The value of γ for the rare-gas crystals can be evaluated from the thermal expansion coefficients in the vicinity of $T = \theta_\infty$. For these crystals γ is, of course, only determined by the acoustic modes. Table I also gives the $K'(\theta_\infty)$ values for the rare-gas crystals. These are to be compared with the measured K at $T = \theta_\infty$ which we call $K(\theta_\infty)$. Figure 2 shows a plot of $K(\theta_\infty)$ vs $K'(\theta_\infty)$ for these rare-gas crystals where n, the number of atoms per primitive crystallographic cell, is $n = 1$. The rare

---

[23] C. Domb and L. Salter, *Phil. Mag.* **43**, 1083 (1952).
[24] F. Clayton and D. N. Batchelder, *J. Phys. C* **6**, 1213 (1973).

TABLE I. OBSERVED AND CALCULATED VALUES FOR THE THERMAL CONDUCTIVITY OF RARE-GAS CRYSTALS ($n = 1$) AT $T = \theta_\infty$ [a]

Crystal	V_M (cm³/mole)	V_M/V_{M0}	θ_0 (°K)	θ_∞ (°K)	γ	K (10^{-3} W/cm deg)	K' (10^{-3} W/cm deg)	ϵ
Ne	13.76	1.028	71	65 (38, 39)	2.76 (44)	1.1[c] (28–30)	0.98	—
Ar[b]	22.53	0.999	94	84 (40, 41)	2.73 (45)	3.9 (24)	3.84	0.96 ± 0.04
Ar	24.03	1.065	79	71 (40, 41)	2.74 (45)	2.9 (28, 31–36)	2.84	2.1
Kr	28.01	1.033	66	60 (40, 42)	2.84 (45)	4.9 (28, 34, 37)	4.09	1.5
Xe	35.40	1.019	59	54 (43)	2.65 (46)	7.4 (34)	6.44	1.3

[a] The literature references are in parenthesis.
[b] K measured at constant volume, all other samples were free-standing.
[c] Extrapolated from 20°K using the free-standing V_M at 20°K assuming $\epsilon = 1.0$.

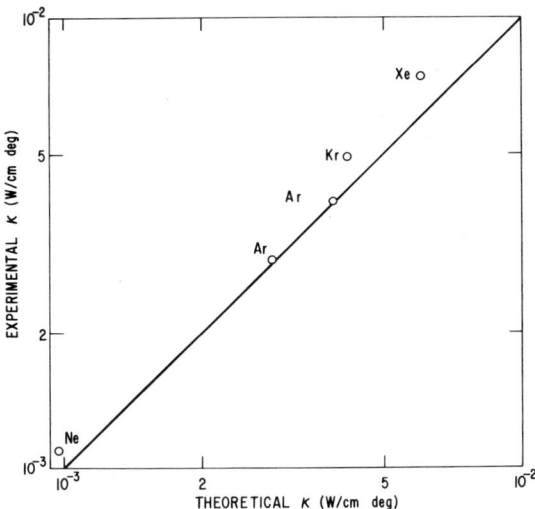

FIG. 2. Experimental versus theoretical values for the thermal conductivity of rare gas crystals at $T = \theta_\infty$. The two points for Ar are for two different molar volumes. The number of atoms per primitive crystallographic unit cell is $n = 1$.

gas crystals have very large thermal expansion coefficients so that it is important to specify the molar volume that was used for both K and K'. The effect of molar volume on the thermal conductivity of solid argon has been carefully studied by Clayton and Batchelder.[24] Similar changes with volume occur for all the rare gas crystals. For a fixed temperature the parameters δ,[25] θ,[26] and γ[27] all vary with molar volume. The largest effect is the variation of θ with volume. From Grüneisen's equation

$$\gamma = -\left(\frac{\partial \ln \theta}{\partial \ln V_M}\right)_T \qquad (2.2)$$

where V_M is the molar volume. One obtains

$$\theta \propto V_M^{-\gamma} . \qquad (2.3)$$

The neutron dispersion curves for the rare gas solids have generally been measured at temperatures near 4.2°K under the equilibrium pressure of the respective rare gas. The θ_0 and θ_∞ values determined from $g(\nu)$ are thus appropriate to the solid at its equilibrium molar volume,

[25] O. G. Peterson, D. N. Batchelder, and R. O. Simmons, *Phys. Rev.* **150**, 703 (1966).
[26] R. Q. Fugate and C. A. Swenson, *J. Low Temp. Phys.* **10**, 317 (1973).
[27] M. S. Anderson, R. Q. Fugate, and C. A. Swenson, *J. Low Temp. Phys.* **10**, 345 (1973).

V_{MO} at $0°K$. Most of the thermal conductivity[24,28-37] measurements on rare-gas crystals, except for Clemans[30] and Clayton and Batchelder,[24] have been made under free-standing conditions where the crystal is allowed to expand as it warms above $0°K$. Thus the θ values are lower than those calculated from $g(\nu)$. This change in θ with temperature has been calculated for all of the crystals from Eq. (2.3), and the appropriate larger molar volume θ values, both θ_0 and θ_∞, are used in Table I. The only exception is for the constant volume data[24] on Ar where the molar volume was held fixed during the thermal conductivity measurements. Table I gives the molar volume, V_M, at a temperature equal to θ_∞. This is the molar volume used for determining both θ_0 and θ_∞ from those given in the literature.[38-43] The γ values in Table I have been chosen from the free-standing thermal expansion data[44-46] near $T = \theta_\infty$. For Ar at constant volume at a temperature of $84°K$ we have used the free-standing value at $T = 0°K$ because the molar volume at $84°K$ was essentially equal to V_{MO}. From Anderson et al.[27] the intrinsic variation of γ with temperature is small, the largest changes in γ are produced by changes in molar volume.

[28] G. K. White and S. B. Woods, *Phil. Mag.* **3**, 785 (1958); see also *Nature (London)* **177**, 851 (1956).
[29] R. M. Kimber and S. J. Rogers, *J. Phys. C* **6**, 2279 (1973).
[30] J. E. Clemans, "Low Temperature Thermal Conductivity Measurements on Solid Neon at Constant Volume", Ph.D. Thesis, Univ. of Illinois, Urbana, 1975.
[31] D. J. Lawrence, A. T. Stewart, and E. W. Guptill, *Can. J. Phys.* **37**, 1069 (1959).
[32] A. Bernè, G. Boato, and M. DePaz, *Nuovo Cimento, B* **46**, 182 (1966).
[33] I. N. Krupskii and V. G. Manzhelii, *Phys. Status Solidi* **24**, K53 (1967).
[34] I. N. Krupskii and V. G. Manzhelii, *Zh. Eksp. Teor. Fiz.* **55**, 2075 (1968) [*Sov. Phys.—J.E.T.P.* **28**, 1097 (1969)].
[35] D. E. Daney, *Cryogenics* **11**, 290 (1971).
[36] D. K. Christen and G. L. Pollack, *Phys. Rev. B* **12**, 3380 (1975).
[37] P. Korpiun, J. Moser, F. J. Pieringer, and E. Luscher, *Proc. Second Int. Conf. on Phonon Scattering in Solids, 1975,* (L. J. Challis, V. W. Rampton, and A. F. G. Wyatt, eds.), p. 377. Plenum, New York, 1976.
[38] J. A. Leake, W. B. Daniels, J. Skalyo, Jr., B. C. Frazer, and G. Shirane, *Phys. Rev.* **181**, 1251 (1969).
[39] J. Behari and B. B. Tripathi, *J. Phys. Soc. Jap.* **31**, 1639 (1971).
[40] G. E. Jelinek, *Phys. Rev. B* **3**, 2716 (1971).
[41] Y. Fujii, N. A. Lurie, R. Pynn, and G. Shirane, *Phys. Rev. B* **10**, 3647 (1974).
[42] J. Skalyo, Jr., Y. Endoh, and G. Shirane, *Phys. Rev. B* **9**, 1797 (1974).
[43] N. A. Lurie, G. Shirane, and J. Skalyo, Jr., *Phys. Rev. B* **9**, 5300 (1974).
[44] J. C. Holste and C. A. Swenson, *J. Low Temp. Phys.* **18**, 477 (1975).
[45] C. R. Tilford and C. A. Swenson, *Phys. Rev. B* **5**, 719 (1972).
[46] V. G. Manzhelii, V. G. Gavrilko, and E. I. Voitovich, *Fiz. Tverd. Tela* **9**, 1483 (1967) [*Sov. Phys.—Solid State* **9**, 1157 (1967)].

The variation in K with temperature is of special interest. Let us define the slopes of the $\ln K$ versus $\ln T$ curve at constant pressure or volume as

$$\epsilon = -\left(\frac{\partial \ln K}{\partial \ln T}\right)_P, \quad m = -\left(\frac{\partial \ln K}{\partial \ln T}\right)_V. \quad (2.4)$$

If V_M is held constant, the $m = 1$ for temperatures $T \geq \theta_\infty$. This was shown by Clayton and Batchelder[24] for Ar. If the crystal is free-standing then generally $\epsilon > 1$ because the crystal expands on heating. The measured values of ϵ at $T = \theta_\infty$ are shown in Table I for Ar, Kr, and Xe. The theoretical calculation of ϵ will be returned to later.

From Table I we see that Eq. (2.1) gives values of $K'(\theta_\infty)$ that agree with the experimental $K(\theta_\infty)$ to within 20%,

$$1.00 \leq [K(\theta_\infty)/K'(\theta_\infty)] \leq 1.20. \quad (2.5)$$

This is better agreement than might be expected considering all of the approximations involved.

3. ADAMANTINE CRYSTALS ($n = 2$)

So far the agreement between $K(\theta_\infty)$ and $K'(\theta_\infty)$ is satisfactory for some nonmetallic crystals that possess only acoustic phonons. An approximate method of proceeding will now be developed for those crystals that possess both acoustic and optic phonon branches. The simplest examples are those with only two atoms per primitive unit cell, i.e., with $n = 2$.

Examples of such crystals are diamond, Si, Ge, GaP, ZnTe, etc. These, like the rare-gas crystals, also have an fcc lattice. The rocksalt-structure crystals like LiF, NaCl, RbI, MgO, etc. also fit this description.

Neutron scattering measurements of the phonon dispersion curves have been made for many of the cubic adamantine crystals. There are experimental results for diamond,[47,48] Si,[47,49] Ge,[50,51] GaP,[52] GaAs,[52] GaSb,[53] InP,[54] InSb,[52] ZnS,[52] ZnSe,[52] ZnTe,[55] and CdTe.[56] Calculated dispersion curves exist for SiC,[52,57] AlSb,[58] and InAs.[59]

[47] G. Dolling and R. A. Cowley, *Proc. Phys. Soc. (London)* **88**, 463 (1966).
[48] J. L. Warren, J. L. Yarnell, G. Dolling, and R. A. Cowley, *Phys. Rev.* **158**, 805 (1967).
[49] G. Dolling and A. D. B. Woods, in "Thermal Neutron Scattering" (P. A. Egelstaff, ed.), ch. 5. Academic Press, New York, 1965.
[50] B. N. Brockhouse and P. K. Iyengar, *Phys. Rev.* **111**, 747 (1958).
[51] G. Nilsson and G. Nelin, *Phys. Rev. B* **3**, 364 (1961).
[52] K. Kunc, *Ann. Phys. (Paris)* **8**, 319 (1973–1974).

In all of the crystals so far studied, the optic branch phonons have much smaller group velocities than those in the acoustic branch. Since $K(\theta)$ varies as the square of the group velocity, we can obtain a good approximation to K by completely ignoring the optic phonons as carriers of the thermal energy and assume that all of the heat is transported by the acoustic phonons in these adamantine crystals. This model has been used by Hamilton and Parrott[17] for their calculations on Si, Ge, and GaAs. Their model even gives the result that the lowest energy branch, the transverse-acoustic phonons, carries more than 80% of the heat. Thus it seems reasonable to assume that if we consider only the acoustic phonons, we should be able to estimate the absolute magnitude of K from a formula similar to Eq. (2.1). This approach has been previously used by Steigmeier and Kudman,[60,61] Slack,[62] Billard and Cabannes,[63] and Roufosse and Klemens.[64] It has been suggested by Wagini[65] that some of these adamantine compounds like AlSb do exhibit a small (15%) contribution to K from optic phonons at high temperatures. This contribution, if present, is ignored for the time being. The possibility of scattering of acoustic phonons by optic phonons is also ignored. The idea is to attempt to generalize Eq. (2.1) to fcc lattices with $n \geq 2$ by assuming that only acoustic phonons carry the heat and interact only with other acoustic phonons. The nature of the quest is to obtain a simple version of Eq. (2.1) that takes n into account. For this we will use a counting procedure to find and include all of the acoustic phonons and their interactions. A more thoroughgoing theory would involve complex calculations that are beyond the scope of this paper. Some attempt at this has been made by Roufosse and Klemens.[66] The results of this

---

[53] M. K. Farr, J. G. Traylor, and S. K. Sinha, *Phys. Rev. B* **11**, 1587 (1975).
[54] P. H. Borcherds, G. F. Alfrey, D. H. Saunderson, and A. D. B. Woods, *J. Phys. C* **8**, 2022 (1975).
[55] N. Vagelatos, D. Wehe, and J. S. King, *J. Chem. Phys.* **60**, 3613 (1974).
[56] J. M. Rowe, R. M. Nicklow, D. L. Price, and K. Zanio, *Phys. Rev. B* **10**, 671 (1974).
[57] R. Banerjee and Y. P. Varshni, *J. Phys. Soc. Jap.* **30**, 1015 (1971).
[58] R. Banerjee and Y. P. Varshni, *Can. J. Phys.* **47**, 451 (1969).
[59] D. N. Talwar and B. K. Agrawal, *Phys. Status Solidi B* **63**, 441 (1974).
[60] E. F. Steigmeier and I. Kudman, *Phys. Rev.* **132**, 508 (1963); *Phys. Rev.* **141**, 767 (1966).
[61] E. F. Steigmeier, in "Thermal Conductivity" (R. P. Tye, ed.), Vol. 2, ch. 4. Academic Press, London, 1969.
[62] G. A. Slack, *J. Appl. Phys.* **35**, 3460 (1964); *J. Phys. Chem. Solids* **34**, 321 (1973).
[63] D. Billard and F. Cabannes, *High Temp.-High Press.* **3**, 201 (1971).
[64] M. Roufosse and P. G. Klemens, *J. Geophys. Res.* **79**, 703 (1974).
[65] H. Wagini, *Z. Naturforsch. A* **21**, 2096 (1966).
[66] M. Roufosse and P. G. Klemens, *Phys. Rev. B* **7**, 5379 (1973).

simple counting scheme have been published previously by Slack.[67] The argument proceeds as follows.

First we take each fcc sublattice with lattice parameter a_0. In each sublattice the atoms will all be identical with a mass $m_i (1 \leq i \leq n)$. For GaAs we have $m_1 = 69.72$ g and $m_2 = 74.92$ g. In the acoustic-phonon branches these i atoms all move more-or-less together, and all have the same Debye temperature, θ, and Grüneisen constant, γ. The $K(\theta_\infty)$ is a sum of the individual contributions $K_i(\theta_\infty)$ from each sublattice. Each K_i' is given by

$$K_i'(\theta_\infty) = B m_i (a_0^3/4)^{1/3} \theta^2 \gamma^{-2} n^{-1}. \tag{3.1}$$

The factor of n^{-1} in Eq. (3.1) is present because each sublattice, i, suffers from phonon–phonon scattering not only from itself, but also from the other $n - 1$ sublattices present. Hence

$$K'(\theta_\infty) = \sum_{i=1}^{n} K_i'(\theta_\infty) = B \left[\frac{a_0^3}{4}\right]^{1/3} \frac{\theta_\infty^2}{\gamma^2} \frac{1}{n} \left[\sum_{i=1}^{n} m_i\right]. \tag{3.2}$$

Note that

$$n\bar{M} = \sum_{i=1}^{n} m_i$$

where $\bar{M}$ is now the average atomic mass of the crystal in grams. If δ^3 is defined as the average volume occupied by one atom of the crystal, then

$$\delta^3 = a_0^3/4n.$$

Hence

$$K'(\theta_\infty) = B(n)^{1/3} \bar{M} \delta \theta_\infty^2 \gamma^{-2}.$$

In these crystals with $n \geq 2$ we must be careful to define θ_∞ to conform to Eq. (1.3) where we now integrate $\nu^2 g(\nu)$ only over the acoustic branches of the phonon spectrum. In the more standard definition[47] of θ_∞, the integration is carried out over the whole phonon spectrum including the optic branches. Hence we shall use $\bar{\theta}_\infty$ to denote the Debye temperature for the three acoustic branches only. Similarly, we really want γ to refer to the acoustic branches only at high temperature and shall call this $\bar{\gamma}_\infty$. The calculation of γ values separately for the acoustic and optic branches has been carried out by Dolling and Cowley.[47] The

---

[67] G. A. Slack, *Proc. Int. Conf. Phonon Scattering in Solids* (H. J. Albany, ed.), p. 24. Centre d'Etudes Nucleaires de Saclay, 1972.

general equation for the acoustic-branch thermal conductivity of an fcc crystal is now

$$K'(\tilde{\theta}_\infty) = Bn^{1/3}\bar{M}\delta(\tilde{\theta}_\infty)^2(\tilde{\gamma}_\infty)^{-2} \qquad (3.3)$$

with the same value of $B = 3.04 \times 10^{-8}$ as used in Eq. (2.1). The value of the Debye temperature at 0°K for the acoustic branch is now $\tilde{\theta}_0$. It can be shown that this is related to the customary value of θ_0 determined from the elastic constants by

$$\tilde{\theta}_0 = \theta_0 n^{-1/3}. \qquad (3.4)$$

(See Anderson[68] and Billard and Cabannes[63] for a discussion of this point.) Note that Eq. (3.3) is rather different from the approaches used previously.[60-63]

We can now apply Eq. (3.3) to the adamantine compounds if $\tilde{\theta}_\infty$ and $\tilde{\gamma}_\infty$ are known. We have calculated $\tilde{\theta}_\infty$ from the $g(\nu)$ curves for diamond, Si, Ge, SiC, AlSb, GaP, GaAs, GaSb, InSb, ZnS, ZnSe, ZnTe, and CdTe by integration over the acoustic phonon part of the spectrum only. These results are shown in Table II, along with the $\tilde{\theta}_0$ values that are obtained from elastic constant or specific heat capacity measurements. For the other crystals in Table II, the value of $\tilde{\theta}_\infty$ was estimated from $\tilde{\theta}_0$ and other data by comparison with the above 13 crystals. The value of $\tilde{\gamma}_\infty$ is more of a problem. The method of handling this has been pointed out by Daniels.[69] Some of the acoustic modes may have negative γ values. The thermal expansion measurements at high temperature give an arithmetic average over all the phonon branches, both acoustic and optic. However, since γ^2 enters into Eq. (3.3), we really want $\bar{\gamma}^2$, the average value of γ^2 over the acoustic branches at high temperature. Calculated values for γ of the acoustic phonons exist for Si,[70,71] Ge,[47,70] and ZnTe.[72] These values are not in very good agreement with the few measurements that have been made.[73-76] Hence the experimental γ

---

[68] O. L. Anderson, *J. Phys. Chem. Solids* **12**, 41 (1959).
[69] W. B. Daniels, in "International Conference on the Physics of Semiconductors", Exeter, 1962, p. 482. Institute of Physics and the Physical Society, London, 1962. See also T. H. K. Barron, *Nature (London)* **178**, 871 (1956).
[70] H. Jex, *Phys. Status Solidi* **B45**, 343 (1971).
[71] I. Ishida, *J. Phys. Soc. Jap.* **39**, 1282 (1975).
[72] J. F. Vetelino, K. V. Namjoshi, and S. S. Mitra, *J. Appl. Phys.* **41**, 5141 (1970); also *Phys. Rev. B* **2**, 967 (1970).
[73] R. T. Payne, *Phys. Rev. Lett.* **13**, 53 (1964).
[74] O. Brafman and S. S. Mitra, *Proc. Second Int. Conf. Light Scattering in Solids*, 1971, (M. Balkanski, ed.), p. 284. Flammarion, Paris, 1971.
[75] W. Richter, J. B. Renucci, and M. Cardona, *Solid State Commun.* **16**, 131 (1975).
[76] B. A. Weinstein and G. J. Piermarini, *Phys. Rev. B* **12**, 1172 (1975).

TABLE II. CALCULATED AND OBSERVED VALUES FOR THE THERMAL CONDUCTIVITY OF ADAMANTINE CRYSTALS ($n = 2$) AT $T = \bar{\theta}_\infty$

Crystal	$\bar{\theta}_0$ (°K)	$\bar{\theta}_\infty$ (°K)	$\gamma_\infty{}^a$	K_1' (W/cm deg)	K (W/cm deg)	ϵ
Diamond	1778 (78, 79)	1450 (47)	0.90 (90)	3.52	~2.2 (63)	~1.4
SiC	857 (63)	740 (52)	0.76 (91)	1.87	1.3 (63)	1.40
Si	512 (80)	395 (47)	0.56 (71)	0.93	1.15 (102)	1.37
Ge	297 (80)	235 (50)	0.76 (92)	0.89	0.83 (102)	1.28
BP	780 (63)	670 (89)	—	1.66	1.1 (63)	~1.4
AlAs	330 (81)	270 (89)	0.66 (93)	0.82	1.1 (103)	—
AlSb	230 (81)	210 (58)	0.60 (94)	0.79	0.90 (60)	1.32
GaP	352 (82)	275 (52)	0.76 (95)	0.81	1.11 (65)	1.22
GaAs	275 (83)	220 (52)	0.75 (94)	0.77	0.81 (104)	1.25
GaSb	214 (82)	165 (53)	0.75 (90)	0.62	0.76 (60, 105)	1.41
InP	240 (84)	220 (54)	0.60 (60)	0.81	1.08 (106–108)	1.44
InAs	199 (83)	165 (59)	0.57 (60)	0.61	0.57 (60, 109)	1.44
InSb	164 (83)	135 (52)	0.56 (96)	0.54	0.48 (110, 111)	1.39
ZnS	269 (85)	230 (52)	0.75 (97)	0.39	0.39 (112)	1.39
ZnSe	215 (85)	190 (52)	0.75 (97)	0.41	0.33 (112)	1.32
ZnTe	179 (86)	155 (55)	0.97 (72)	0.39	0.44 (112)	1.36
CdTe	127 (85)	120 (56)	0.52 (98)	0.31	0.24 (112)	1.34
HgSe	120 (87)	110 (89)	0.17 (100)	0.29	0.055 (113)	1.79
HgTe	112 (88)	100 (89)	0.46 (101)	0.29	0.12 (114)	1.71

^{a} See text for value of γ used in computing K_1'.

value as determined from thermal expansion data at high temperatures, γ_∞, has been recorded in Table II. The actual value of γ used in Eq. (3.3) for calculating $K'(\bar{\theta}_\infty)$ was 0.70 for diamond, Si, Ge, SiC, and the III–V compounds, while 0.83 was used for the II–VI compounds. These were chosen so as to lie somewhere between the extreme limits of the experimental values of γ_∞. Clearly, better data on γ values for the various phonon branches are needed.

We note that $\bar{\gamma}_\infty$ cannot be evaluated just from the third-order elastic constants. These constants give $\bar{\gamma}_0$, which is just the thermal value of γ_0 at absolute zero[77] for the lowest energy phonons. Hamilton and Parrott[17] have essentially used γ_0 in their calculations, a procedure which is not really correct for the high temperature region that we are concerned with.

In Table II we give the various parameters used for calculating $K'(\bar{\theta}_\infty)$ from Eq. (3.3). The $\bar{\theta}_0$ values have been calculated from literature

[77] P. W. Sparks and C. A. Swenson, *Phys. Rev.* **163**, 779 (1967).

values[63,78-88] of θ_0 using Eq. (3.4) with $n = 2$. The $\bar{\theta}_\infty$ values have already been explained except for BP, AlAs, HgSe, and HgTe. For these crystals $\bar{\theta}_\infty$ was estimated[89] using the same method as Steigmeier.[81] The γ_∞ values are also taken from the literature.[60,71,72,90-101] The fifth column in Table II is labeled K_1'. The subscript "1" on these calculated values of thermal conductivity means that only the first set of branches, the acoustic phonon, were used. The inclusion or exclusion of the next higher or optic phonon branches becomes more of a question in the rocksalt structure compounds, and is discussed later.

The observed values of K at $T = \bar{\theta}_\infty$ are given in the sixth column of Table II. The seventh column gives the measured slope, ϵ, of the K

---

[78] J. C. Phillips, *Phys. Rev.* **113**, 147 (1959).
[79] M. H. Grimsditch and A. K. Ramdas, *Phys. Rev. B* **11**, 3139 (1975).
[80] P. Flubacher, A. J. Leadbetter, and J. A. Morrison, *Phil. Mag.* **4**, 273 (1959).
[81] E. F. Steigmeier, *Appl. Phys. Lett.* **3**, 6 (1963).
[82] W. F. Boyle and R. J. Sladek, *Phys. Rev. B* **11**, 2933 (1975).
[83] J. C. Holste, *Phys. Rev. B* **6**, 2495 (1972).
[84] Calculated from elastic constant data at 300°K of F. S. Hickernell and W. R. Gayton, *J. Appl. Phys.* **37**, 462 (1966).
[85] J. A. Birch, *J. Phys. C* **8**, 2043 (1975).
[86] B. H. Lee, *J. Appl. Phys.* **41**, 2984 (1970).
[87] A. Lehoczky, D. A. Nelson, and C. R. Whitsett, *Phys. Rev.* **188**, 1069 (1969).
[88] R. I. Cottam and G. A. Saunders, *J. Phys. Chem. Solids* **36**, 187 (1975).
[89] Estimated from a log-log plot of $\bar{\theta}_\infty$ versus $(\bar{M})^{1/2}\delta^{3/2}$ for the adamantine crystals.
[90] D. Gerlich, *J. Phys. Chem. Solids* **35**, 1026 (1974).
[91] Calculated from the thermal expansion coefficient measurements of A. Taylor and R. M. Jones, in "Silicon Carbide" (J. R. O'Connor and J. Smiltens, eds.), p. 147. Pergamon Press, New York, 1960, and the elastic constants of ref. 57.
[92] R. R. Reeber, *Phys. Status Solidi A* **32**, 321 (1975).
[93] Calculated from the thermal expansion data of M. Ettenberg and R. J. Paff, *J. Appl. Phys.* **41**, 3926 (1970) and the bulk modulus of GaP from R. Weil and W. O. Groves, *J. Appl. Phys.* **39**, 4049 (1968).
[94] S. I. Novikova, *Fiz. Tverd. Tela* **7**, 2683 (1965) [*Sov. Phys.—Solid State* **7**, 2170 (1966)].
[95] Calculated from the thermal expansion data of E. D. Pierron, D. L. Parker, and J. B. McNeely, *J. Appl. Phys.* **38**, 4669 (1967) and I. Kudman and R. J. Paff, *J. Appl. Phys.* **43**, 3670 (1972) and the bulk modulus of GaP (see ref. 93).
[96] Average of values from ref. 60 and 90.
[97] T. F. Smith and G. K. White, *J. Phys. C* **8**, 2031 (1975).
[98] Calculated from the thermal expansion coefficient measurements of S. I. Novikova, *Fiz. Tverd. Tela* **2**, 2341 (1960) [*Sov. Phys.—Solid State* **2**, 2087 (1961)] and the elastic constant measurements of ref. 99.
[99] Yu. Kh. Vekilov and A. P. Rusakov, *Fiz. Tverd. Tela* **13**, 1157 (1971) [*Sov. Phys.—Solid State* **13**, 956 (1971)].
[100] Calculated from the thermal expansion data of V. V. Zhdanova, V. I. Lukina, and S. I. Novikova, *Phys. Status Solidi* **13**, K19 (1966) and the elastic constants of ref. 87.
[101] Calculated from the thermal expansion data of S. I. Novikova and N. Kh. Abrikosov, *Fiz. Tverd. Tela* **5**, 2138 (1963) [*Sov. Phys.—Solid State* **5**, 1558 (1964)] and the elastic constants of ref. 88.

versus T curve. The K and ϵ values have been taken from the recent literature.[60,63,65,102-114] The only K values that have been obtained by dubious extrapolation are those for diamond and BP. For this extrapolation an average slope $\epsilon = 1.4$ was assumed. An isotope scattering correction has been applied to the raw data[102] for Si and Ge. For all other crystals in Table II this isotope correction is negligible. The K and K_1' values from Table II have been plotted in Fig. 3.

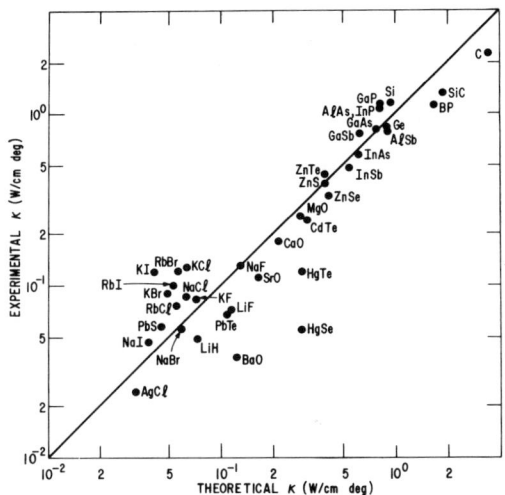

FIG. 3. Experimental versus theoretical values for the thermal conductivity of adamantine and rocksalt structure crystals at $T = \bar{\theta}_\infty$. The theoretical values are K_1'; in all cases $n = 2$, see text.

[102] C. J. Glassbrenner and G. A. Slack, *Phys. Rev.* **134**, A1058 (1964). The $K(\bar{\theta}_\infty)$ values have been corrected for isotope scattering. This correction is particularly large for Ge.

[103] M. A. Afromowitz, *J. Appl. Phys.* **44**, 1292 (1973), some small extrapolation used with $\epsilon = 1.25$ assumed.

[104] R. O. Carlson, G. A. Slack, and S. J. Silverman, *J. Appl. Phys.* **36**, 505 (1965).

[105] G. LeGuillou and H. J. Albany, *J. Phys. (Paris)* **31**, 495 (1970).

[106] I. Kudman and E. F. Steigmeier, *Phys. Rev.* **133**, A1665 (1964).

[107] M. Boutard and P. Pinard, *J. Phys. (Paris)* **33**, 787 (1972).

[108] S. A. Aliev, A. Ya. Nashelskii, and S. S. Shalyt, *Fiz. Tverd. Tela* **7**, 1590 (1965) [*Sov. Phys.—Solid State* **7**, 1287 (1965)].

[109] G. LeGuillou and H. J. Albany, *Phys. Rev. B* **5**, 2301 (1972).

[110] M. G. Holland, *Phys. Rev.* **134**, A471 (1964).

[111] M. I. Aliev, R. M. Dzhabbarov, D. G. Arasly, and M. A. Alieva, *Fiz. Tekh. Poluprov.* **7**, 427 (1973) [*Sov. Phys.—Semicond.* **7**, 311 (1973)].

[112] G. A. Slack, *Phys. Rev. B* **6**, 3791 (1972).

[113] C. R. Whitsett, D. A. Nelson, J. G. Broerman, and E. C. Paxhia, *Phys. Rev. B* **7**, 4625 (1973).

[114] C. R. Whitsett and D. A. Nelson, *Phys. Rev. B* **5**, 3125 (1972).

For the first 17 crystals

$$0.65 \leq [K(\tilde{\theta}_\infty)/K_1'(\tilde{\theta}_\infty)] \leq 1.4 \tag{3.5}$$

and

$$1.20 \leq \epsilon \leq 1.45. \tag{3.6}$$

For HgSe and HgTe the observed values of ϵ appear to be too large, and the values of the lattice K too small. It appears reasonable to suppose that some extra phonon scattering by the free carriers[113,114] is present which is not accounted for by the theory used here that considers only phonon–phonon scattering. The anomalously low K values of HgSe and HgTe have been noted previously.[112]

The agreement between theory and experiment in Eq. (3.5) is ±40%. This is not quite as good as the ±20% found for the rare-gas crystals.

We have, so far, ignored the possibility that the observed K is lower than the calculated one because the acoustic phonons are also scattered by interaction with optic phonons. This possibility for crystals with $n \geq 2$ was first suggested by Blackman[115] and has been mentioned by Gurevich.[116] A review of the theoretical problem was later given by Ziman.[117] The first experimental work in this area appears to be that of Devyatkova and Smirnov,[118] on rocksalt structures, followed by that of Steigmeier and Kudman,[60,106] Leroux-Hugon and Veyssie,[119] and Wagini[65] on adamantine structure crystals. Some calculations[120,121] on two-dimensional lattices may help to solve some problems. Many recent papers have been concerned with this question.[66,122–125]

The interaction between acoustic and optic phonons can become very strong in special crystals like $SrTiO_3$, $KTaO_3$, and SbSI, as shown by

---

[115] M. Blackman, *Phil. Mag.* **19**, 989 (1935).
[116] V. L. Gurevich, *Fiz. Tverd. Tela* **1**, 1474 (1959) [*Sov. Phys.—Solid State* **1**, 1351 (1960)].
[117] J. M. Ziman, in "Electrons and Phonons," p. 144. Oxford Univ. Press, London, 1960.
[118] E. D. Devyatkova and I. A. Smirnov, *Fiz. Tverd. Tela* **4**, 2507 (1962) [*Sov. Phys.—Solid State* **4**, 1836 (1963)].
[119] P. Leroux-Hugon and J. J. Veyssie, *Phys. Status Solidi* **8**, 561 (1965).
[120] I. V. Kirillova, I. L. Korobova, and B. Ya. Moizhes, *Fiz. Tverd. Tela* **9**, 1281 (1967) [*Sov. Phys.—Solid State* **9**, 1003 (1967)].
[121] Yu. A. Logachev, B. Ya. Moizhes, and A. S. Skal, *Fiz. Tverd. Tela* **12**, 2791 (1970) [*Sov. Phys.—Solid State* **12**, 2251 (1971)].
[122] Yu. A. Logachev and M. S. Yur'ev, *Fiz. Tverd. Tela* **14**, 3336 (1972) [*Sov. Phys.—Solid State* **14**, 2826 (1973)].
[123] Yu. A. Logachev and L. N. Vasil'ev, *Fiz. Tverd. Tela* **15**, 1612 (1973) [*Sov. Phys.—Solid State* **15**, 1081 (1973)].
[124] A. V. Petrov, N. S. Tsypkina, and Yu. A. Logachev, *Fiz. Tverd. Tela* **16**, 65 (1974) [*Sov. Phys.—Solid State* **16**, 39 (1974)].
[125] J. P. Moore, R. K. Williams, and R. S. Graves, *Phys. Rev. B* **11**, 3107 (1975).

the work of Steigmeier et al.:a1:a2:a6:x0:a1:a2:a9 and Barrett and Holland.[130] These cases, in which there are very-low-energy optic–phonon branches, lead naturally into the effects of solid–solid phase changes on the thermal conductivity.

For the crystals in Table II there appears to be very little correlation between the $[K/K_1']$ ratio and the energy separation between the acoustic and optic branches. Even for AlSb where the mass ratio of the two elements is 4.50, the separation between branches is still sufficiently small so that two acoustic phonons can interact to produce an optic phonon. Thus the scatter of the $[K/K_1']$ ratio, caused mainly by uncertainties in the calculation of K', obscures any real evidence in these adamantine crystals of acoustic–optic scattering or of a small amount of heat transfer by optic phonons. The neat dependence of $[K/K']$ on mass ratio originally found by Steigmeier and Kudman[60] may be caused mostly by their use of θ instead of $\bar{\theta}$ in calculating K'. The question of scattering by optic phonons is considered further in Section IV.

4. Rocksalt Structure Crystals ($n = 2$)

The rocksalt or sodium chloride crystal structure is also fcc with $n = 2$, and K measurements have been made on many nonmetallic crystals with this structure. Also many neutron scattering studies of the phonon dispersion curves have been carried out. Hence we are in a good position to test Eq. (3.3). A large difference between these crystals and the adamantine structures, however, is that some of the optic phonons have group velocities as large as the velocities of the acoustic phonons, hence we may find optic phonons contributing to K in a substantial manner. The first theoretical paper dealing with this possibility seems to have been published by Gurevich,[116] with later theoretical contributions by Kirillova et al.,[120] and Logachev et al.[121–124] Some experimental evidence, which depends on the deviation of ϵ from the value $\epsilon = 1$ for $T \sim \theta$, has been presented by Devyatkova and Smirnov,[118] Farag et al.,[131] and Wagini[65] for crystals of KBr, NaI, PbS, PbSe, and AlSb. The collected data for NaF also appear to exhibit[62] this same behavior, although when plotted differently[132] the conclusion was drawn that

[126] E. F. Steigmeier and R. Klein, Helv. Phys. Acta 39, 594 (1966).
[127] E. F. Steigmeier, Phys. Rev. 168, 523 (1968).
[128] E. F. Steigmeier, Helv. Phys. Acta 41, 406 (1968).
[129] E. F. Steigmeier and J. W. Merz, Helv. Phys. Acta 41, 1206 (1968).
[130] H. H. Barrett and M. G. Holland, Phys. Rev. B 2, 3441 (1970).
[131] B. S. Farag, I. A. Smirnov, and Y. L. Yousef, Physica 31, 1673 (1965).
[132] I. A. Smirnov, Fiz. Tverd. Tela 9, 1845 (1967) [Sov. Phys.—Solid State 9, 1454 (1967)].

enhanced optic–phonon scattering of acoustic phonons rather than a contribution from optic–phonon thermal conductivity was present. The experimental evidence in the published literature for optic–phonon heat transport is not yet clear.

We can calculate the expected contribution to K from acoustic phonons by using Eq. (3.3). The parameters used and the results are shown in Table III. The $\tilde{\theta}_0$ values have been calculated from θ_0 values[133–140] using Eq. (3.4).

TABLE III. CALCULATED AND OBSERVED VALUES FOR THE THERMAL CONDUCTIVITY OF ROCKSALT STRUCTURE CRYSTALS ($n = 2$) AT $T = \tilde{\theta}_\infty$

Crystal	$\tilde{\theta}_0$ (°K)	$\tilde{\theta}_\infty$ (°K)	γ_∞	K_1' (10^{-3} W/cm deg)	K (10^{-3} W/cm deg)	ϵ	ν_{max} (THz)
LiH	944 (133)	615 (141)	1.28 (158)	72	49 (171)	1.42	17.0
LiF	583 (134)	500 (154)	1.50 (159)	111	71 (172)	1.22	11.5
NaF	392 (134)	395 (155)	1.50 (159)	129	130 (171)	1.30	9.0
NaCl	256 (134)	220 (142)	1.56 (160)	63	86 (124)	1.24	5.19
NaBr	179 (134)	150 (143)	1.50 (159)	59	56 (124)	1.20	3.15
NaI	133 (134)	100 (144)	1.56 (161)	38	47 (173)	~2.0	2.35
KF	267 (135)	235 (145)	1.52 (162)	71	83 (174)	≥1.0	5.73
KCl	187 (134)	172 (142)	1.45 (163)	63	128 (175)	1.21	4.11
KBr	137 (134)	117 (146)	1.45 (164)	49	91 (176)	1.21	2.80
KI	104 (134)	87 (147)	1.45 (159)	41	120 (177)	1.60	2.06
RbCl	135 (134)	124 (142)	1.45 (165)	56	76 (125)	1.23	3.11
RbBr	109 (134)	105 (148)	1.45 (165)	57	122 (125)	1.13	2.50
RbI	86 (134)	84 (149)	1.41 (161)	53	99 (125)	1.15	1.98
AgCl	128 (136)	124 (150)	1.90 (166)	32	24 (178)	1.0	3.25
MgO	750 (137)	600 (156)	1.44 (167)	282	250 (179)	~1.0	16.6
CaO	490 (137)	450 (151)	1.57 (168)	212	180 (180)	~1.0	13.0
SrO	318 (137)	270 (152)	1.52 (168)	162	110 (180)	~1.0	7.2
BaO	184 (138)	183 (153)	1.50 (169)	121	38 (180)	0.6	4.89
PbS	175 (139)	115 (139)	2.0 (170)	45	57 (181)	0.86	3.00
PbTe	127 (140)	105 (157)	1.45 (170)	109	68 (181)	0.80	2.85

[133] B. Yates, G. H. Wostenholm, and J. L. Bingham, *J. Phys. C* **7**, 1769 (1974).
[134] C. R. Cleavelin, D. O. Pederson, and B. J. Marshall, *Phys. Rev. B* **5**, 3193 (1972).
[135] J. T. Lewis, A. Lehoczky, and C. V. Briscoe, *Phys. Rev.* **161**, 877 (1967).
[136] J. Donecker, *Phys. Status Solidi* **37**, 275 (1970).
[137] E. Gmelin, *Z. Naturforsch. A* **25**, 887 (1970).
[138] E. Gmelin, *Z. Naturforsch. A* **24**, 1794 (1969).
[139] M. E. Elcombe, *Proc. Roy. Soc. (London)* **300**, 210 (1967).
[140] M. Saint-Paul and K. H. Gobrecht, *Phys. Status Solidi A* **18**, K111 (1973).

The $\bar{\theta}_\infty$ values[139,141–153] for all of the crystals except for LiF, NaF, MgO, and PbTe in Table III were obtained from Eq. (1.3) by direct integration of g(ν) over the acoustic portion of the phonon spectrum. The last column in Table III gives the maximum frequency, ν_M, in terahertz of the acoustic phonons considered during the integration. For the remaining four crystals the overlap between the acoustic and optic phonon branches was so great that the published g(ν) could not readily be separated into acoustic and optic parts. Thus $\bar{\theta}_\infty$ values were estimated from the available dispersion curves[154–157] by comparison with the other 16 crystals. The γ_∞ values are derived from the thermal expansion, elastic constant, and specific heat capacity data at high temperatures.[158–170] These thermal values were used in computing both $K_1{}'$ and

[141] J. L. Verble, J. L. Warren, and J. L. Yarnell, *Phys. Rev.* **168**, 980 (1968).
[142] G. Raunio and S. Rolandson, *Phys. Rev. B* **2**, 2098 (1970).
[143] J. S. Reid, T. Smith, and W. J. L. Buyers, *Phys. Rev. B* **1**, 1833 (1970).
[144] R. A. Cowley, W. Cochran, B. N. Brockhouse, and A. D. B. Woods, *Phys. Rev.* **131**, 1030 (1963).
[145] W. Bührer, *Phys. Status Solidi* **41**, 789 (1970).
[146] W. Bluthardt, W. Schneider, and M. Wagner, *Phys. Status Solidi* **56**, 453 (1973).
[147] G. Dolling, R. A. Cowley, C. Schittenhelm, and I. M. Thorson, *Phys. Rev.* **147**, 577 (1966).
[148] S. Rolandson and G. Raunio, *J. Phys. C* **4**, 958 (1971).
[149] G. Raunio and S. Rolandson, *Phys. Status Solidi* **40**, 749 (1970).
[150] P. R. Vijayaraghavan, R. M. Nicklow, H. G. Smith, and M. K. Wilkinson, *Phys. Rev. B* **1**, 4819 (1970).
[151] D. H. Saunderson and G. E. Peckham, *J. Phys. C* **4**, 2009 (1971).
[152] K. H. Rieder, R. Migoni, and B. Renker, *Phys. Rev. B* **12**, 3374 (1975).
[153] S. S. Chang, C. W. Tompson, E. Gürmen, and L. D. Muhlestein, *J. Phys. Chem. Solids* **36**, 769 (1975).
[154] G. Dolling, H. G. Smith, R. M. Nicklow, P. R. Vijayaraghavan, and M. K. Wilkinson, *Phys. Rev.* **168**, 970 (1968).
[155] W. J. L. Buyers, *Phys. Rev.* **153**, 923 (1967).
[156] M. J. L. Sangster, G. Peckham, and D. H. Saunderson, *J. Phys. C* **3**, 1026 (1970).
[157] W. Cochran, R. A. Cowley, G. Dolling, and M. M. Elcombe, *Proc. Roy. Soc. (London), Ser. A* **293**, 433 (1966).
[158] D. Gerlich and C. S. Smith, *J. Phys. Chem. Solids* **35**, 1587 (1974).
[159] J. E. Rapp and H. D. Merchant, *J. Appl. Phys.* **44**, 3919 (1973).
[160] Average of data from refs. 159 and 162.
[161] H. D. Merchant, K. K. Srivastava, and H. D. Pandey, *Crit. Rev. Solid State Science* **3**, 451 (1973).
[162] R. W. Roberts and C. S. Smith, *J. Phys. Chem. Solids* **31**, 619 (1970).
[163] R. R. Rao and M. Peter, *Helv. Phys. Acta* **47**, 705 (1974).
[164] C. H. Panter, *J. Phys. C* **7**, 4483 (1974).
[165] G. K. White and J. G. Collins, *Proc. Roy. Soc. (London), Ser. A* **333**, 237 (1973).
[166] K. F. Loje and D. E. Schuele, *J. Phys. Chem. Solids* **31**, 2051 (1970).
[167] G. K. White and O. L. Anderson, *J. Appl. Phys.* **37**, 430 (1966).

K_2'. The values for K_1' as computed from Eq. (3.3) using $\bar{\theta}_\infty$ are shown in the fifth column of Table III. The quantity K_2' is explained later.

The experimental values of K and ϵ are taken from numerous literature references.[124,125,171–181] Not all of the available data are cited. Those given are believed to be the most accurate at $T = \bar{\theta}_\infty$. No corrections to the data are made for the phonon scattering caused by the natural isotopes. This correction is quite small and is generally smaller than the uncertainty in the experimental numbers themselves. For PbS and PbTe the experimental K in Table III is the lattice component only, the electronic contribution has already been subtracted.

The results in Table III show that for the rocksalt structure crystals

$$0.31 \leq [K(\bar{\theta}_\infty)/K_1'(\bar{\theta}_\infty)] \leq 2.9$$

and

$$0.6 \leq \epsilon \leq 2.0.$$

These values should be compared with Eqs. (3.5) and (3.6) for the adamantine crystals. If we eliminate the points for PbS and PbTe where there may be free-carrier scattering of the phonons, and BaO where the low K and ϵ values make it appear that the sample studied was very impure, then the more representative ranges are

$$0.64 \leq [K(\bar{\theta}_\infty)/K_1'(\bar{\theta}_\infty)] \leq 2.9 \qquad (4.1)$$

[168] R. Ruppin, *Solid State Commun.* **10**, 1053 (1972).
[169] Estimated by comparison with MgO, CaO, and SrO.
[170] S. I. Novikova and N. Kh. Abrikosov, *Fiz. Tverd. Tela* **5**, 1913 (1963) [*Sov. Phys.—Solid State* **5**, 1397 (1964)].
[171] G. A. Slack, *J. Phys. Chem. Solids* **34**, 321 (1973).
[172] A. A. Men, A. Z. Chechelnitskii, V. A. Sokolov, and E. N. Simun, *Fiz. Tverd. Tela* **15**, 2773 (1973) [*Sov. Phys.—Solid State* **15**, 1844 (1974)].
[173] S. J. Rogers, *Phys. Rev.* **B3**, 1440 (1971); and T. F. McNelly, Ph.D. Thesis, Cornell Univ., Ithaca, New York, 1974.
[174] A. Eucken and G. Kuhn, *Z. Phys. Chem.* **134**, 193 (1928).
[175] E. D. Devyatkova and I. A. Smirnov, *Fiz. Tverd. Tela* **4**, 1972 (1962) [*Sov. Phys.—Solid State* **4**, 1445 (1963)] and also ref. 124.
[176] F. C. Baumann and R. O. Pohl, *Phys. Rev.* **163**, 843 (1967).
[177] C. T. Walker, *Phys. Rev.* **132**, 1963 (1963); R. L. Rosenbaum, C. K. Chau, M. V. Klein, *Phys. Rev.* **186**, 852 (1969); and ref. 176.
[178] C. K. Chau and M. V. Klein, *Phys. Rev. B* **1**, 2642 (1970); and W. Bausch, F. Guckenbiehl, and W. Waidelich, *Phys. Lett. A* **28**, 38 (1968).
[179] G. A. Slack, *Phys. Rev.* **126**, 427 (1962); see also ref. 180.
[180] N. N. Kovalev, A. V. Petrov, and O. V. Sorokin, *Fiz. Tverd. Tela* **13**, 291 (1971) [*Sov. Phys.—Solid State* **13**, 232 (1971)].
[181] S. S. Shalyt, V. M. Muzhdaba, and A. D. Galetskaya, *Fiz. Tverd. Tela* **10**, 1277 (1968) [*Sov. Phys.—Solid State* **10**, 1018 (1968)].

and
$$1.0 \leq \epsilon \leq 2.0 \tag{4.2}$$

for the rocksalt structure.

5. Heat Transport by Optic Phonons

The most obvious difference between Eq. (3.5) and Eq. (4.1) is the upper limit. The observed values of K tend to be higher than the calculated ones. The K versus K' plot is given in Fig. 3. However, if we plot K/K_1' versus the mass ratio for these compounds as in Fig. 4 we see that, except for the strange case of KI, the high K/K_1' values tend to occur for mass ratios near unity. It is just near unity mass ratios that the optic branches, when plotted in an extended zone plot, appear to be a smooth continuation of the acoustic branches. Hence when $M_1 = M_2$, we can consider that Eq. (3.3) should be modified by replacing $\tilde{\theta}_\infty$ by θ_∞, where θ_∞ is given by Eq. (1.3) integrated over both the acoustic and optic parts of the phonon spectrum. Hence we define K_2' as the calculated thermal conductivity when both acoustic and optic branches are included. Thus

$$K_2'(\theta_\infty) = B n^{1/3} \bar{M} \delta (\theta_\infty)^2 (\gamma_\infty)^{-2}. \tag{5.1}$$

So that we may compare the observed and calculated values at a temperature of $T = \tilde{\theta}_\infty$, we really want $K_2'(\tilde{\theta}_\infty)$. The extrapolation from $T = \theta_\infty$ to $T = \tilde{\theta}_\infty$ is made by assuming that the computed K' will have the same slope, ϵ, as the experimental curve. Hence

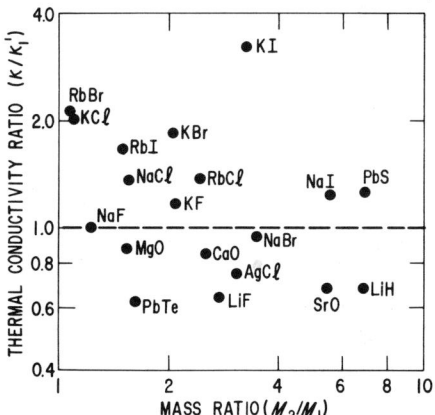

Fig. 4. The ratio of the experimental to theoretical value of thermal conductivity for rocksalt structure crystals at $T = \tilde{\theta}_\infty$ as a function of the mass ratio. K_1' is computed by assuming that only the acoustic phonons carry the heat. In all cases $n = 2$.

$$K_2{}'(\bar{\theta}_\infty) = Bn^{1/3}\bar{M}\delta(\theta_\infty)^2(\gamma_\infty)^{-2}(\theta_\infty/\bar{\theta}_\infty)^\epsilon \qquad (5.2)$$

and

$$[K_2{}'(\bar{\theta}_\infty)/K_1{}'(\bar{\theta}_\infty)] = (\theta_\infty/\bar{\theta}_\infty)^{2+\epsilon}. \qquad (5.3)$$

We have assumed that $\gamma_\infty = \bar{\gamma}_\infty$ since we are using the thermal expansion value for γ_∞ anyway. Several additional references[182–184] are needed to obtain θ_∞ values for all of the crystals. For the last three crystals in Table IV a value of $\epsilon = 1$ was used in Eq. (5.2).

In Fig. 5 the $K/K_2{}'$ values are plotted versus the mass ratio. For mass ratios near unity $K_2{}'$ is a good approximation to K, see RbBr and KCl. If it were a good approximation for all of the crystals the points would all fall near $K/K_2{}' = 1$. Clearly they do not. As the mass ratio increases, the approximation becomes progressively worse. By comparison with

TABLE IV. CALCULATED VALUES FOR THE THERMAL CONDUCTIVITY OF ROCKSALT STRUCTURE CRYSTALS ($n = 2$) AT $T = \bar{\theta}_\infty$ IF BOTH ACOUSTIC AND OPTIC BRANCHES ARE INCLUDED [see Eq. (5.2)]

Crystal	θ_∞ (°K)	$K_2{}'$ (10^{-3} W/cm deg)	$K/K_2{}'$	M_1/M_2
LiH	1325 (182)	991	0.050	6.88
LiF	630 (154)	234	0.30	2.74
NaF	445 (183)	191	0.68	1.21
NaCl	286 (142)	147	0.59	1.54
NaBr	220 (143)	200	0.28	3.48
NaI	187 (144)	250	0.10	5.52
KF	320 (145)	180	0.46	2.06
KCl	232 (142)	165	0.78	1.10
KBr	183 (144)	206	0.44	2.04
KI	158 (147)	327	0.37	3.25
RbCl	191 (142)	225	0.34	2.41
RbBr	144 (148)	153	0.80	1.07
RbI	121 (149)	167	0.59	1.48
AgCl	210 (150)	157	0.15	3.04
MgO	767 (156)	590	0.42	1.52
CaO	575 (184)	443	0.41	2.51
SrO	450 (152)	748	0.15	5.48
BaO	350 (153)	845	0.045	8.58
PbS	230 (139)	360	0.16	6.46
PbTe	130 (157)	206	0.33	1.62

[182] J. Neuberger, *J. Nonmetals* **1**, 271 (1973).
[183] A. Gosh, A. N. Basu, and S. Sengupta, *Proc. Roy. Soc. (London), Ser. A* **340**, 199 (1974).
[184] K. S. Upadhyaya and R. K. Singh, *J. Phys. Chem. Solids* **36**, 293 (1975).

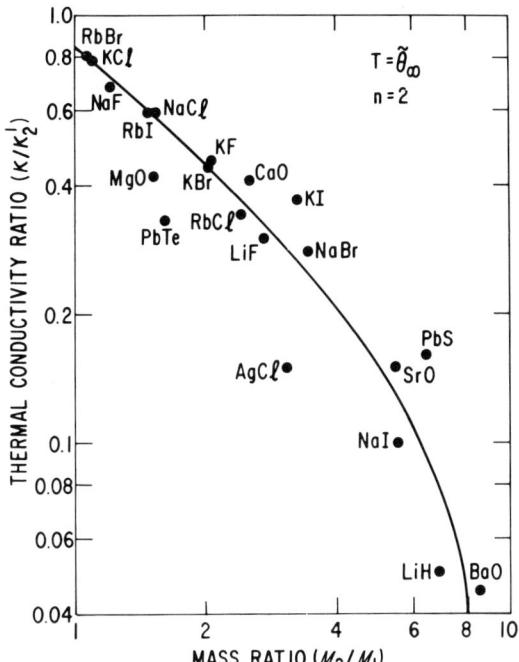

FIG. 5. The ratio of the experimental to theoretical value of the thermal conductivity for rocksalt structure crystals at $T = \tilde{\theta}_\infty$ as a function of the mass ratio. K_2' is computed by assuming that both acoustic and optic phonons carry the heat. In all cases $n = 2$.

Fig. 4 it is clearly a better approximation than K_1' for RbBr and KCl. For the range $M_2/M_1 \geq 1.5$ the K_1' is the better approximation. This agrees with the predictions of Logachev et al.[121] So it is clear that optic phonons can carry heat in crystals if the acoustic–optic gap is quite small and if the optic phonon velocity is large. This is the case for RbBr[148] and KCl[142] where at $T = \tilde{\theta}_\infty$ maybe 20% to 50% of the heat is carried by optic phonons, see Fig. 4. For NaF there is probably less, but some, heat transport by optic phonons. For the rest of the crystals in Table III almost all of the heat is carried by acoustic phonons.

6. Other Crystals ($n \geq 3$)

The next more complex cases are the fcc crystals with $n = 3$. Some examples are CaF_2, Mg_2Si, and ThO_2. The phonon disperson curves have been measured for CaF_2,[185] SrF_2,[186] BaF_2,[187] UO_2,[188] and Mg_2Sn,[189]

[185] M. M. Elcombe and A. W. Pryor, *J. Phys. C* **3**, 492 (1970).
[186] M. M. Elcombe, *J. Phys. C* **5**, 2702 (1972).
[187] J. P. Hurrell and V. J. Minkiewicz, *Solid State Commun.* **8**, 463 (1970).

and calculated for Mg_2Si^{190} and $Mg_2Ge.^{191}$ One of the consistent features of these dispersion curves is that the six optic-phonon branches have very low group velocities compared to the acoustic branches. Thus Eq. (3.3) might work well for predicting the K of these crystals. Other fcc structures with $n \geq 4$ for which good thermal conductivity data exist are few. Spinel, $MgAl_2O_4$, is one. Here $n = 14$. No neutron scattering data exist for it, but there are some[192] data for the related crystal Fe_3O_4. A normal mode analysis of the spinel structure has been made by Lutz,[193] and optical measurements[194,195] give the energies of some of the phonons at the zone center. Using these results and the measured elastic constants,[196] a schematic phonon dispersion curve has been constructed, only 22 branches are shown because some of the 42 possible ones are degenerate. This dispersion curve is shown in Fig. 6 together with dispersion curves for Ar,[40,41] GaP,[52] KCl,[142] and CaF_2.[185] The value of $\bar{\theta}_\infty = 352°K$ has been estimated for $MgAl_2O_4$ from the elastic constants,[196] and is used as the scaling parameter in Fig. 6.

For crystals with $n > 14$ we go to β-boron and the various boron polymorph structures.[197] These are fcc or slightly distorted fcc structures. With these crystals we can go up to $n = 414$ for YB_{68}.[198] Their relevant properties are given in Table VI. No phonon dispersion curve measurements exist for these crystals, so that we will use the simplified model which assumes that only the three acoustic phonon branches contribute to K, and all the optic ones can be ignored. We ignore them both as possible contributors to K and, also, as possible scatterers for the acoustic phonons. Then we can apply Eq. (3.3) and see how K' compares to K.

The $n = 3$ crystals are treated in Table V where the literature references for the phonon dispersion curves,[185−191,199] θ_0 val-

---

[188] G. Dolling, R. A. Cowley, and A. D. B. Woods, *Can. J. Phys.* **43**, 1397 (1965).

[189] R. J. Kearney, T. G. Worlton, and R. E. Schmunk, *J. Phys. Chem. Solids* **31**, 1085 (1970).

[190] W. B. Whitten, P. L. Chung, and G. C. Danielson, *J. Phys. Chem. Solids* **26**, 49 (1965).

[191] P. L. Chung, W. B. Whitten, and G. C. Danielson, *J. Phys. Chem. Solids* **26**, 1753 (1965).

[192] H. Watanabe and B. N. Brockhouse, *Phys. Lett.* **1**, 189 (1962).

[193] H. D. Lutz, *Z. Naturforsch. A* **24**, 1417 (1969).

[194] G. A. Slack, *Phys. Rev.* **134**, A1268 (1964).

[195] M. P. O'Horo, A. L. Frisillo, and W. B. White, *J. Phys. Chem. Solids* **34**, 23 (1973).

[196] M. F. Lewis, *J. Acoust. Soc. Am.* **40**, 728 (1966); see also Ref. 220.

[197] G. A. Slack, D. W. Oliver, and F. H. Horn, *Phys. Rev. B* **4**, 1714 (1971).

[198] G. A. Slack, D. W. Oliver, G. D. Brower, and J. D. Young, *J. Phys. Chem. Solids* **38**, 45 (1977).

[199] The $\bar{\theta}_\infty$ value for ThO_2 was calculated from the data in ref. 188 by assuming that $\theta \sim \bar{M}^{-1/2}$.

TABLE V. CALCULATED AND OBSERVED VALUES FOR THE THERMAL CONDUCTIVITY OF FCC CRYSTALS ($n = 3$) AT $T = \theta_\infty$

Crystal	$\tilde{\theta}_0$ (°K)	$\tilde{\theta}_\infty$ (°K)	γ_∞	K_1' (10^{-3} W/cm deg)	K (10^{-3} W/cm deg)	ϵ	ν_{max} (THz)
CaF_2	354 (185)	345 (185)	1.89 (205)	91	85 (211–213)	~1.0	7.50
SrF_2	262 (186)	260 (186)	1.65 (206)	115	105 (213)	1.25	6.00
BaF_2	196 (200)	190 (187)	1.57 (205)	102	120 (213)	~1.0	4.65
ThO_2	273 (201)	230 (199)	1.97 (207)	131	200 (214)	1.28	5.33
Mg_2Si	401 (202)	330 (190)	1.32 (208)	193	105 (215)	1.12	8.32
Mg_2Ge	333 (203)	230 (191)	1.38 (209)	137	130 (215)	1.12	5.70
Mg_2Sn	234 (204)	155 (189)	1.27 (210)	122	160 (215, 216)	1.19	3.67

ues,[185,186,200–204] γ_∞ values,[205–210] and experimental K and ϵ values[211–216] are given. For CaF_2, SrF_2, BaF_2, and UO_2 the $\tilde{\theta}_\infty$ values were calculated from $g(\nu)$ using Eq. (1.3). For the other crystals various approximations were used based on the published phonon dispersion curves. The γ_∞ values are, again, derived from the thermal expansion at high temperature. The ν_{max} column in Table V gives the maximum frequency of the acoustic phonons.

[200] D. Gerlich, *Phys. Rev. A* **135**, 1331 (1964).
[201] M. Ali and P. Nagels, *Phys. Status Solidi* **21**, 113 (1967), also from ref. 188.
[202] B. C. Gerstein, F. J. Jelinek, M. Habenschuss, W. D. Shickell, J. R. Mullaly, and P. L. Chung, *J. Chem. Phys.* **47**, 2109 (1967).
[203] B. C. Gerstein, P. L. Chung, and G. C. Danielson, *J. Phys. Chem. Solids* **27**, 1161 (1966).
[204] F. J. Jelinek, W. D. Shickell, and B. C. Gerstein, *J. Phys. Chem. Solids* **28**, 267 (1967).
[205] C. Wong and D. E. Schuele, *J. Phys. Chem. Solids* **29**, 1309 (1968).
[206] Calculated from data of A. C. Bailey and B. Yates, *Proc. Phys. Soc.* **91**, 390 (1967) and D. Gerlich, *Phys. Rev. A* **136**, 1366 (1964).
[207] G. H. Winslow, *High Temp. Sci.* **3**, 361 (1972).
[208] Calculated from data of ref. 190 and S. I. Novikova, *Fiz. Tverd. Tela* **10**, 3141 (1968) [*Sov. Phys.—Solid State* **10**, 2481 (1968)].
[209] Calculated from data of ref. 191 and I. Ya. Dutchak and Y. P. Yamolyuk, *Izv. Vyssh. Ucheb. Zaved., Fiz.* no. 11, 146 (1973) [*Sov. Phys. J.* **16**, 1606 (1973)].
[210] Calculated from data of Novikova in ref. 208 and L. C. Davis, W. B. Whitten, and G. C. Danielson, *J. Phys. Chem. Solids* **28**, 439 (1967).
[211] G. A. Slack, *Phys. Rev.* **122**, 1451 (1961).
[212] L. S. Parfenjeva, I. A. Smirnov, and V. V. Tikhonov, *Fiz. Tverd. Tela* **13**, 1509 (1971) [*Sov. Phys.—Solid State* **13**, 1267 (1971)].
[213] B. M. Mogilevskii and V. F. Tumpurova, *Fiz. Tverd. Tela* **16**, 1786 (1974) [*Sov. Phys.—Solid State* **16**, 1161 (1974)].
[214] Y. S. Touloukian, R. W. Powell, C. Y. Ho, and P. G. Klemens, in "Thermophysical Properties of Matter," Vol. 2. Plenum, New York, 1970.
[215] J. J. Martin, *J. Phys. Chem. Solids* **33**, 1139 (1972).
[216] V. K. Zaitsev and E. N. Nikitin, *Fiz. Tverd. Tela* **12**, 357 (1970) [*Sov. Phys.—Solid State* **12**, 289 (1970)].

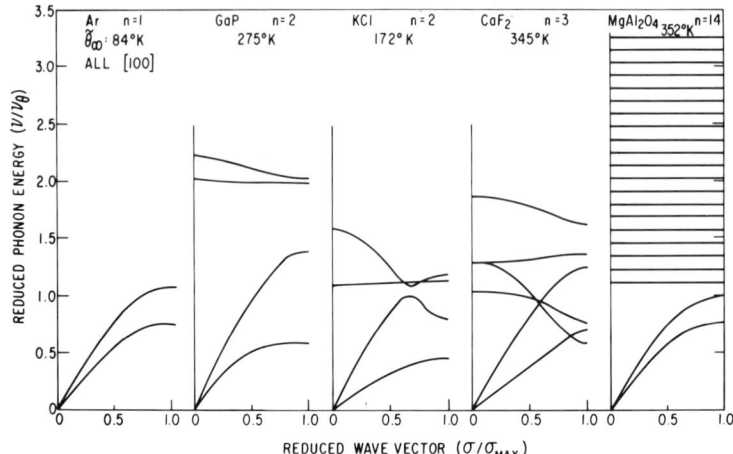

FIG. 6. The phonon dispersion curves in the [100] direction for several different crystals of increasing n value. The zone boundary occurs at $\sigma = \sigma_{\max}$. The phonon frequencies are scaled to ν_θ, where $h\nu_\theta = k\tilde{\theta}_\infty$. The results for Ar, GaP, KCl, and CaF$_2$ are taken from data in the literature. The results for MgAl$_2$O$_4$ are only schematic in order to show the increasing complexity of the phonon spectrum as n increases. Here all of the optic phonons are assumed to have zero group velocity.

We can calculate K_2' and a K_3' values for the $n = 3$ crystals in a manner analagous to Eq. (5.1) for the $n = 2$ crystals. This yields the following ratios at $T = \tilde{\theta}_\infty$:

$$[K_3'(\tilde{\theta}_\infty)/K_1'(\tilde{\theta}_\infty)] = (\tilde{\tilde{\theta}}_\infty/\tilde{\theta}_\infty)^{2+\epsilon} \tag{6.1}$$

where $\tilde{\tilde{\theta}}_\infty$ is obtained from Eq. (1.3) by integrating over the whole phonon spectrum, and is equal to the normally used value of θ_∞. Similarly,

$$[K_2'(\tilde{\theta}_\infty)/K_1'(\tilde{\theta}_\infty)] = (\tilde{\theta}_\infty/\tilde{\theta}_\infty)^{2+\epsilon} \tag{6.2}$$

where $\tilde{\theta}_\infty$ is obtained from Eq. (1.3) by integrating over the acoustic plus the first three optic branches. These calculations have been made for ThO$_2$ as an example using the $g(\nu)$ curves[199] for UO$_2$ with a small shift to account for the slight mass difference. The results are $\tilde{\theta}_\infty = 230°$K, $\tilde{\theta}_\infty = 455°$K, $\tilde{\tilde{\theta}}_\infty = 642°$K, and at $T = 230°$K one calculates $K_1' = 0.131$ W/cm, $K_2' = 1.23$ W/cm deg, and $K_3' = 3.80$ W/cm deg. The measured value at $230°$K is $K = 0.20$ W/cm deg, which is much closer to K_1' than to either of the others. Thus it appears that optic phonons are of very little importance in carrying heat in ThO$_2$ at $230°$K, in agreement with our general model. Note also that the thorium-to-oxygen mass ratio of 14.5 is also quite large. A similar calculation for CaF_2 where the mass ratio is only 2.5 yields essentially the same conclusion. In Table V and Fig. 7 we find

$$0.5 \leq [K(\tilde{\theta}_\infty)/K_1'(\tilde{\theta}_\infty)] \leq 1.5 \qquad (6.3)$$

and

$$1.0 \leq \epsilon \leq 1.3. \qquad (6.4)$$

Thus the agreement of K and K_1' for $n = 3$ is at least as good as for $n = 2$ [see Eqs. (3.5) and (4.1)].

Proceeding to crystals where $n \geq 4$, the data in Table VI have been collected. Here not all of the crystals are cubic, so that our understanding is less certain. However, for the noncubic crystals we have taken an experimental value of K averaged over the various crystallographic directions. The anisotropy in K is very small, i.e., less than 20%, for BeO, Al_2O_3, and β-boron; it is about 2 in crystals of SiO_2. In Table VI the $\tilde{\theta}_0$ values[198,217-223] are either from the literature or have been calculated from the elastic constants. The $\tilde{\theta}_\infty$ values have been estimated from the phonon dispersion curves[224-226] by using the following approximation. Assume that $g(\nu)$ is proportional to ν^2 over the whole zone for each phonon branch. Then sum $\int \nu^2 g(\nu) d\nu$ for the various branches in the high symmetry directions according to their degeneracy. For the a-axis in hexagonal structures this degeneracy is 3, while it is 1 for the c-axis. Then

$$(\tilde{\theta}_\infty)^2 = \frac{h^2}{k^2} \frac{(\Sigma \nu_B^5)}{(\Sigma \nu_B^3)} \qquad (6.5)$$

where ν_B is the phonon frequency at the boundary of the Brillouin zone.

[217] G. A. Slack and S. B. Austerman, *J. Appl. Phys.* **42**, 4713 (1971).
[218] O. L. Anderson, E. Schreiber, R. C. Liebermann, and N. Soga, *Rev. Geophys.* **6**, 491 (1968).
[219] W. R. Manning, O. Hunter, Jr., and B. R. Powell, Jr., *J. Am. Ceram. Soc.* **52**, 436 (1969).
[220] Z. P. Chang and G. R. Barsch, *J. Geophys. Res.* **78**, 2418 (1973); θ_0 calculated from the elastic constants.
[221] Estimated from β-boron $\theta_0 = 1370°K$.
[222] G. A. Slack and D. W. Oliver, *Phys. Rev. B* **4**, 592 (1971).
[223] I. M. Silvestrova, L. M. Belayev, Y. V. Pisarevski, and T. Niemyski, *Mat. Res. Bull.* **9**, 1101 (1974).
[224] M. A. Nusimovici, *Ann. Phys. (Paris)* **4**, 97 (1969).
[225] M. E. Striefler and G. R. Barsch, *Phys. Rev. B* **12**, 4553 (1975).
[226] W. Kappus, *Z. Phys. B* **21**, 325 (1975).

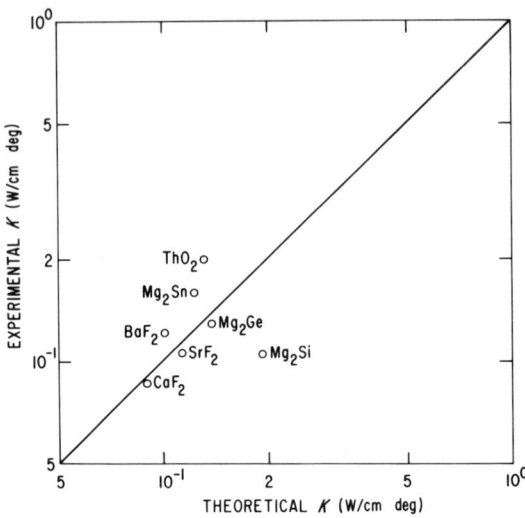

FIG. 7. Experimental versus theoretical values for the thermal conductivity of flourite structure crystals at $T = \bar{\theta}_\infty$. The theoretical values are K_1'. In all cases $n = 3$.

Equation (6.5) is only an approximation for the acoustic branches, but it gives $\bar{\theta}_\infty$ values in moderate agreement with the $\bar{\theta}_0$ values for BeO, SiO$_2$, and Al$_2$O$_3$. For the other crystals in Table VI no phonon dispersion curves have been measured or calculated so it has been assumed[227] that $\bar{\theta}_\infty$ is just equal to $\bar{\theta}_0$. The values[218,220,228–233] of γ_∞ are taken from thermal expansion data. The measured K values in Table VI have been collected from the literature.[179,194,197,217,222,234,235] The K versus K_1' plot is shown in Fig. 8.

[227] $\bar{\theta}_\infty$ is assumed to be equal to $\bar{\theta}_0$ because no data on the phonon dispersion curves are available.
[228] Calculated from the data of G. A. Slack and S. F. Bartram, *J. Appl. Phys.* **46**, 89 (1975) and C. F. Cline, H. L. Dunegan, and G. W. Henderson, *J. Appl. Phys.* **38**, 1944 (1967).
[229] D. A. Gerlich, *J. Phys. Chem. Solids* **31**, 1188 (1970).
[230] W. R. Manning and O. Hunter, Jr., *J. Am. Ceram. Soc.* **52**, 492 (1969).
[231] Assumed to be the same as for β-boron.
[232] Calculated from the data of S. Geller, G. P. Espinosa, and P. B. Crandall, *J. Appl. Cryst.* **2**, 86 (1969) and W. J. Alton and A. J. Barlow, *J. Appl. Phys.* **38**, 3023 (1967).
[233] Calculated from the thermal expansion data of E. Dupuy and L. Hackspill, *Compt. Rend.* **197**, 229 (1933); F. N. Tavadze, I. A. Bairamashvili, G. V. Tsagareishvili, and D. V. Khantadze, *Stud. Cercet. Met.* **10**, 49 (1965); and the elastic constants of ref. 223.
[234] See ref. 214, p. 182.
[235] P. H. Klein and W. J. Croft, *J. Appl. Phys.* **38**, 1603 (1967), and ref. 214, p. 240.

TABLE VI. CALCULATED AND OBSERVED VALUES FOR THE THERMAL CONDUCTIVITY OF COMPLEX CRYSTALS ($n \geq 4$) AT $T = \bar{\theta}_\infty$

Crystal	Structure	n	$\bar{\theta}_0$ (°K)	$\bar{\theta}_\infty$ (°K)	γ_∞	K_1' (W/cm deg)	K (W/cm deg)	ϵ	ν_{max} (THz)
BeO	Hex.	4	806 (217)	810 (224)	1.38 (228)	0.40	0.70 (217)	1.7	19.4
α-SiO$_2$	Hex.	9	275 (218)	180 (225)	0.70 (218)	0.20	0.13 (234)	1.2	3.9
α-Al$_2$O$_3$	Rhombo.	10	478 (218)	390 (226)	1.32 (229)	0.24	0.24 (179)	1.6	10.0
Y$_2$O$_3$	bcc	10	228 (219)	228 (227)	1.09 (230)	0.32	0.40 (235)	1.4	—
MgAl$_2$O$_4$	fcc	14	352 (220)	352 (227)	1.41 (220)	0.20	0.20 (194)	1.0	—
B$_{12}$As$_2$	Rhombo.	14	390 (221)	390 (227)	1.0 (231)	0.47	0.92 (197)	2.1	—
Y$_2$O$_3$	bcc	40	144 (219)	144 (227)	1.09 (230)	0.20	0.76 (235)	1.4	—
Y$_3$Al$_5$O$_{12}$	bcc	80	175 (222)	175 (227)	1.3 (232)	0.16	0.18 (222)	1.3	—
β-Boron	Rhombo.	105	290 (223)	290 (227)	1.0 (233)	0.25	0.27 (197)	1.9	—
YB$_{68}$	fcc	414	170 (198)	170 (227)	1.0 (231)	0.16	0.020 (197)	0.0	—

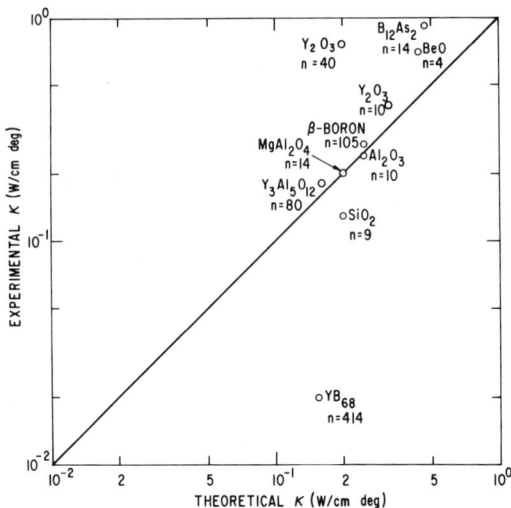

FIG. 8. Experimental versus theoretical values for the thermal conductivity of various crystals at $T = \tilde{\theta}_\infty$. The theoretical values are K_1'. In all cases $n \geq 4$.

An examination of Table VI and Fig. 8 shows that, except for YB_{68}, at $T = \tilde{\theta}_\infty$

$$0.6 \leq K/K_1' \leq 3.8 \tag{6.6}$$

and

$$1.0 \leq \epsilon \leq 2.1. \tag{6.7}$$

This agreement between K and K_1' is not quite as good as found previously. The main difficulty is a poor knowledge of the spectrum for the optic phonons. In SiO_2[225] and Al_2O_3[226] the spectra show that all of the optic modes have almost zero group velocity. Thus K_1' is a good approximation to K for these two crystals. In BeO[224] the lowest three optic modes have substantial group velocities. Thus, as in KCl and RbBr, we appear to have some heat conduction by optic modes in BeO. The mass ratio in BeO is 1.78, again rather small. So we expect $K > K_1'$ in BeO, as is found. For $B_{12}As_2$ and Y_2O_3 the phonon spectrum is not known, but optic phonon transport may be present. In the case of Y_2O_3 the primitive crystallographic unit cell has $n = 40$, the regular cubic unit cell with $a_0 = 10.604$ Å has 80 atoms. This 80-atom cell is actually composed of 8 unit cubes of $a_0 = 5.302$ Å with different spatial orientations. Thus each of these simpler units has 10 atoms. As shown in Table VI, if we assume that Y_2O_3 has 10 atoms per unit cell we obtain better agreement between K_1' and K than when we assume $n = 40$. So a knowledge of both the crystal structure and the phonon spectrum is needed in order to correctly estimate the thermal conductivity.

Note that in the calculation of K_1' for Table VI we have used Eq. (3.3), even though not all of the crystals have the fcc structure. This is somewhat justified by the observed agreement between K_1' and K. In reality the value of B should change somewhat with a change in crystal structure. This variation has been ignored.

7. GENERALIZATIONS

If it is assumed that none of the optic phonons carry heat, we can make the following generalizations about complex crystals with large n. For temperatures $T \geq \tilde{\theta}_\infty$ we can use Eq. (3.3) and a value of $\epsilon \cong 1$ to give

$$K_1'(T) = Bn^{1/3}\bar{M}\delta(\tilde{\theta}_\infty)^3 T^{-1}(\tilde{\gamma}_\infty)^{-2}. \tag{7.1}$$

We observe that, to a rough approximation, in Tables I–III, V, and VI

$$\tilde{\theta}_0 \cong \tilde{\theta}_\infty.$$

From Eq. (34)

$$(\tilde{\theta}_\infty)^3 = (\theta_0)^3 n^{-1}.$$

Then Eq. (7.1) can be written as

$$K_1'(T) = B\bar{M}\delta(\theta_0)^3 n^{-2/3} T^{-1}(\gamma)^{-2}. \tag{7.2}$$

The values of θ_0 and γ depend mostly on $\bar{M}$, δ, and on the interatomic bonding. They do not depend on n. Hence for a series of similar crystals that differ mainly in the complexity of their crystal structures, the K values at some fixed temperature will decrease with increasing n as $n^{-2/3}$. This is essentially the origin of the crystal complexity factor introduced by Oliver and Slack[236] to account for the decrease in K as the crystal structure becomes more complex. The $n^{-2/3}$ dependence was derived by Slack,[67] and has been mentioned by Roufosse and Klemens.[66] The decrease in K with increasing n was noticed by Eucken and Kuhn[174] in 1928, but no quantitative theory was given. It was elaborated on by Missenard,[237] who found empirically an $n^{-1/2}$ dependence of K.

An interesting series for studying this effect might be the one consisting of Ge, GaAs, ZnSe, CuBr, ZnGeAs$_2$, Cu$_2$GeSe$_3$, and

---

[236] D. W. Oliver and G. A. Slack, *J. Appl. Phys.* **37**, 1542 (1966).
[237] A. Missenard, "Conductivité Thermique des Solides, Liquides, Gaz et de Leurs Mélanges," ch. 1. II. Eyrolles, Paris, 1965.

Cu_3AsSe_4. These crystals have adamantine structures with $2 \leq n \leq 16$. Some K data exist[238] for these crystals, but not enough information is available for computing K'. Some phonon dispersion curves have been published[239] for $n = 8$. Other complex crystals for which K has been measured recently are fluorapatite[240] ($Ca_5P_3O_{12}F$, $n = 42$), topaz[241] ($Al_2SiO_4F_2$, $n = 36$), and sulfur[242] (S, $n = 32$). What happens at very large values of n? In the boron series we have $n^{2/3} = 55.5$ for YB_{68}. Thus the effect on K of increasing n can be as large as a factor of about 50. The extreme limit to increasing n is that for a glass where $n \to \infty$. If we take room temperature as a modestly high temperature (i.e., $> \tilde{\theta}_\infty$) for a complex crystal, and if we realize that the range of K values for nonmetallic solids at 300°K is about

$$10 \text{ W/cm°K} \geq K(300°K) \geq 10^{-2} \text{ W/cm°K},$$

then an upper limit on n can be set by

$$n^{2/3} \cong 10^3 \text{ or } n \cong 30{,}000.$$

This means that K will no longer depend on n when $n \geq 30{,}000$. For $\infty \geq n \geq 3 \times 10^4$ the material behaves essentially like a glass. At $n = 3 \times 10^4$ the crystal lattice constant will be approximately 100 Å. In YB_{68} one has $a_0 = 23.4$ Å. It would be interesting to find crystals with $a_0 \cong 100$ Å where the atoms are all well ordered in a periodic three-dimensional array. Some organic crystals may approach this. The K of complex organic crystals has been calculated in a somewhat different manner by Keyes[243] and Bondi.[244]

Another conclusion from the present analysis of complex crystals is that the important parameter is $\tilde{\theta}_\infty$, not θ_∞ or θ_0. From Eq. (3.4) we see that $\tilde{\theta} = \theta_0 n^{-1/3}$, and hence is much less than θ_0 for large n. The exponential increase of K with decreasing T that is characteristic of Umklapp processes[217] will not occur until T drops to $\sim(1/2)\tilde{\theta}$, which is $T \cong (1/2)\theta_0 n^{-1/3}$. For large n this Umklapp behavior may never be seen, as in the case of $Y_3Al_5O_{12}$ and other garnets[222] where $n = 80$. This feature

---

[238] D. P. Spitzer, *J. Phys. Chem. Solids* **31**, 19 (1970).

[239] A. S. Poplavnoi and V. G. Tjuterev, *J. Phys. (Colloq.) C* **3**, 169 (1975).

[240] R. H. Hopkins, D. H. Damon, P. Piotrowski, M. S. Walker, and J. H. Uphoff, *J. Appl. Phys.* **42**, 272 (1971).

[241] M. V. Romanova, I. A. Smirnov, and V. V. Tikhonov, *Fiz. Tverd. Tela* **13**, 1812 (1971) [*Sov. Phys.—Solid State* **13**, 1515 (1971)].

[242] G. A. Slack, *Phys. Rev.* **139**, A507 (1965).

[243] R. W. Keyes, *J. Chem. Phys.* **31**, 452 (1959); and "Organic Semiconductors," p. 123. (T. T. Brophy and T. W. Buttrey, eds.). Macmillan, New York, 1962.

[244] A. Bondi, *J. Appl. Phys.* **37**, 4648 (1966).

of the acoustic phonon branches was pointed out for silicon by Holland[110,245] where $n = 2$.

8. Longitudinal and Transverse Phonons

In the derivation of Eq. (3.3) it was assumed that the longitudinal and transverse acoustic phonons contribute to K in similar fashions and that there is no need to distinguish between them. It was assumed that their relaxation times are similar and, hence, their γ^2 values are similar in magnitude. It is possible to separate the contributions of the longitudinal and transverse phonons to $g(\nu)$ and, hence, define a transverse Debye temperature as

$$(\tilde{\theta}_{\infty T})^2 = \frac{5h^2}{3k^2} \frac{\int \nu^2 g_T(\nu) d\nu}{\int g_T(\nu) d\nu} \tag{8.1}$$

where $g_T(\nu)$ is the phonon density of states for transverse acoustic phonons. A similar definition of $\tilde{\theta}_{\infty L}$ can be made. Thus $\tilde{\theta}_\infty$ is just equal to

$$(\tilde{\theta}_\infty)^2 = [2(\tilde{\theta}_{\infty T})^2 + (\tilde{\theta}_{\infty L})^2]/3. \tag{8.2}$$

If it is assumed that the Grüneisen constants for T and L phonons are nearly identical, then the sum

$$K' = K_T' + K_L'$$

can be easily separated and the ratio of K_L' to K_T' becomes

$$(K_L'/K_T') = (1/2)(\tilde{\theta}_{\infty L}/\tilde{\theta}_{\infty T})^2. \tag{8.3}$$

If Eq. (8.3) is applied to KCl the result is $(K_L'/K_T') = 1.0$. For GaAs the result is about 3.0. Thus one concludes that the longitudinal and transverse acoustic phonons contribute about equally to the total thermal conductivity. Holland[245] was one of the first to consider this problem, and considered Si and Ge specifically. His ideas were elaborated on by Bhandari and Verma[246,247] and later by others.[105,107] A summary of the discussion about the relevant relaxation times was published in three letters.[248-250] The present state of knowledge is uncertain. Both phonon polarizations do contribute to the total heat

---

[245] M. G. Holland, *Phys. Rev.* **132**, 2461 (1963).
[246] C. M. Bhandari and G. S. Verma, *Phys. Rev.* **138**, A288 (1965).
[247] C. M. Bhandari and G. S. Verma, *Phys. Rev.* **140**, A2101 (1965).
[248] G. L. Guthrie, *Phys. Rev. B* **3**, 3573 (1971).
[249] G. S. Verma, C. M. Bhandari, and Y. P. Joshi, *Phys. Rev. B* **3**, 3574 (1971).
[250] M. G. Holland, *Phys. Rev. B* **3**, 3575 (1971).

transport; their relative percentages are unknown. Some recent work on this problem has been published by Logachev et al.[121,122] Equation (8.3) is clearly an oversimplification of the problem because the scattering mechanisms of longitudinal and transverse phonons are different.

III. Volume Dependence of the Thermal Conductivity

9. Hydrostatic Stress

In Eq. (3.3) it was shown how K depends on various crystal parameters for the temperature range $T \geq \tilde{\theta}_\infty$. In order to find the volume dependence of K we can differentiate K with respect to volume. If we apply a hydrostatic pressure, the crystal volume decreases and the Debye temperature increases. Hence we expect an increase in K with increasing pressure. The first experimental measurements of this effect in nonmetals were done by Bridgman.[251] He also measured glass, some rocks, and various nonmetallic liquids. The theory of the volume dependence of K has been studied by Lawson,[252] Rodionov,[253] Klemens,[254] Slack,[67] and others.[255–259] Let us use the following definitions for various derivatives at constant temperature:

$$g = -\left(\frac{\partial \ln K}{\partial \ln V}\right)_T \tag{9.1}$$

$$q = +\left(\frac{\partial \ln \gamma}{\partial \ln V}\right)_T \tag{9.2}$$

$$\gamma = -\left(\frac{\partial \ln \theta}{\partial \ln V}\right)_T. \tag{9.3}$$

The second and third equations are the common definitions of q and γ.

---

[251] P. W. Bridgman, "Physics of High Pressure," ch. 11. Macmillan, New York, 1931; see also *Proc. Amer. Acad. Art. Sci.* **59**, 141 (1923) and *Amer. J. Sci.* **7**, 81 (1924).
[252] A. W. Lawson, *J. Phys. Chem. Solids* **3**, 155 (1957).
[253] K. P. Rodionov, *Fiz. Metal. Metalloved.* **6**, 745 (1958) [*Phys. Metals Metallog. USSR* **6**(4), 160 (1958)].
[254] P. G. Klemens, *Proc. Ninth Conf. Thermal Conductivity*, 1969, (H. R. Shanks, ed.). U.S. Atomic Energy Commission, Conf. 691002, 1970.
[255] D. L. Mooney and R. G. Steg, *High Temp.-High Press.* **1**, 237 (1969).
[256] R. E. Barker and R. Y. S. Chen, *J. Chem. Phys.* **53**, 2616 (1970).
[257] L. Bohlin, *High Temp.-High Press.* **5**, 581 (1973).
[258] L. Bohlin and P. Andersson, *Solid State Commun.* **14**, 711 (1974).
[259] P. G. Klemens and D. J. Ecsedy, *in* "Phonon Scattering in Solids," (L. J. Challis, V. W. Rampton, and A. F. G. Wyatt, eds.). Plenum Press, New York, 1976.

Then in the temperature range $T \geq \tilde{\theta}_\infty$ where

$$K' = Bn^{1/3}\bar{M}\delta(\tilde{\theta}_\infty)^3/\gamma^2 T,$$

we have

$$g = 3\gamma + 2q - 1/3. \qquad (9.4)$$

Note that δ is proportional to $V^{1/3}$. This yields the $-1/3$ in Eq. (9.4). The factor B does not change with volume (or temperature).

Since Bridgman's work there have been a few other measurements of g. Wilkinson and Wilks[260] measured the K of solid ^{4}He for a series of different molar volumes. Since then there have been many measurements on solid helium. Measurements have also been made for solid neon.[30] However, the He and Ne data are for temperatures low compared to $\tilde{\theta}_\infty$. The data of Clayton and Batchelder[24] on solid argon are for $T \geq \tilde{\theta}_\infty$ as well as lower temperatures, and are pertinent to the present discussion. Other data exist in the range $T \geq \tilde{\theta}_\infty$ for several different solids.[256,261-268] Some data are for plastics,[256,264,268] some are for Bi[262] and Te.[265] The present interest is in those crystals that are cubic and have a simple crystal structure. These are KCl,[261,266] NaCl,[267] KBr,[261] and AgCl.[263,268]

The pressure dependence of K, which is generally the measured quantity, can be converted to the volume dependence of K by using

$$g = -\left(\frac{\partial \ln K}{\partial \ln V}\right)_T = B_T\left(\frac{\partial \ln K}{\partial P}\right)_T \qquad (9.5)$$

where B_T is the isothermal bulk modulus. For NaCl and KCl[269] these B_T values have been determined. For AgCl the elastic constants[270] have been used to compute B_T. The values of g so obtained are listed in Table VII as the experimentally measured ones. For Ar the value of g was determined[24] at 75°K at an average molar volume of 23.2 cm³/mole from a plot of $\ln K$ versus $\ln V_M$.

---

[260] K. R. Wilkinson and J. Wilks, *Proc. Phys. Soc. A* **64**, 89 (1951).
[261] D. S. Hughes and F. Sawin, *Phys. Rev.* **161**, 861 (1967).
[262] T. Rosander and G. Backstrom, *Phys. Lett. A* **29**, 517 (1969).
[263] T. Rosander and G. Backstrom, *Phys. Scripta* **1**, 269 (1970).
[264] P. Andersson and G. Backstrom, *High Temp.-High Press.* **4**, 101 (1972).
[265] Kh. I. Amirkhanov, Ya. B. Magomeddov, and S. N. Emirov, *Fiz. Tverd. Tela* **15**, 1512 (1973). [*Sov. Phys.—Solid State* **15**, 1015 (1973)].
[266] O. Alm and G. Backstrom, *J. Phys. Chem. Solids* **35**, 421 (1974).
[267] O. Alm and G. Backstrom, *High Temp.-High Press.* **7**, 235 (1975).
[268] P. Andersson and G. Backstrom, *Rev. Sci. Instrum.* **47**, 205 (1976).
[269] G. R. Barsch and Z. P. Chang, *Phys. Status Solidi* **19**, 139 (1967).
[270] W. Hidshaw, J. T. Lewis, and C. V. Briscoe, *Phys. Rev.* **163**, 876 (1967).

TABLE VII. THE VOLUME DEPENDENCE OF K FOR SEVERAL CUBIC CRYSTALS

Crystal	T (°K)	R^a	$B_T{}^b$	$K_0{}^c$	q	g^d theory	g^d meas.
Ar	75	—	—	~3	1.0 (272)	9.89	6 ± 2
NaCl	298	2.88 (267)	2.34 (269)	60	1.40 (273)	7.15	6.7 ± 0.7
KCl	298	3.30 (266)	1.74 (269)	66	1.77 (273)	7.56	5.7 ± 0.5
AgCl	300	2.53 (268)	4.20 (270)	7.9	2.14 (273)	9.65	10.6 ± 0.6

$^a R = [\partial(\ln K)/\partial P]_T$ as measured, in units of 10^{-11} cm^3/erg.

$^b B_T$ = Isothermal bulk modulus, in units of 10^{+11} erg/cm^3.

$^c K_0$ = Measured thermal conductivity at given temperature at zero pressure, in units of 10^{-3} W/cm deg.

^{d} The measured value is from Eq. (9.5), the uncertainty is mainly from R, and the theoretical value is from Eq. (9.4).

The computed value of g from Eq. (9.4) depends on q. For Ne it has been experimentally determined[27] that $q = 1$, and this value seems reasonable[271,272] for Ar. For the alkali halides values have been calculated[273-275] in the range $1 \leq q \leq 2$, and the values from Roberts and Ruppin[273] have been used. The q for AgCl has been taken to be equal to that of RbCl.[273]

From Table VII it can be seen that there is a fairly good agreement between the observed and calculated values of g. The experimental results of Hughes and Sawin[261] have been neglected because of the strange behavior of their K versus P curves and the very low values of K at zero pressure. Another calculation of g values has been made by Bohlin and Andersson.[258] They assume $q = 2\gamma$, which gives erroneously large q values, hence they obtain computed values of g that are too large.

10. Uniaxial Stress

So far we have dealt with the effects of hydrostatic pressure on K. For cubic crystals the lattice symmetry does not change under hydrostatic pressure unless an abrupt phase change occurs.[262] However, the symmetry does change under uniaxial stress. It appears that there is only one experiment that has measured these effects on K in the range $T \geq \tilde{\theta}_\infty$ where the stresses have been kept within the elastic limits of the crystal.

[271] G. L. Pollack, Rev. Mod. Phys. 36, 748 (1964), see p. 786.
[272] C. A. Swenson, J. Phys. Chem. Solids 29, 1337 (1968).
[273] R. W. Roberts and R. Ruppin, Phys. Rev. B 4, 2041 (1971).
[274] H. H. Demarest, Jr., J. Phys. Chem. Solids 35, 1393 (1974).
[275] R. R. Rao, J. Phys. Soc. Jap. 38, 1080 (1975).

This is the work of Smirnov and Moizhes[276] on alkali halides from 100°K to 376°K. In these crystals phonon–phonon scattering is dominant, and the effect of stress on K was very large. Increases in K were seen that were 100 times larger than the effect of comparable hydrostatic pressures. The authors believed that the allowed phonon–phonon scattering processes were decreased in number as the symmetry of the crystal was lowered. Such effects in pure crystals need much more work before a clear understanding is obtained.

IV. Temperature Dependence of the Thermal Conductivity

11. Thermal Expansion Effects

Clayton and Batchelder[24] showed that there are significant differences for Ar crystals when K is measured at constant volume rather than at constant pressure. Since most K measurements are made at constant pressure, it would be convenient to be able to convert the results to constant volume. If we consider the temperature range $T \geq \tilde{\theta}_\infty$, we have seen that the value of K, measured at constant pressure, is of the form

$$K_P = A/T^\epsilon \tag{11.1}$$

where A depends on pressure, but is independent of T. [See Eq. (2.4) for the definition of ϵ.] Now if we measure K at constant volume the result is

$$K_V = G/T^m \tag{11.2}$$

where G depends on volume, but is independent of T. If only acoustic three-phonon scattering processes are present in the crystal, then $m = 1$. Ranninger[277] has pointed out that if the crystal volume depends on temperature, then $\epsilon > m$. If we take derivatives of Eqs. (11.1) and (11.2), we obtain

$$\epsilon = m + gT \left(\frac{\partial \ln V}{\partial T}\right)_P ,$$

where the subscript P means constant pressure. If α is the linear thermal expansion coefficient, then $3\alpha = (\partial \ln V/\partial T)_P$ and

$$\epsilon = m + \eta_{th} = m + 3\alpha gT. \tag{11.3}$$

---

[276] I. A. Smirnov and B. Ya. Moizhes, *Fiz. Tverd. Tela* **5**, 1958 (1963) [*Sov. Phys.—Solid State* **5**, 1430 (1964)].

[277] J. Ranninger, *Phys. Rev.* **140**, A2031 (1965).

Here η_{th} is the contribution to ϵ from thermal expansion. Equation (11.3) is reminiscent of the expression for the ratio of the specific heat at constant pressure to the specific heat at constant volume, i.e.,

$$(C_p/C_V) = 1 + 3\alpha\gamma T.$$

Equation (11.3) states that the slope of the thermal conductivity curve measured at constant pressure is steeper than the slope measured at constant volume. This can be checked.

For the rare gas crystals the results are shown in Table VIII. For all of the crystals it has been assumed that $q = 1$ and $m = 1$. The α values have been taken from the recent literature.[25,46,278] The agreement between observed and calculated values of ϵ is very good. The trend in ϵ from Ar to Xe reflects the rapid change in α through the series. Since a value of $m = 1$ works well for all three crystals, it is possible to conclude that only three-phonon scattering processes are occurring in the range $T \geq \theta_\infty$. This is in agreement with Clayton and Batchelder[24] and contrary to the supposition of Krupskii and Manzhelii[34] that four-phonon processes were responsible for the rapid drop in K observed with increasing temperature. If it is assumed that g in Eq. (11.3) is independent of temperature, then ϵ versus T can be computed from the measured thermal expansion. Using the data of Perterson et al.[25] for α of Ar, the ϵ versus T curve in Fig. 9 was computed with $g = 9.89$. The experimental ϵ values for constant volume ($V_M = 22.14$ cm³/mole) were taken from Clayton and Batchelder,[24] those for constant pressure were from Krupskii and Manzhelii.[34] Notice that above 30°K, that is above about $\theta/2$, the calculated and measured ϵ values for constant pressure are in good agreement. The constant volume results are closer to $\epsilon \cong 1$.

We can now look at the rocksalt structure compounds in Table III. Values of ϵ for all of them can be computed from Eq. (11.3) using $m = 1$ and the known values of α[159] and g.[273] Some results are shown in Table IX. Of all of the crystals in Table III from LiH to SrO, the worst agreement in ϵ values is found for crystals with high mass ratios such as

TABLE VIII. THE TEMPERATURE DEPENDENCE OF K MEASURED AT CONSTANT PRESSURE FOR RARE GAS CRYSTALS COMPARED WITH THEORY

Crystal	T (°K)	V_M (cm³/mole)	γ_∞	α (10^{-4}/deg)	g (calc.)	ϵ calc.	ϵ obs.
Ar	71	24.03	2.74	5.77 (25)	9.89	2.22	2.0
Kr	60	28.01	2.84	3.01 (278)	10.19	1.55	1.5
Xe	54	35.40	2.65	1.90 (46)	9.62	1.30	1.3

[278] P. Korpiun and H. J. Coufal, *Phys. Status Solidi* **6**, 187 (1971).

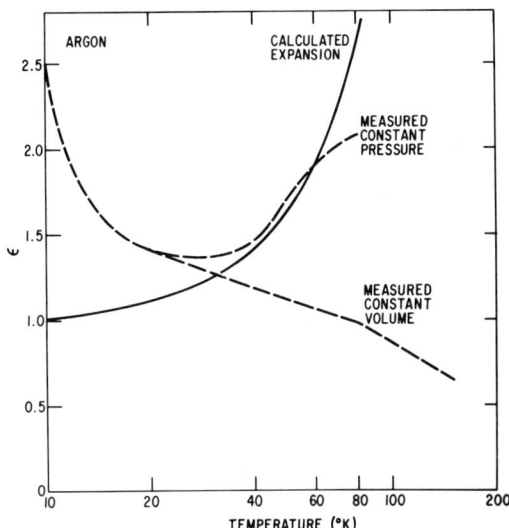

Fig. 9. The exponential factor, ϵ, versus temperature for Ar measured at constant volume and at constant pressure. The solid line shows the calculated value for constant pressure. Below 30°K the measured ϵ rises as the umklapp process die out exponentially with decreasing temperature. This dying is not included in the calculation.

KI and NaI. Generally the best agreement is for crystals like NaF, KCl, and RbBr where the mass ratio is close to unity. Plots of ϵ versus T for KCl and KI are shown in Figs. 10 and 11. For KCl the agreement is good for 50°K to 400°K. Above 400°K the photon radiation makes the measured ϵ values smaller than the calculated ones. The experimental ϵ values for KCl are taken from Rosenbaum et al.[177] on single crystals and Petrov et al.[124] on polycrystalline aggregates. For KI the measured ϵ values are from Walker[177] and Petrov et al.[124] For KI the measured ϵ value for $T \leq 340°K$ is always larger than that calculated from the thermal expansion. The same is true for NaI (see Table IX). In the case of the adamantine crystals in Table II, the thermal expansion with $m = 1$ in Eq. (11.3) does not appear to account for the measured ϵ values even at the highest temperatures. Perhaps some other mechanism is operating in these cases.

12. Scattering by Optic Phonons

One possible cause of the large measured values of ϵ is that $m > 1$. Originally Pomeranchuk[279] suggested that four-phonon processes are

[279] See ref. 5.

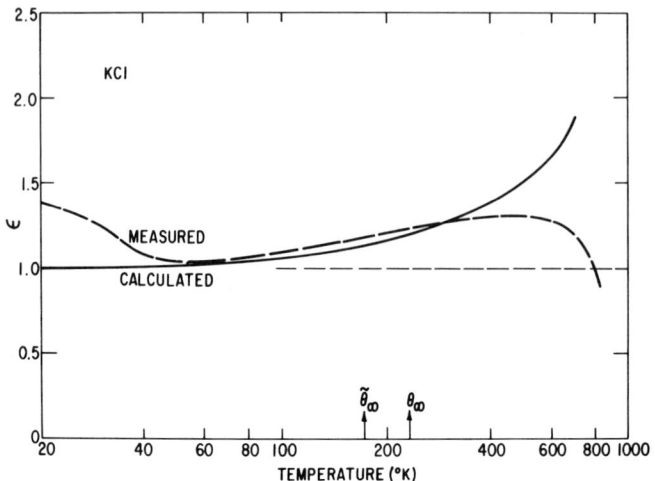

FIG. 10. The exponential factor, ϵ, versus temperature for KCl measured at a constant pressure of one atmosphere. The solid line shows the calculated value for constant pressure assuming only thermal expansion effects.

important and he showed that such processes could produce m values $1 \leq m \leq 2$. Ecsedy and Klemens[259,280] have shown that the strength of four-phonon processes is probably too weak compared to three-phonon processes to produce m values significantly greater than unity. They believe that three-phonon processes involving an optic mode are responsible for some of the $\epsilon > 1$ values observed. Some quantitative estimates of the effects of optic phonons have been given by Ziman,[117] Leroux-Hugon and Veyssie,[119] and Logachev and Vasilev.[123] It is possible to construct a simple model for the scattering of acoustic by optic phonons, and this is done as follows. Let us confine the argument to temperatures $T \geq \tilde{\theta}_\infty/2$. In this region the thermal resistivity, $1/K$, produced by three-phonon processes where acoustic phonons interact with each other is

$$W = 1/K = AT. \tag{12.1}$$

Now let us assume[117] that the crystal has $n = 2$ and has an optic phonon branch at a single frequency ν_{op}. This optic branch is essentially an Einstein oscillator, while the acoustic branch is represented by Debye oscillators with a whole spectrum of frequencies up to a maximum frequency of $\tilde{\theta}_\infty k/h$ (see Ziman[117]). Let us define a temperature $\theta_{op} = h\nu_{op}/k$. Then in this model $\theta_{op} > \tilde{\theta}_\infty$. The assumption is now made that

[280] D. J. Ecsedy and P. G. Klemens, *Bull. Am. Phys. Soc.* **20**, 356 (1975).

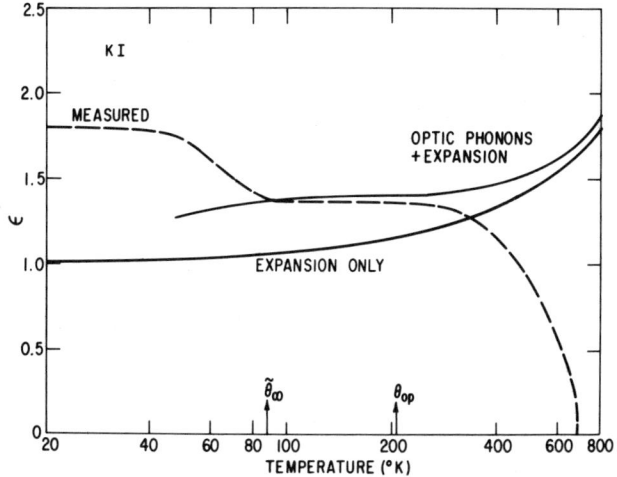

FIG. 11. The exponential factor, ϵ, versus temperature for KI measured at a constant pressure of one atmosphere. The solid lines show the calculated values when both thermal expansion and scattering of acoustic phonons by optic phonons are considered.

the thermal resistance offered to the acoustic phonons as carriers is that produced by the sum of the acoustic and optic scattering, and that the scattering is proportional to the average thermal energy present in each phonon branch. No energy is carried by the optic phonons in this model. Thus the two terms sum to

$$W \cong \frac{AT}{(1+S_{\rm op})}\left[1 + S_{\rm op}\left(\frac{x}{e^x - 1}\right)\right] \qquad (12.2)$$

where $x = h\nu_{\rm op}/hT$, and $S_{\rm op}$ = relative strength of the optic–acoustic

TABLE IX. THE TEMPERATURE DEPENDENCE OF K MEASURED AT CONSTANT PRESSURE FOR SOME ROCKSALT STRUCTURE CRYSTALS COMPARED WITH THEORY AT $T = \tilde{\theta}_\infty$

Crystal	$\tilde{\theta}_\infty$ (°K)	q	α (10^{-6}/deg)	g	ϵ calc.[a]	meas.
KF	235	1.57	30	7.37	1.16	≥1.0
KCl	172	1.77	32	7.56	1.12	1.21
KBr	117	1.37	32	6.76	1.08	1.21
KI	87	1.26	30	6.54	1.05	1.50
NaI	100	1.77	34	7.89	1.08	~2.0

[a] From Eq. (11.3) with $m = 1$.

phonon interaction. If $S_{op} = 1$, then both phonon branches have equal coupling coefficients to the acoustic phonons. In the limit of high temperatures as $x \to 0$ note that Eq. (12.2) reduces to Eq. (12.1) for any value of S_{op}. The logarithmic derivative of Eq. (12.2) is just

$$m = + \left(\frac{\partial \ln W}{\partial \ln T}\right) = 1 + \eta_{op} \qquad (12.3)$$

where η_{op} is the contribution to ϵ from the scattering of acoustic phonons by optic phonons. The value of η_{op} is given by

$$\eta_{op} = S_{op}Z(Z + x - 1)/(S_{op}Z + 1) \qquad (12.4)$$

where

$$x = (\theta_{op}/T) \quad \text{and} \quad Z = x/(e^x - 1).$$

Note that $\eta_{op} \to 0$ when $S_{op} \to 0$ and also when $T \to \infty$. A plot of η_{op} versus (T/θ_{op}) is shown in Fig. 12. The η_{op} versus T curve is rather broad, values of η_{op} within 5% of the maximum occur over the range $0.30 \leq (T/\theta_{op}) \leq 0.75$.

How can we apply Eq. (12.3) to specific crystals? For θ_{op} we shall take the maximum frequency in the optical spectrum, and shall restrict the discussion to crystals where $n = 2$. Furthermore, because we have made the assumption that no heat is carried by the optic phonons, we need a large gap between the acoustic and optic branches. In the alkali halides this means a large mass ratio such as found in NaI and KI. In the adamantine crystals we will require $\theta_{op} \geq 1.5\tilde{\theta}_\infty$. This covers all of the crystals in Table II except for diamond. Then for these crystals the

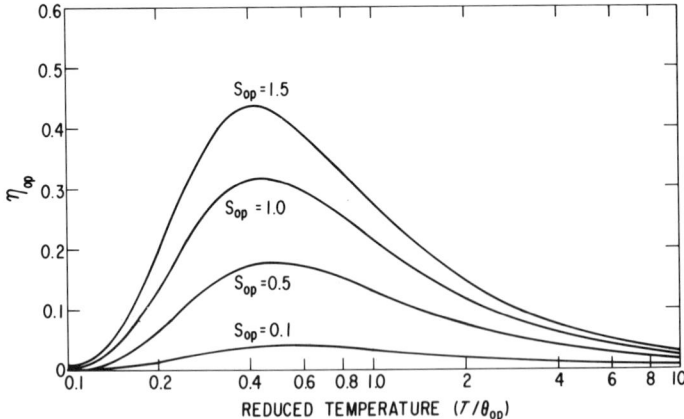

FIG. 12. The calculated value of the optic phonon scattering contribution, η_{op}, to ϵ plotted versus reduced temperature. The factor S_{op} gives the relative scattering strength of the optic compared to the acoustic phonons.

measured value of ϵ is

$$\epsilon = m + \eta_{th}$$

where $m = 1$ if only acoustic three-phonon processes are important. Thus

$$\epsilon = 1 + \eta_{th} + \eta_{op}. \quad (12.5)$$

We have applied Eq. (12.5) to KI, Si, and CdTe. The results are shown in Figs. 11, 13, and 14. The experimental ϵ values for Si are from Glassbrenner and Slack,[102] and for CdTe from Slack[112] and Horowitz and Wurst.[281] The α values for Si are from Slack and Bartram,[282] the q value from Rao.[275] For CdTe the α values are from Novikova[98] and later references.[283–285] A value of $q = 1.50$ was assumed for CdTe, this yields $g = 4.23$. For all three crystals it was assumed that $S_{op} = 1.2$.

In Figs. 11, 13, and 14 for $T \geq \tilde{\theta}_\infty$ the agreement between experiment and the theory from Eq. (12.5) is moderately good. If only Eq. (11.3) is used with $m = 1$ the agreement is rather poor. Note that in Fig. 11 for

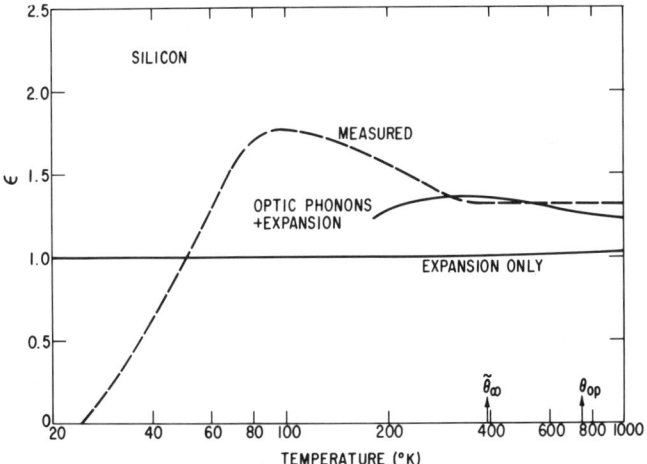

FIG. 13. The exponential factor, ϵ, versus temperature for Si. The calculated solid line shows that the dominant contribution to $\epsilon - 1$ above 300°K is from scattering by optic phonons. Below 300°K the value of ϵ is determined by umklapp, isotope, and boundary scattering effects.

[281] N. C. Horowitz and J. C. Wurst, *J. Am. Ceram. Soc.* **58**, 462 (1975).
[282] G. A. Slack and S. F. Bartram, *J. Appl. Phys.* **46**, 89 (1975).
[283] M. G. Williams, R. D. Tomlinson, and M. J. Hampshire, *Solid State Commun.* **7**, 1831 (1969).
[284] V. M. Glazov, S. N. Chizhevskaya, and S. B. Evgen'ev, *Zh. Fiz. Khim.* **43**, 373 (1969) [*Russ. J. Phys. Chem.* **43**, 201 (1969)].
[285] R. D. Greenough and S. B. Palmer, *J. Phys. D* **6**, 587 (1973).

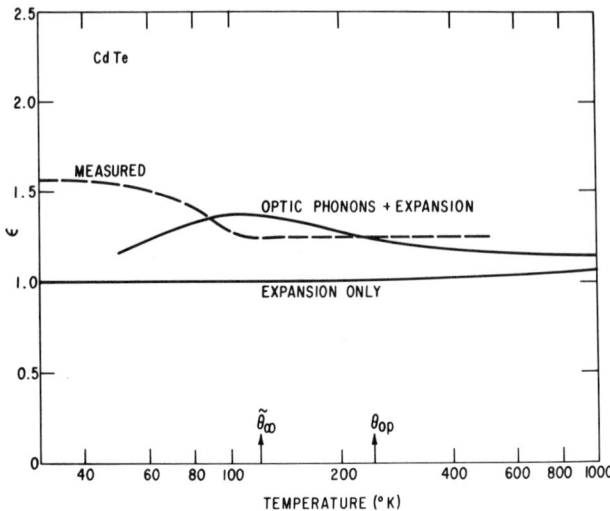

FIG. 14. The exponential factor, ϵ, versus temperature for CdTe. The calculated values are shown by the solid lines.

KI for $T > 300°K$ the internal photon radiation is causing ϵ to drop rapidly. This radiation is not present for Si and CdTe. Thus the present simple model, which assumes only three-phonon processes, thermal expansion, and scattering by optic phonons in special cases, can account for the temperature dependence of ϵ for $T \geq \tilde{\theta}_\infty$. For example, all of the ϵ values in Table II (except for HgSe and HgTe) can be accounted for with a range of $1.4 \geq S_{0p} \geq 0.6$. Further tests of the model used for Eq. (12.2) should be made in order to determine whether a more complex model is actually needed.

The calculation of ϵ versus T for crystals such as NaBr and KBr that possess only moderate separations between the acoustic and optic branches is more difficult because both heat transport and scattering may be produced by the optic phonons. The extension of Eq. (12.2) to cases where $n > 2$ is also difficult. However, from the observed ϵ values in Tables V and VI it does not seem that any new phenomena need be taken into account.

V. Thermal Conductivity at the Melting Point

13. Theory for Liquids

In previous sections we have shown how to understand the magnitude, volume, and temperature dependence of the thermal conductivity of nonmetallic crystals at temperatures above $\bar{\theta}_\infty$. We can extrapolate these results to the melting point of the solid. How does the thermal conductivity of the solid compare with that of the liquid at the melting point? Can the behavior of the liquid be related to that of the solid? This is the main question of this section. The discussion will be limited to those nonmetallic solids that are also nonmetallic in the liquid state. This restriction eliminates many semiconductors in Tables II and III because they become metallic on melting. In particular Si, Ge, GaSb, InSb, and PbTe have been studied[286] and found to possess this type of behavior.

Since phonons transport heat in nonmetallic crystals, it is worthwhile to consider the question of phonons in liquids.[287-289] Phonon-like excitations do exist at all frequencies for longitudinal waves and at high frequencies for transverse waves. Neutron scattering studies have verified some of these models and actual phonon dispersion curves for liquids have been measured.[291-293] Both the diffusive and vibratory modes[286,290,294] of a liquid contribute to the thermal conductivity as

$$K_{tot} = K_{diff} + K_{vib}. \tag{13.1}$$

For liquids near the melting point the density is sufficiently high so that the diffusive mode contribution is very small. Hence it will be neglected in the following discussion, and K_{vib} is the only important term. The "lattice" disorder in a liquid is generally sufficiently large so that the mean-free-path, l, of a phonon is of the order of a few intermolecular distances. Then the model of Debye[2] can be used to estimate K_L for a liquid assuming that l is fixed by the fluid density. That is, at constant volume

$$K_L = vl\rho C_v /3M \tag{13.2}$$

where v = velocity of sound, ρ = liquid density, C_v = specific heat capacity per mole at constant volume, and M = molecular weight. If we let $l = (M/\rho N_A)^{1/3}$, where N_A = Avogadro's number, we obtain

$$K_L = v\rho^{2/3} C_v /3 M^{2/3} N_A^{1/3}. \tag{13.3}$$

[286] A. R. Regel, I. A. Smirnov, and E. V. Shadrichev, *Phys. Status Solidi A* **5**, 13 (1971).
[287] S. J. Cocking, *Adv. Phys.* **16**, 189 (1967).
[288] P. A. Egelstaff, "An Introduction to the Liquid State," pp. 159, 169. Academic Press, London, 1967.
[289] P. Bratby, T. Gaskell, and N. H. March, *Phys. Chem. Liquids* **2**, 53 (1970).
[290] C. A. Croxton, "Introduction to Liquid State Physics," p. 57. Wiley, New York, 1975.
[291] D. G. Henshaw and A. D. B. Woods, *Phys. Rev.* **121**, 1266 (1961).
[292] K. Sköld and K. E. Larsson, *Phys. Rev.* **161**, 102 (1967).
[293] J. K. Wilmshurst and J. M. Bracker, in "Molten Salts," p. 291. (G. Mamantov, ed.). Decker, New York, 1969.
[294] E. McLaughlin, *Chem. Rev.* **64**, 390 (1964).

This equation, or a variant of it, has been used by Bridgman[251] and by Horrocks and McLaughlin[294,295] to explain the thermal conductivity of dense liquids. The relationship of Eq. (13.3) to other theoretical models for liquids is explained in several texts.[237,294,296,297] The variation in K_L with volume at constant temperature defined as

$$g_L \equiv -\left(\frac{\partial \ln K_L}{\partial \ln V}\right)_T \tag{13.4}$$

can be obtained from Eq. (13.3) as

$$g_L = -\left(\frac{\partial \ln v}{\partial \ln V}\right)_T + \frac{2}{3}.$$

This converts to the equation of Kamal and McLaughlin:[298]

$$g_L = \gamma_L + (1/3)$$

where $\tag{13.5}$

$$\gamma_L = -\left(\frac{\partial \ln \nu}{\partial \ln V}\right)_T.$$

Here ν is the vibration frequency of the molecules, and γ_L is the Grüneisen constant of the liquid. Equation (13.5) depends on the relationship

$$\left(\frac{\partial \ln \nu}{\partial \ln V}\right)_T = \left(\frac{\partial \ln v}{\partial \ln V}\right)_T - \frac{1}{3}.$$

If it is assumed that the γ_L can be evaluated for a liquid in the same way as for a solid, namely

$$\gamma_L = 3\alpha B_T M/\rho C_v \tag{13.6}$$

where 3α = volume thermal expansion coefficient of the liquid and B_T = isothermal bulk modulus of the liquid, then γ_L can be computed separately from g_L. One expects for dense liquids that $1 \leq \gamma_L \leq 3$, and is similar in magnitude to the γ values for solids.

---

[295] J. Horrocks and E. McLaughlin, *Trans. Faraday Soc.* **56**, 206 (1960).
[296] N. V. Tsederberg, "Thermal Conductivity of Gases and Liquids." M.I.T. Press, Cambridge, 1965.
[297] Y. S. Touloukian, P. E. Liley, and S. C. Saxena, *in* "Thermophysical Properties of Matter," Vol. 3. Plenum, New York, 1970.
[298] I. Kamal and E. McLaughlin, *Trans. Faraday Soc.* **60**, 809 (1964).

14. VOLUME DEPENDENCE FOR LIQUIDS

The relationship of Eq. (13.5) to Eq. (9.4) which gives g for the solid should be noted; clearly $g_L < g_s$.

If the value of g_L is known, then the measured thermal conductivity of a liquid as a function of density at constant temperature is given by

$$K_L \sim (\rho)^{g_L}. \tag{14.1}$$

When a solid melts the liquid and solid densities at the melting point, ρ_L and ρ_s, are generally different. What would K_L be at the melting point if no change in density had occurred? Define a fictitious thermal conductivity of the liquid, K_F, as

$$K_F = K_L(\rho_s/\rho_L)^{g_L}. \tag{14.2}$$

The question of interest is the relationship of K_F to K_s at the melting point.

It is expected that $K_F \leq K_s$ at the melting point because the mean-free-path, l, of the phonons in the solid is still temperature dependent and is being determined by three-phonon collisions. In the liquid the mean-free-path has a value comparable to the intermolecular distances. This difference in the mechanisms that determine l is primarily responsible for the difference between g_L and g_s. Some further reduction in K is caused by the fact that $v_s > v_L$. A further reduction in K is caused by some shifting in the phonon density-of-states to diffusive modes at low phonon energies and wavenumbers.[290]

The value of g_L for some liquids has been determined experimentally. Table X shows some of the results. Let us consider the case of liquid argon. The pressure dependence of K has been measured by Ziebland and Burton,[299] Ikenberry and Rice,[300] and Bailey and Kellner.[301] The pressure dependence of the liquid density has been measured by various authors.[302-305] From these data the volume dependence of K at fixed temperature can be determined. The result is that at high liquid densities over the range 1.35 g/cm³ $\leq \rho_L \leq$ 1.46 g/cm³ and over the temperature range 84°K $\leq T \leq$ 120°K the value of g_L is essentially constant at $g_L = 2.80 \pm 0.10$ for liquid argon. At lower liquid densities (and higher

---

[299] H. Ziebland and J. T. A. Burton, *Brit. J. Appl. Phys.* **9**, 52 (1958).
[300] L. D. Ikenberry and S. A. Rice, *J. Chem. Phys.* **39**, 1561 (1963).
[301] B. J. Bailey and K. Kellner, *Physica* **39**, 444 (1968).
[302] A. VanItterbeek, O. Verbeke, and K. Staes, *Physica* **29**, 742 (1963).
[303] W. vanWitzenburg and J. C. Stryland, *Can. J. Phys.* **46**, 811 (1968).
[304] R. K. Crawford and W. B. Daniels, *J. Chem. Phys.* **50**, 3171 (1969).
[305] W. B. Streett and L. A. K. Staveley, *J. Chem. Phys.* **50**, 2302 (1969).

TABLE X. THE VOLUME DEPENDENCE OF K FOR SOME LIQUIDS FROM EQ. (13.4)

Liquid	ρ_L (g/cm³)	g_L	References K_L	$\rho_{L'}$ ρ_s
³He	0.135	~0.7	(306)	(317, 318)
⁴He	0.183	1.15	(306)	(319, 320)
H₂	0.077	1.90	(307)	(307, 321, 322)
Ne	1.247	—	(308, 309)	(271, 323)
N₂	0.869	1.90	(299)	(324, 325)
Ar	1.425	2.80	(300, 301)	(302–305)
CH₄	0.452	2.85	(300)	(302, 325, 326)
Kr	2.455	2.60	(300)	(327, 328)
Xe	2.963	2.70	(300)	(40, 329–332)
NH₃	0.734	2.15	(310)	(310, 333)
H₂O	1.000	1.50	(311)	(334, 335)
Toluene	0.954	3.60	(312–314)	(336–338)
Benzene	0.895	3.60	(315, 316)	(336, 337, 339)

temperatures) g_L is smaller. For example, at $\rho_L = 1.19$ g/cm³ and $T = 120°$K its value is $g_L = 2.25$.

If it is assumed that $g_L = 2.80$ for argon liquid densities as high as $\rho_L = 1.621$ g/cm³, which is ρ_s at the triple point, then the K_F value at the triple point is given by

$$K_F(83.8°\text{K}) = 1.26 \left(\frac{1.621}{1.425}\right)^{2.80} \text{mW/cm deg}$$

$$K_F(83.8°\text{K}) = 1.81 \text{ mW/cm deg}.$$

In Fig. 15 is shown a plot of K_s[24] at $\rho_s = 1.621$ g/cm³ and K_L at $\rho_L = 1.425$ g/cm³ versus temperature from the data of Ikenberry and Rice.[300] The K_s curve varies as T^{-1}, whereas the K_L curve is nearly independent of temperature. The K_F versus T curve is shown as computed for a liquid density of 1.621 g/cm³. Note that K_s, if extrapolated, would meet K_F at a temperature of 1.10 times the triple point temperature. So the thermal conductivity of the fictious liquid at the triple point is very close to that of the solid, the ratio at 83.8°K for argon is

$$K_F/K_s = 0.86.$$

Many other liquids behave in the same way.

For a number of liquids values of g_L have been calculated from measurements of K_L under pressure[299–301,306–316] combined with data for ρ_L

[306] J. F. Kerrisk and W. E. Keller, *Phys. Rev.* **177**, 341 (1969).
[307] H. M. Roder and D. E. Diller, *J. Chem. Phys.* **52**, 5928 (1970).
[308] E. Löchterman, *Cryogenics* **3**, 44 (1963).
[309] L. Bewilogua and T. Yoshimura, *J. Low Temp. Phys.* **8**, 255 (1972).

versus pressure.[302-305,307,310,317-339] In some cases no κ_L versus pressure data exist, so g_L values must be estimated by comparison with liquids of a similar structure and melting point. Since g_L depends mainly on the liquid density and very little on temperature, the g_L values in Table X refer to liquid densities at the triple point or melting point and to temperatures in this region.

15. Volume Dependence for Disordered Solids

Note that the g_L values in Table X are considerably smaller than the g_s values in Table VII. In order to show that these small values are a

[310] D. P. Needham and H. Ziebland, *Int. J. Heat Mass Trans.* **8**, 1387 (1965).
[311] V. J. Castelli and E. M. Stanley, *J. Chem. Eng. Data* **19**, 8 (1974), and B. LeNeindre, P. Bury, R. Tufeu, and B. Vodar, *J. Chem. Eng. Data* **21**, 265 (1975).
[312] E. McLaughlin and J. F. T. Pittman, *Phil. Trans. Roy. Soc. (London)* **270**, 579 (1971).
[313] V. Z. Geller and G. V. Zaporozhan, *Izv. Vyssh. Ucheb. Zaved., Neft Gaz* **17**(10), 69 (1974).
[314] Yu. L. Rastorguev, B. A. Grigor'ev, and G. F. Bogatov, *Inzh.-Fiz. Zh.* **17**, 847 (1969) [*J. Eng. Phys.* **17**, 1370 (1969)].
[315] Yu. L. Rastorguev and V. V. Pugach, *Izv. Vyssh. Ucheb. Zaved., Neft Gaz* **13**(8), 69 (1970).
[316] T. S. Akhundov, *Izv. Vyssh. Ucheb. Zaved., Neft. Gaz* **17**(2), 78 (1974).
[317] R. H. Sherman and F. J. Edeskuty, *Ann. Phys. (N.Y.)* **9**, 522 (1960).
[318] E. R. Grilly, *J. Low Temp. Phys.* **4**, 615 (1971).
[319] P. H. Keesom, "Helium," p. 240. Elsevier, Amsterdam.
[320] C. A. Swenson, *Phys. Rev.* **79**, 626 (1950).
[321] R. F. Dwyer, G. A. Cook, O. E. Berwaldt, and H. E. Nevins, *J. Chem. Phys.* **43**, 801 (1965).
[322] H. M. Roder, L. A. Weber, and R. D. Goodwin, U.S. Nat. Bur. Stand., Monograph 94 (1965).
[323] J. C. Holste and C. A. Swenson, *J. Low Temp. Phys.* **18**, 477 (1975).
[324] F. Din, "Thermodynamic Functions of Gases," Vol. 3. Butterworths, London, 1961.
[325] V. M. Cheng, W. B. Daniels and R. K. Crawford, *Phys. Rev. B* **11**, 3972 (1975).
[326] R. D. Goodwin and R. Prydz, *J. Res. U.S. Nat. Bur. Stand., A* **76**, 81 (1972).
[327] W. B. Streett and L. A. K. Staveley, *J. Chem. Phys.* **55**, 2495 (1971).
[328] D. L. Losee and R. O. Simmons, *Phys. Rev.* **172**, 944 (1968).
[329] M. J. Terry, J. T. Lynch, M. Bunclark, K. R. Mansell, and L. A. K. Staveley, *J. Chem. Therm.* **1**, 413 (1969).
[330] W. B. Streett, L. S. Sagan, and L. A. K. Staveley, *J. Chem. Therm.* **5**, 633 (1973).
[331] J. C. Stryland, J. E. Crawford, and M. A. Mastoor, *Can. J. Phys.* **38**, 1546 (1960).
[332] J. R. Packard and C. A. Swenson, *J. Phys. Chem. Solids* **24**, 1405 (1963).
[333] V. G. Manzhelii and A. M. Tolkachev, *Fiz. Tverd. Tela* **5**, 3413 (1963) [*Sov. Phys.— Solid State* **5**, 2506 (1964)].
[334] R. W. Powell, *Adv. Phys.* **7**, 276 (1958).
[335] G. S. Kell and E. Whalley, *Phil. Trans. Roy. Soc. (London), Ser. A*, **258**, 565 (1965).
[336] G. Egloff, "Physical Constants of Hydrocarbons." Reinhold, New York, 1940.
[337] A. R. Ubbelhode, "Melting and Crystal Structure." Oxford Univ. Press, New York and London, 1965.
[338] See ref. 314.
[339] G. A. Holder and E. Whalley, *Trans. Faraday Soc.* **58**, 2095 (1952).

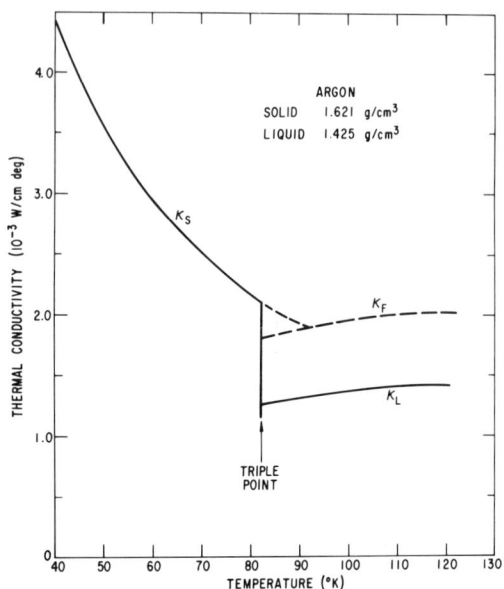

FIG. 15. The thermal conductivity versus temperature for solid and liquid argon at constant density in each phase. The solid at 1.621 g/cm³ is in equilibrium with the liquid at 1.425 g/cm³ at the triple point. The curve K_F is for a fictitious liquid equal in density to the solid.

consequence of a small mean-free-path for the phonons that is caused by disorder, it is instructive to look at g values for some glasses, plastics, and rocks. Table XI shows some data collected from the literature[251,340–342] for pressures near one atmosphere and temperatures near room temperature. For the clearly noncrystalline materials in Table XI, the range of g values is $0.64 \leq g_s \leq 2.58$; very similar to the range for liquids in Table X. The last three entries in Table XI are for solids that are impure or partially crystalline, and the g values are approaching those found in Table VII. Thus there is a distinct difference in g values for liquids as compared to pure, undoped nonmetallic solids. However, impure and partially crystalline solids form intermediate cases. Klemens[343] has discussed the case of impure crystals.

[340] S. W. Kieffer, I. C. Getting, and G. C. Kennedy, *J. Geophys. Res.* **81**, 3018 (1976).
[341] P. Andersson and B. Sundqvist, *J. Polym. Sci.* **13**, 243 (1975).
[342] P. Andersson, "Specific Heat, Thermal Conductivity, and Thermal Diffusivity of Solids at High Pressures," Ph.D. Thesis, Umeå University, Sweden, 1973, unpublished; see also ref. 258.
[343] See ref. 254.

TABLE XI. THE VOLUME DEPENDENCE OF K FOR SOME SOLIDS

Solid	T (°K)	ρ_s (g/cm³)	g_s
Quartz glass	313	2.21	0.65 (340)
Pyrex glass	303	2.234	1.25 (251)
Polystyrene	303	1.05	1.84 (341)
Isotactic polypropylene	303	0.90	1.85 (341)
Low-density polyethylene	303	0.93	2.16 (342)
Polymethylmethacrylate	303	1.18	2.43 (341)
Teflon	303	2.15	2.58 (342)
Basalt	303	2.924	3.02 (251)
Limestone	348	2.602	4.78 (251)
High-density polyethylene	303	0.97	4.96 (342)

16. GRÜNEISEN PARAMETER FOR LIQUIDS

The volume dependence of K_L has not been measured for all of the liquids of interest. Thus g_L values must be estimated by some other method. From Eqs. (13.5) and (13.6) one can use other thermodynamic properties to find g_L. In Table XII are some γ_L values computed from data in the literature.[305,327,330,334–348] For the rare-gas liquids Ar, Kr, and Xe, the agreement between g_L, as measured directly, and $\gamma_L + (1/3)$ is fairly good. Hence it seems permissible to use a value of $g_L = 2.80$ for Ne. For benzene the agreement between g_L and $\gamma_L + (1/3)$ is poor. Benzene is not a liquid composed of simple spheres for which the heat capacity is only due to the translational motion of the molecules. Hence the measured C_v is too large and γ_L is too small. So Eq. (13.5) is not generally applicable to liquids composed of complex molecules.

TABLE XII. RELATION OF g_L AND γ_L FOR SEVERAL LIQUIDS NEAR THE TRIPLE POINT

Liquid	g_L	$\gamma_L + 1/3$	γ_L	References
Ne	—	3.08	2.75	(344)
Ar	2.80	2.90	2.57	(305, 345)
Kr	2.60	3.17	2.84	(327, 346)
Xe	2.70	3.07	2.74	(330)
Benzene	3.60	1.58	1.25	(347, 348)

[344] C. Gladun, *Cryogenics* **6**, 27 (1966).
[345] J. Stephenson, *Can. J. Phys.* **53**, 1367 (1975).
[346] W. B. Streett, H. I. Ringermacher, and J. L. Burch, *J. Chem. Phys.* **57**, 3829 (1972).
[347] L. A. K. Staveley, W. I. Tupman, and K. R. Hart, *Trans. Faraday Soc.* **51**, 323 (1955).
[348] J. S. Burlew, *J. Am. Chem. Soc.* **62**, 696 (1940).

17. Solid–Liquid Transition

With the g_L values from Table X and $g_L = 2.80$ for Ne, the K_F values for a number of simple liquids at the liquid–solid transition can now be calculated. The results are shown in Table XIII. The ρ_L and ρ_s values are taken from the references in Table X. The K_L and K_s values have been collected from the literature. Some references[349-355] have not been mentioned previously. The K_F values have been computed from Eq. (14.2).

One of the striking results from Table XIII is the K_s to K_F ratio for ^{4}He. A plot of K_L, K_s, and K_F for ^{4}He is shown in Fig. 16. Both K_L and K_s are values for constant densities, these are $\rho_L = 0.1829$ g/cm^3 and $\rho_s = 0.1961$ g/cm^3. The liquid–solid transition takes place at a pressure of 68.5 atmospheres[320] at 1.9°K. The ^{4}He solid has a hexagonal-close-packed crystal structure and K_s is anisotropic. However the anisotropy has been measured and it is very small[350] close to the melting point. Hence the K_s curve in Fig. 16 is independent of orientation. Note that the extrapolated K_s line crosses K_F at a temperature of 1.8 times the melting temperature. This same factor for Ar in Fig. 15 is 1.1. The ^{4}He solid melts prematurely.

The measured g_L values for toluene and benzene together with some estimated values for other substances containing complex molecular groups are given in Table XIV. The solid and liquid density values have been measured previously.[336-338,356-357] The K_L and K_s values at the melting point are from the published literature.[312-316,357-365] The K_F

---

[349] B. Bertman, H. A. Fairbank, C. W. White, and M. J. Crooks, *Phys. Rev.* **142**, 74 (1966).
[350] R. Berman, C. R. Day, D. P. Goulder, and J. E. Vos, *J. Phys. C* **6**, 2119 (1973).
[351] R. W. Hill and B. Schneidmesser, *Z. Phys. Chem.* **16**, 257 (1958).
[352] L. A. Koloskova, I. N. Krupskii, V. G. Manzhelii, and B. Ya. Gorodilov, *Fiz. Tverd. Tela* **15**, 1913 (1973) [*Sov. Phys.—Solid State* **15**, 1278 (1973)].
[353] V. G. Manzhelii and I. N. Krupskii, *Fiz. Tverd. Tela* **10**, 284 (1968) [*Sov. Phys.—Solid State* **10**, 221 (1968)].
[354] I. N. Krupskii, V. G. Manzhelii, and L. A. Koloskova, *Phys. Status Solidi* **27**, 263 (1968).
[355] E. H. Ratcliffe, *Phil. Mag.* **7**, 1197 (1962).
[356] B. Meyer, "Elemental Sulfur." Interscience, New York, 1965.
[357] A. G. Turnbull, *Z. Phys. Chem. (Frankfurt)* **42**, 243 (1964).
[358] B. M. Mogilevskii and V. G. Surin, *Fiz. Tverd. Tela* **12**, 3118 (1970) [*Sov. Phys.—Solid State* **12**, 2522 (1971)].
[359] J. E. S. Venart, *J. Chem. Eng. Data* **10**, 239 (1965).
[360] A. Eucken and E. Schröder, *Ann. Phys. (Paris)* **36**, 609 (1939).
[361] A. F. Chudnovskii, B. M. Mogilevskii, and V. G. Surin, *Inzh.-Fiz. Zh.* **19**, 705 (1970) [*J. Eng. Phys.* **19**, 1295 (1970)].

TABLE XIII. COMPARISON OF K_L AND K_s AT THE MELTING POINT FOR SOME SIMPLE SUBSTANCES

Substance	T (°K)	P (atmos.)	g_L	ρ_L (gm/cm³)	ρ_L/ρ_s	K_L	K_F	K_s
						(10^{-3} W/cm deg)		
³He	1.82	68.5	~0.7	0.135	0.958	0.16 (306)	0.17	1.0 (349)[b]
⁴He	1.90	33.0	1.15	0.183	0.933	0.16 (306)	0.17	2.33 (349, 350)[c]
H₂ (para)	13.80[a]	0.070	1.90	0.077	0.889	0.84 (307)	1.05	4.4 (351)
Ne	24.54[a]	0.427	2.80	1.247	0.864	1.05 (308, 309)	1.58	3.1 (28)
N₂	63.15[a]	0.124	1.90	0.869	0.922	1.62 (299)	1.89	1.9 (352)
Ar	83.82[a]	0.679	2.80	1.425	0.879	1.26 (300, 301)	1.80	2.1 (34)
Ar	88	170.	2.80	1.435	0.885	1.27 (299)	1.79	1.78 (299)
Ar	91	290.	2.80	1.444	0.889	1.38 (300)	1.92	2.20 (300)
Ar	95.1	480.	2.80	1.466	0.892	1.39 (301)	1.91	2.07 (301)
CH₄	90.67[a]	0.115	2.85	0.452	0.925	2.19 (300)	2.73	3.00 (353)
	99	340	2.85	0.461	0.933	2.37 (300)	2.89	2.88 (300)
Kr	115.95[a]	0.722	2.60	2.455	0.877	0.92 (300)	1.29	1.40 (34)
	125.5	350.	2.60	2.507	0.878	1.00 (300)	1.40	1.30 (300)
Xe	161.4[a]	0.806	2.70	2.963	0.873	0.72 (300)	1.04	1.11 (34)
	170.3	240.	2.70	3.017	0.883	0.77 (300)	1.07	1.06 (300)
NH₃	195.4[a]	0.060	2.15	0.734	0.895	7.2 (310)	9.1	8.6 (354)
H₂O	273.2[a]	0.006	1.50	1.000	1.091	5.66 (311)	4.97	22.5 (355)

[a] Triple point.
[b] Body-centered cubic ³He.
[c] Hexagonal close-packed ⁴He.

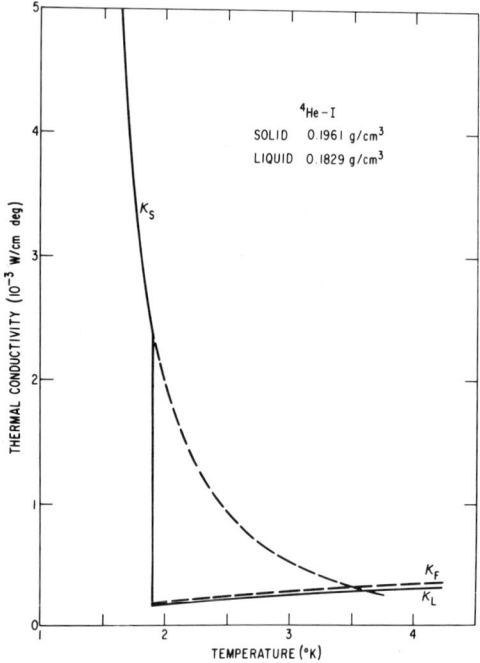

FIG. 16. The thermal conductivity versus temperature for solid and liquid helium under pressure. The pressure varies with temperature in such a way that the liquid and solid densities are held constant at the given values. The melting point, under pressure, is 1.9°K. The K_F curve is for a fictitious liquid, see text.

values are from Eq. (14.2). The K_s/K_F ratios from Tables XIII and XIV are plotted versus melting temperature in Fig. 17. All of the points except for ^{3}He and H_2O lie on a smooth curve. In ^{4}He, H_2, and Ne the solid melts before the phonon–phonon scattering reaches its maximum value. This "premature" melting is a quantum effect.[366] The substances with the larger values of the quantum parameter,[367] Λ^*, have the larger K_s/K_F ratios. For classical substances $\Lambda^* = 0$ and $K_s/K_F \cong 1$. The anomalous behavior of ^{3}He may be associated with the fact that it is a

[362] B. M. Mogilevskii, V. G. Surin, and A. F. Chudnovskii, *Fiz. Tverd. Tela* **12**, 1556 (1970) [*Sov. Phys.—Solid State* **12**, 1228 (1970)].
[363] R. Bachmann, *Wärme und Stoffübertragung* **2**, 129 (1969).
[364] B. M. Mogilevskii and V. G. Surin, *Fiz. Tverd. Tela* **13**, 2471 (1971) [*Sov. Phys.—Solid State* **13**, 2071 (1972)].
[365] A. Sugawara, *J. Appl. Phys.* **36**, 2375 (1965).
[366] See ref. 377, p. 52.
[367] J. DeBoer, *Physica* **14**, 139 (1948).

TABLE XIV. COMPARISON OF K_L AND K_s AT THE MELTING POINT AT ONE ATMOSPHERE PRESSURE FOR SOME COMPLEX SUBSTANCES

Substance	T (°K)	g_L	ρ_L (gm/cm³)	ρ_L/ρ_s	K_L (10⁻³ W/cm deg)	K_F	K_s
Toluene	178.2	3.60	0.954	0.899	1.59 (312–314)	2.34	2.24 (358)
Benzene	278.7	3.60	0.895	0.882	1.52 (315, 316, 359)	2.39	2.60 (360, 361)
Cyclohexane	279.8	2.00^a	0.791	0.951	1.29 (362)	1.43	1.38 (362)
Napthalene	353.2	3.60^a	0.978	0.857	1.31 (363)	2.28	2.31 (364)
White phosphorus	317.2	3.00^a	1.745	0.969	1.86 (365)	2.04	2.26 (365)
Orthorhombic sulfur	387	3.60^a	1.809	0.868	1.28 (357)	2.13	2.41 (357)

^{a} Assumed value.

Fermi liquid whereas all the others are Bose–Einstein liquids, or it may be that the reported K_s is less than the intrinsic value due to sample impurities or experimental difficulties. The anomalous behavior of H_2O is real, but not understood. The solid contracts on melting in contrast to all of the other substances studied. Hence $K_F < K_L$. This means that the density correction in Eq. (14.2) goes the wrong way on melting. The measured sound velocity of longitudinal waves in solid H_2O is 2.25 times

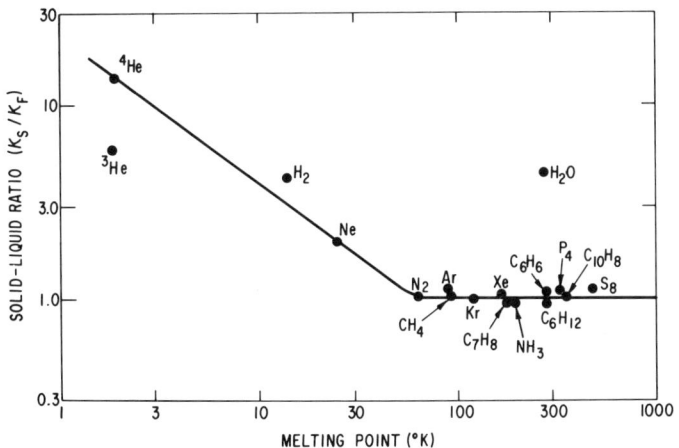

FIG. 17. The ratio of the thermal conductivity of the solid phase to that of the fictitious liquid phase of equal density at the melting point. The solid line indicates the trend of the data.

that in liquid water.[368] This factor explains part of the value of $K_s/K_L = 4.0$, but not all. Bridgman[369] pointed out that the value of K_L for H_2O is readily understandable in comparison with other liquids. Hence K_s at the melting point is abnormally large for H_2O.

The results in Fig. 17 for classical liquids can be extended to even higher temperatures by considering the results for rocksalt structure crystals. In Table XV are listed those substances for which both K_s and K_L at the melting point are available. The ρ_s and ρ_L values are taken from current literature,[370,371] the K_s values are generally extrapolated from the data[124,171] used in Table III, except for AgBr,[372] and are assumed to contain no contribution from photon heat transport. The K_L values have been collected from recent work.[373-376] In these measurements the problem of measuring K_L without error caused by photon heat transport is quite difficult and may not have been accomplished in all cases. The temperatures at the melting points are quite high and photon transport can be nonnegligible. The behavior of K_s and K_L at the melting point for KCl is shown in Fig. 1. For all of the rocksalt structure solids the value of K_F was computed from Eq. (14.2) using an assumed value of

TABLE XV. COMPARISON OF K_L AND K_s AT THE MELTING POINT AT ONE ATMOSPHERE PRESSURE FOR SOME SUBSTANCES CRYSTALLIZING WITH THE ROCKSALT STRUCTURE

Substance	T (°K)	ρ_L (gm/cm³)	ρ_L/ρ_s	K_L (10^{-3} W/cm deg)	K_F (10^{-3} W/cm deg)	K_s (10^{-3} W/cm deg)	K_s/K_F
NaF	1268	1.945	0.794	12.5 (373)	25	28 (124)	1.12
NaCl	1073	1.556	0.812	9.9 (373, 374)	18	12 (124, 171)	0.67
NaBr	1020	2.341	0.822	3.1 (375)	5.6	5.4 (124)	0.96
KCl	1043	1.527	0.851	8.0 (373, 374)	13	12 (124)	0.92
KBr	1007	2.127	0.860	2.9 (375)	4.6	4.7 (124)	1.02
KI	954	2.441	0.866	2.3 (375)	3.5	3.5 (124)	1.00
AgBr	707	5.575	0.924	2.8 (376)	3.5	4.1 (372)	1.17

[368] N. E. Dorsey, "Properties of Ordinary Water-Substance." Reinhold, New York, 1940.
[369] P. W. Bridgman, "Physics of High Pressures," p. 318. MacMillan, New York, 1931.
[370] M. Blander, "Molten Salt Chemistry," Interscience, New York, 1964.
[371] H. Schinke and F. Sauerwald, Z. Anorg. Allg. Chem. 287, 313 (1956).
[372] T. E. Popchapsky, J. Chem. Phys. 21, 1539 (1953).
[373] P. V. Polyakov and V. M. Mozhaev, Teplofiz. Vys. Temp. 13, 661 (1975) [High Temp. 13, 600 (1975)].
[374] V. I. Fedorov and V. I. Machuev, Teplofiz. Vys. Temp. 8, 912 (1970) [High Temp. 8, 858 (1970)].
[375] J. McDonald and H. T. Davis, Phys. Chem. Liquids 2, 119 (1971).
[376] K. Cornwell, J. Phys. D 4, 441 (1971).

$g_L = 3.00$. Neither experimental values of g_L nor the pressure dependence of K_L could be found in the literature. The best agreement between K_F and K_s was found by assuming that $g_L = 3.00$ for these rocksalt structure crystals.

18. Summary

From Fig. 17 and Table XV it is concluded that for classical liquids with melting temperatures between 60°K and 1268°K the ratio of the solid to liquid thermal conductivity at the melting point is approximately given by

$$\frac{K_s}{K_L} \cong \left(\frac{\rho_s}{\rho_L}\right)^{g_L} \qquad (18.1)$$

where g_L is defined by Eq. (13.4) and depends only on the properties of the liquid. Measured values of g_L fall in the range $1 \le g_L \le 4$.

A somewhat different analysis of the effect of melting on K has been given by Rao[377] and later by Keyes[378] and Mogilevskii et al.[379] This model involves knowing the entropy of fusion. The agreement between theory and experiment is either not very good[378] or else some unknown coefficients are used.[379] So this theory was not employed here.

VI. Minimum Thermal Conductivity

The minimum mean free path of a phonon in a crystal must be of the order of one phonon wavelength. Shorter mean free paths are meaningless if we wish to discuss the problem in phonon terminology. This problem has been discussed by Roufosse and Klemens.[64] What is the magnitude and temperature dependence of K if all phonons are so frequently scattered that their mean free path is this minimum value? How does this value of K compare with the measured value at the melting point? There are the questions dealt with in this section.

19. Rare-Gas Crystals ($n = 1$)

For crystals with $n = 1$ only acoustic phonons are present. A Debye model is used for the heat capacity and it is assumed that all of the phonons have the same group velocity v. For phonons of frequency ν

---

[377] M. R. Rao, *Indian J. Phys.* **16**, 155 (1942).
[378] R. W. Keyes, *J. Chem. Phys.* **31**, 452 (1959).
[379] B. M. Mogilevskii, V. G. Surin, and A. F. Chudnovskii, *Inzh.-Fiz. Zh.* **21**, 702 (1971) [*J. Eng. Phys.* **21**, 1297 (1971)].

the minimum mean free path is just

$$l = v/\nu. \quad (19.1)$$

In the limit of temperatures high compared to the Debye temperature the heat capacity per unit volume is just

$$C_v = 3k/\delta^3. \quad (19.2)$$

The expression for the minimum K_s equivalent to Eq. (13.2) in the high-temperature limit for these acoustic phonons is

$$K'_{SMINA\infty} = 3kv^2/2\delta^3\nu_A. \quad (19.3)$$

The prime denotes a theoretical value. Here the highest frequency for the acoustic phonons is

$$\nu_A = k\tilde{\theta}_\infty/h. \quad (19.4)$$

The temperature dependence of K'_{SMINA} is

$$[K'_{SMINA}/K'_{SMINA\infty}] = \frac{2}{x_A^2} \int_0^{x_A} \frac{x^3 e^x dx}{(e^x - 1)^2} \quad (19.5)$$

where $x = h\nu/kT$ and $x_A = h\nu_A/kT$. The transport integral in Eq. (19.5) has been tabulated.[380]

The value of $K'_{SMINA\infty}$ has been evaluated in Table XVI for the rare-gas crystals with $n = 1$. The average sound velocity for the solid at the triple point was evaluated from the measured elastic constants and density. The longitudinal and transverse velocities in the [100] direction were calculated from $\sqrt{C_{11}/\rho}$ and $\sqrt{C_{44}/\rho}$. The average, v, is

$$v = \frac{1}{3}\left[\sqrt{\frac{C_{11}}{\rho}} + 2\sqrt{\frac{C_{44}}{\rho}}\right]. \quad (19.6)$$

These v values[381-384] are very close to those measured[385-387] on polycrystalline samples. The values of ν_A were calculated from the θ_∞ values in Table I.

[380] W. M. Rogers and R. L. Powell, U. S. Nat. Bur. Stand. Circular 595, July 3, 1958.
[381] S. Gewurtz, H. Kiefte, D. Landheer, R. A. McLaren, and B. P. Stoicheff, *Phys. Rev. Lett.* **29**, 1454 (1972).
[382] S. Gewurtz and B. P. Stoicheff, *Phys. Rev. B* **10**, 3487 (1974).
[383] D. Landheer, H. E. Jackson, R. A. McLaren, and B. P. Stoicheff, *Phys. Rev. B* **13**, 888 (1976).
[384] W. S. Gornall and B. P. Stoicheff, *Phys. Rev. B* **4**, 4518 (1971).
[385] R. Balzer, D. S. Kupperman, and R. O. Simmons, *Phys. Rev. B* **4**, 3636 (1971).
[386] G. J. Keeler and D. N. Batchelder, *J. Phys. C* **3**, 510 (1970).
[387] D. S. Kupperman and R. O. Simmons, *J. Phys. C* **4**, L5 (1971).

TABLE XVI. THE MINIMUM THERMAL CONDUCTIVITY FOR RARE GAS CRYSTALS ($n = 1$) AT THE TRIPLE POINT

Crystal	T (°K)	δ (10^{-8} cm)	ν_A (THz)	v (10^5 cm/sec)	$K'_{\text{SMINA}\infty}$ (10^{-3} W/cm deg)
Ne	24.54	2.852	1.28	0.730 (381)	3.51
Ar	83.82	3.445	1.40	0.956 (382)	3.13
Kr	115.95	3.676	1.25	0.773 (383)	1.99
Xe	161.4	4.005	1.12	0.751 (384)	1.62

A comparison of $K'_{\text{SMINA}\infty}$ from Table XVI and the measured K_S at the triple point from Table XIII shows fairly good agreement, although the calculated $K'_{\text{SMINA}\infty}$ is about 50% larger than the measured K_S. If the phonon dispersion had been included in this model, the K' values would have been lower and the agreement probably better. However, the main conclusion is that the measured value at the triple point for all four rare-gas solids is very close to the minimum possible thermal conductivity. The phonon–phonon scattering is at its maximum effectiveness and no further reduction in K is possible.

20. ADAMANTINE AND ROCKSALT STRUCTURE CRYSTALS ($n = 2$)

The technique for calculating the minimum thermal conductivity for crystals with $n = 2$ must involve a scheme for handling the optic modes. The specific heat capacity at high temperatures for a general crystal with n atoms per primitive unit cell has $(1/n)$ of the total assignable to the three acoustic modes and $[(n - 1)/n]$ to the $3(n - 1)$ optic modes. Hence for the acoustic modes the minimum thermal conductivity, based on Eq. (19.3), is

$$K'_{\text{SMINA}\infty} = 3kv^2/2n\delta^3\nu_A. \tag{20.1}$$

For an optic mode of frequency ν_o the minimum mean-free-path is δ and the effective propagation velocity is a distance δ in the time period of one vibration; that is

$$v = \delta/\nu_o. \tag{20.2}$$

Thus, for a single set of three optic branches

$$K'_{\text{SMINO}\infty} = \frac{k\nu_o}{n\delta}.$$

For $n \geq 3$ we have to sum over all the optic branches and

$$K'_{\text{SMINO}\infty} = \frac{k}{n\delta} \sum_{i=1}^{n-1} \nu_{oi}. \tag{20.3}$$

The temperature dependence of K' can be approximated by assuming that the specific heat capacity can be represented by an Einstein oscillator of frequency ν_0. Thus

$$K'_{\text{SMINO}} = \frac{k}{n\delta} \sum_{i=1}^{n-1} \nu_{oi} \frac{x^2 e^x}{(e^x - 1)^2} \tag{20.4}$$

where $x = h\nu_{oi}/kT$ for each value of i.

For the crystals with $n = 2$ it has been assumed for simplicity that ν_0 is the highest frequency of the phonon spectrum measured at the zone center. It might be more accurate to take the frequency at the peak of the phonon-density-of-states for the optic branches, but this was not done. The values of ν_A were calculated using Eq. (19.4). Both ν_0 and $\bar{\theta}_\infty$ were taken from the references for the phonon dispersion curves used in Tables II and III. The only exception is AgBr where the recent work of Dorner et al.[388] was used. The average sound velocities were calculated from Eq. (19.6) by extrapolating the measured elastic constants[389-399] to the melting point. The values of δ were calculated from the molar volumes of the solids at their melting points. The total thermal conductivity at the melting point is the sum of

$$K'_{\text{SMIN}\infty} \text{ (total)} = K'_{\text{SMINA}\infty} + K'_{\text{SMINO}\infty} . \tag{20.5}$$

This value is given in the next-to-last column of Table XVII. The measured values of K_S for these crystals extrapolated to the melting point are given in the last column. Some values are taken from Table XV, the rest are taken from other literature sources.[102,112,281,400,401] Note

---

[388] B. Dorner, W. vonderOsten, and W. Bührer, *J. Phys. C* **9**, 723 (1976).
[389] S. P. Nikanorov, Yu. A. Burenkov, and A. V. Stepanov, *Fiz. Tverd. Tela* **13**, 3001 (1972) [*Sov. Phys.—Solid State* **13**, 2516 (1972)].
[390] Yu. A. Burenkov, S. P. Nikaronov, and A. V. Stepanov, *Fiz. Tverd. Tela* **12**, 2428 (1970) [*Sov. Phys.—Solid State* **12**, 1940 (1971)].
[391] R. F. Potter, *Phys. Rev.* **103**, 47 (1956).
[392] R. D. Greenough and S. B. Palmer, *J. Phys. D* **6**, 587 (1973).
[393] C. Susse, *Compt. Rend.* **247**, 1174 (1958).
[394] A. V. Sharko and A. A. Botaki, *Fiz. Tverd. Tela* **12**, 1559 (1970) [*Sov. Phys.—Solid State* **12**, 1232 (1970)].
[395] F. D. Enck, *Phys. Rev.* **119**, 1873 (1960).
[396] A. V. Sharko and A. A. Botaki, *Fiz. Tverd. Tela* **12**, 2247 (1970) [*Sov. Phys.—Solid State* **12**, 1796 (1971)].
[397] A. V. Sharko and A. A. Botaki, *Fiz. Tverd. Tela* **12**, 2702 (1970) [*Sov. Phys.—Solid State* **12**, 2171 (1971)].
[398] D. S. Tannhauser, L. J. Brunner, and A. W. Lawson, *Phys. Rev.* **102**, 1276 (1956).
[399] B. Houston, R. E. Strakna, and H. S. Belson, *J. Appl. Phys.* **39**, 3913 (1968).
[400] H. Wagini, *Z. Naturforsch., A* **19**, 1541 (1964).
[401] E. D. Devyatkova and I. A. Smirnov, *Fiz. Tverd. Tela* **3**, 2298 (1961) [*Sov. Phys.—Solid State* **3**, 1666 (1962)].

TABLE XVII. THE MINIMUM THERMAL CONDUCTIVITY AT THE MELTING POINT FOR CRYSTALS WITH $n = 2$

Crystal	δ (10^{-8} cm)	ν_A (THz)	ν_0 (THz)	v (10^5 cm/sec)	$K'_{SMIN A\infty}$	$K'_{SMIN 0\infty}$	K'_{SMIN} (total)	K_s	
						(10^{-3} W/cm deg)			
Si	2.73	8.23	15.3	6.36 (389)	25.01	3.87	28.88	139	(102)
Ge	2.85	4.90	9.00	3.75 (390)	12.87	2.18	15.05	100	(102)
InSb	3.25	2.81	5.71	2.53 (391)	6.88	1.21	8.09	57	(400)
CdTe	3.26	2.50	5.08	2.26 (392)	6.12	1.08	7.20	11	(112, 281)
LiF	2.09	10.4	19.7	4.21 (393)	19.35	6.51	25.86	27	(171)
NaBr	2.77	3.13	6.22	1.89 (394)	5.59	1.55	7.14	5.4	(124)
KCl	3.26	3.58	4.10	2.19 (395)	4.02	0.87	4.89	12	(124)
KBr	3.42	2.44	5.00	1.67 (396)	2.97	1.01	3.98	4.7	(124)
KI	3.66	1.81	4.3	1.37 (397)	2.19	0.81	3.10	3.5	(124)
AgBr	2.96	2.29	4.10	1.11 (398)	2.16	0.96	3.12	4.1	(372)
PbTe	3.29	2.19	3.4	1.39 (399)	2.57	0.71	3.28	4.8	(401)

that these values are for the lattice K; any free carrier or photon K has been subtracted. The case of KCl deserves some comment. The K' values in Table XVII for KCl were calculated by assuming that there was a distinct separation into acoustic and optic branches. This model does not work well for KCl because its mass ratio is nearly unity, as was pointed out in Section 5. The minimum K for KCl is calculated more accurately by using Eqs. (19.3) and (19.4) with $\theta_\infty = 232°K$ from Table IV. The result is

$$K'_{SMIN\infty} \text{ (total)} = 6.0 \times 10^{-3} \text{ W/cm deg}$$

for KCl. This is somewhat larger than the value given in Table XVII and is to be preferred. For crystals such as NaF and NaCl the behavior will be somewhere between that of KCl and KI. For KI the acoustic–optic phonon separation is substantial and Eq. (20.5) applies.

A comparison of the last two columns of Table XVII shows that for the rocksalt structure crystals the measured K_S at the melting point is 1 to 2 times the calculated minimum. For the adamantine crystals this ratio is 1.5 to 7 times the minimum. So, for at least some crystals, as the temperature is raised the phonon–phonon scattering has not yet reached a maximum "possible" rate when the crystal melts.

The minimum K produced by the optic phonons is given in the seventh column of Table XVII. This optic mode contribution is 15–30% of the total value in the last column. The small absolute values of this minimum optic mode contribution are the reason why it was possible to neglect the heat transport by optic phonons for most of the crystals in Sections 3–7. If the temperature dependence of this optic mode contribution as calculated from Eq. (20.4) is included, then at $T = \tilde{\theta}_\infty$ the minimum optic contribution is even less. For example, in Si at $T = \tilde{\theta}_\infty$ the values are

$$K'_1 = 930 \times 10^{-3} \quad \text{W/cm deg}$$
$$K'_{SMINO} = 2.9 \times 10^{-3} \quad \text{W/cm deg.} \tag{20.6}$$

The minimum optic contribution is 0.3% of the total. Many other crystals behave similarly.

How close to these calculated minimum K values have any of the measurements come, other than at the melting point? Studies on mixed-crystals of alkali halides[174,402,403] have not given values of K_S at all close to K'_{SMIN}. Studies on Ge–Si mixed crystals[61] at 300°K and above show

---

[402] W. S. Williams, *Phys. Rev.* **119**, 1021 (1960).
[403] B. D. Nathan, L. F. Low, and R. H. Tait, *Solid State Commun.* **19**, 615 (1976).

that even in the most heavily doped crystals of $Ge_{0.5}Si_{0.5}$ the measured K is about twice the calculated minimum at 300°K. Similar studies on mixed crystals of other elements with $PbTe$,[404,405] $InSb$,[406,407] and $CdTe$[408,409] show that the lowest thermal conductivities achieved are higher than the minimum values by factors of 2 or more. Clearly there is some room for improvement in these thermoelectric materials by producing further decreases in the thermal conductivity. It is conceivable that the thermal conductivity of an amorphous material might be close to the minimum value. The results reported[410] for amorphous Ge show K values much higher, a factor of about 9 times, than the calculated minimum. The reason for this large difference is not known. This discrepancy is particularly interesting with respect to the results on vitreous SiO_2, which are considered later.

21. Fluorite Structure Crystals ($n = 3$)

For these crystals we have three acoustic branches and six optic branches. We shall describe the three acoustic branches by a Debye temperature $\bar{\theta}_\infty$. This corresponds to an acoustic frequency given by Eq. (19.4). The optic branches will be approximated by two optic frequencies ν_{01} and ν_{02}, where ν_{01} is chosen to occur at some peak in the phonon density of states such that

$$\nu_A < \nu_{01} < \nu_{02},$$

and ν_{02} is taken, as before, to be the highest phonon frequency of any optic phonon. With this model we can calculate the minimum K for all of the crystals listed in Table V. However, the only crystal for which there are appreciable data at high temperatures is ThO_2. Some data for ZrO_2 are also available. The phonon spectrum for ThO_2 is assumed to be very similar to that of UO_2.[188] The phonon spectrum for cubic ZrO_2 is unknown, but the frequency ranges for the cubic, monoclinic, and tetragonal forms are all similar.[411] The elastic constants of ThO_2[412] and

---

[404] V. A. Saakyan and V. V. Tikhonov, *Fiz. Tverd. Tela* **10**, 920 (1968) [*Sov. Phys.—Solid State* **10**, 724 (1968)].

[405] G. T. Alekseeva, B. A. Efimova, L. M. Ostrovskaya, O. S. Serebryannikova, and M. I. Tyspin, *Fiz. Tekh. Poluprov.* **4**, 1322 (1970) [*Sov. Phys.—Semicond.* **4**, 1122 (1971)].

[406] I. Kudman, L. Ekstrom, and T. Seidel, *J. Appl. Phys.* **38**, 4641 (1967).

[407] A. G. Briggs, L. J. Challis, and F. W. Sheard, *J. Phys. C* **3**, 687 (1970).

[408] A. D. Stuckes and R. P. Chasmar, *J. Phys. Chem. Solids* **25**, 469 (1964).

[409] A. D. Stuckes and G. Farrell, *J. Phys. Chem. Solids* **25**, 477 (1964).

[410] P. Nath and K. L. Chopra, *Phys. Rev. B* **10**, 3412 (1974).

[411] C. M. Phillippi and K. S. Mazdiyasni, *J. Am. Ceram. Soc.* **54**, 254 (1971).

[412] P. M. Macedo, W. Capps, and J. B. Wachtman, Jr., *J. Am. Ceram. Soc.* **47**, 651 (1964).

UO_2[413] and Young's modules of ZrO_2[414,415] are used to estimate the average sound velocities. The derived values are given in Table XVIII. No attempt is made to estimate how these values in Table XVIII vary with temperature. The calculated minimum thermal conductivity values in the high temperature limit ($T \gg h\nu_{02}/k$) are also given in the table. Note that the total optic mode contribution is about 30% of the total at high temperatures.

The temperature dependence of the minimum thermal conductivity was calculated using Eqs. (19.5) and (20.4) and is shown in Figs. 18 and 19 as the dashed lines. The measured curve for pure polycrystalline ThO_2 is a composite of values taken from several sources[416–418] corrected for porosity to the values expected at theoretical density (9.98 g/cm^3). Note that above 1500°K the curve does appear to flatten and be limited by the minimum thermal conductivity up to 2600°K. The occurrence of this effect in ThO_2 depends on the large ratio of the melting temperature to $\bar{\theta}_\infty$, a factor of 15.5. This ratio is higher than for any of

TABLE XVIII. MINIMUM THERMAL CONDUCTIVITY FOR SOME OXIDE CRYSTALS WITH $n \geq 3$

	α-Al_2O_3	α-SiO_2	ThO_2	ZrO_2
δ (10^{-8} cm)	2.04	2.32	2.45	2.21
v (10^5 cm/sec)	7.9 (422)	4.4 (68)	3.90 (412)	4.20 (414, 415)
$\bar{\theta}_\infty$ (°K)	390.0 (226)	180.0 (225)	230.0 (199)	380.0
MP (°K)	2345.0	1995.0	3573.0	2973.0
ν_A (THz)	8.13	3.75	4.79	7.92
n	10.0	9.0	3.0	3.0
ν_{0max} (THz)	31.0	35.0	19.0	19.1
$K'_A{}^a$	18.7	9.5	14.9	14.2
$\Sigma K'_0{}^a$	10.6	10.0	6.0	6.4
K' (total)^a	29.3	19.5	20.9	20.6
Other ν_0 (THz)	10, 12, 13, 14, 15, 17, 20, 25, 31	5, 8, 12, 14, 21, 23, 33, 34	13.0	11.4

^{a} These are the minimum K values in the high-temperature limit. The total value is the sum of acoustic and optic contributions. The units are 10^{-3} W/cm deg.

[413] J. B. Wachtman, Jr., M. L. Wheat, H. J. Anderson, and J. L. Bates, *J. Nucl. Mater.* **16**, 39 (1965).
[414] F. H. Norton, "Refractories" (3rd ed.), p. 319. McGraw-Hill, New York, 1949.
[415] H. H. Sturhahn, W. Dawihl, and G. Thamerus, *Ber. Deut. Keram. Gesell.* **52**, 59 (1975).
[416] See ref. 214, p. 195.
[417] J. R. MacEwan and R. L. Stoute, *J. Am. Ceram. Soc.* **52**, 160 (1969).
[418] M. Faucher, F. Cabannes, A. M. Anthony, B. Piriou, and J. Simonato, *Rev. Int. Hautes Temp. Refract.* **7**, 290 (1970).

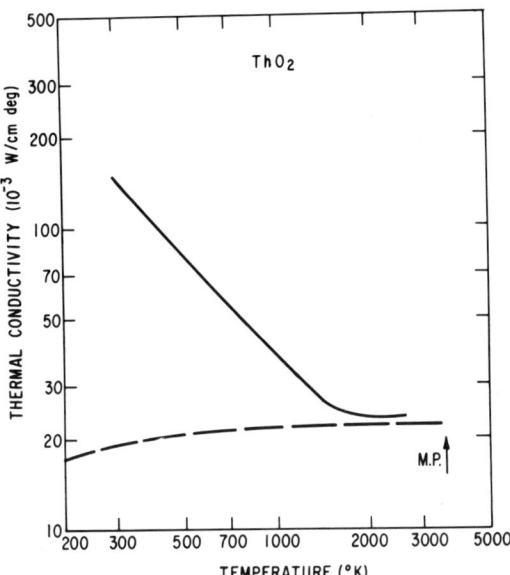

FIG. 18. The thermal conductivity versus temperature for ThO_2 up to the melting point. The measured value (—) approaches the calculated minimum (---) above 1700°K.

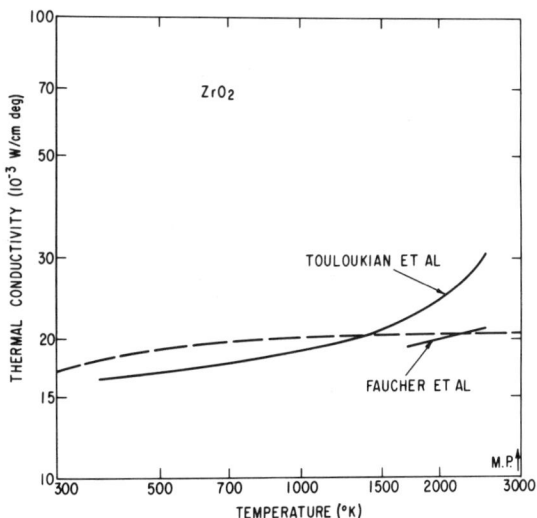

FIG. 19. The thermal conductivity versus temperature for ZrO_2 up to the melting point. The measured values (—) are close to the calculated minimum (---).

the other crystals in Tables I–V. Thus the flattening is not expected to occur in very many materials because they melt before the phonon-mean-free path reaches its minimum value. Apparently TiO_2 and $Si_2Al_6O_{13}$ exhibit similar behavior,[419] and Kingery has explained the observed temperature independent K as the occurrence of the minimum K. The same conclusion has been reached by Schatz and Simmons[420] for natural minerals at high temperatures.

The behavior of ZrO_2 is shown in Fig. 19. The measured K values were taken from several sources.[418,421] The values have been corrected to zero porosity. The upturn in K above 1500°K is believed to be caused by internal photon radiation, and the data of Faucher et al.[418] are probably a better measure of the behavior of the phonon contribution to the thermal conductivity. The data in Fig. 19 are for both pure ZrO_2 and for stabilized material that contains several percent of CaO, MgO or Y_2O_3. In either case the ZrO_2 sample may be a mixture of several different crystal forms of ZrO_2. Hence the phonon-impurity scattering can be very severe[419] and large enough to reduce the measured K to values very near to the minimum calculated ones. In view of the uncertainties in the parameters for ZrO_2 in Table XVIII, the agreement is considered quite satisfactory.

22. Other Crystals ($n \geq 4$)

As n increases the number of optic modes increases and the fraction of the minimum K produced by the optic modes should also increase. Let us consider α-Al_2O_3 and α-SiO_2 from Table VI. The quest is to calculate the minimum K from the known phonon spectra.[225,226] The relevant parameters are given in Table XVIII and are assumed to be independent of temperature. The sound velocities[68,422] are averages over all directions and polarizations in these noncubic crystals. The numerous optic modes are approximated by the several frequencies, $n - 1$ in number, given in Table XVIII. These approximate frequencies were taken from the calculated dispersion curves.[225,226] In the high temperature limit, Eqs. (20.1) and (20.3), the optic modes account for 36% of the total minimum K in α-Al_2O_3 and 51% in α-SiO_2. The temperature dependence of this minimum K has been calculated and is shown in Figs. 20 and 21.

---

[419] W. D. Kingery, *J. Am. Ceram. Soc.* **38**, 251 (1955).
[420] J. F. Schatz and G. Simmons, *J. Geophys. Res.* **77**, 6966 (1972).
[421] See ref. 214, p. 246.
[422] J. B. Wachtman, Jr., W. E. Tefft, D. G. Lam, and R. P. Stinchfield, *J. Res. U. S. Nat. Bur. Stand., A,* **64**, 213 (1960).

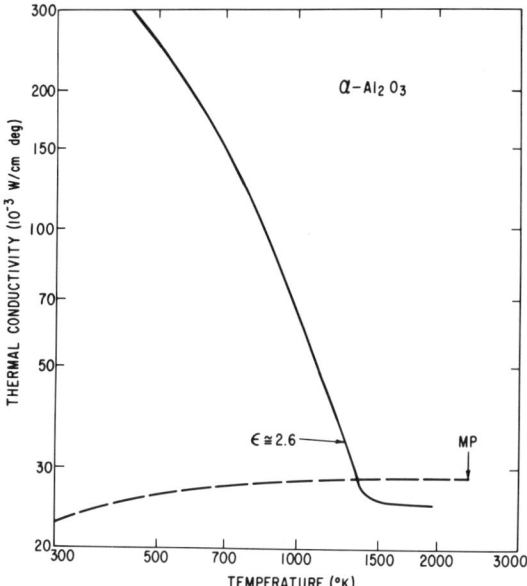

FIG. 20. The thermal conductivity versus temperature for $\alpha - Al_2O_3$ up to the melting point, showing both the measured (—) and the calculated minimum (---) value. The slope of curve at 1250°K is −2.6.

The experimental results for α-Al_2O_3 are from Touloukian et al.[423] up to 700°K. Above 700°K they are from Schatz and Simmons,[424] and are considerably less accurate but do represent the lattice K corrected for photon transport. The measured values above 1500°K are somewhat uncertain, but the values do decrease to the vicinity of the calculated minimum K. Better data are clearly needed. Note that if the measured curve is correct, the value of the slope ϵ at the highest temperatures is about 2.6. This large value may be a consequence of the large number of optic modes in this crystal.

The experimental results for crystal α-SiO_2 are shown in Fig. 21, and are an average of the two principal values from Touloukian et al.[425] Since α-SiO_2 is unstable above 856°K, no data are available at higher temperatures. However, SiO_2 does exist at a glass and the measured K

---

[423] See ref. 214, p. 97.
[424] J. F. Schatz and G. Simmons, *J. Appl. Phys.* **43**, 2586 (1972).
[425] See ref. 214, pp. 182, 183.
[426] J. H. McTaggart and G. A. Slack, *Cryogenics* **9**, 384 (1969).

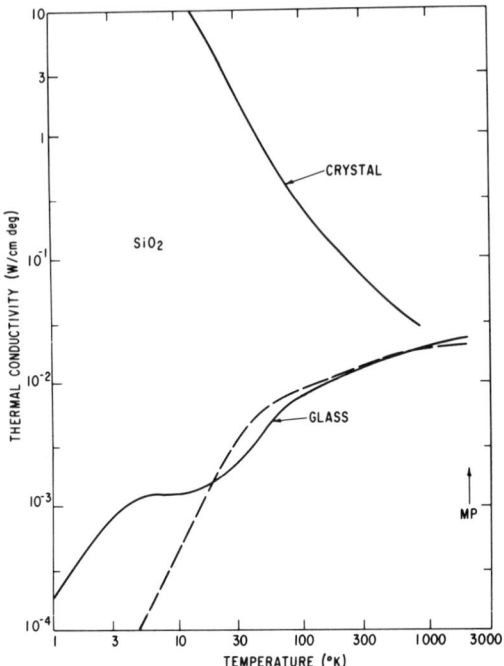

FIG. 21. The thermal conductivity versus temperature for crystalline $\alpha - SiO_2$ and SiO_2 glass. The values (—) for $\alpha - SiO_2$ are an average over the various crystallographic directions. The calculated minimum (---) thermal conductivity is close to that of the glass from 20°K to 2000°K.

of this glass[425-431] is shown in Fig. 21. The agreement between the calculated minimum K for α-SiO_2 using the parameters in Table XVIII and measured K of SiO_2 glass is remarkably good from 20°K to 2000°K. Again the solid curve in Fig. 21 represents the lattice thermal conductivity only, the photon component having been corrected for. The calculated temperature dependence from Eqs. (19.5) and (20.4) fits the measured curve quite well. Note that Eq. (19.5) gives $K \propto T^2$ at very low temperatures, not T^3. This T^2 temperature dependence is a direct consequence of the assumption that the minimum mean-free-path of a

[427] K. L. Wray and T. J. Connolly, *J. Appl. Phys.* **30**, 1702 (1959).
[428] A. G. Romashin, *Teplofiz. Vys. Temp.* **7**, 659 (1969) [*High Temp.* **7**, 604 (1969)].
[429] A. A. Men and A. Z. Chechelnitskii, *Teplofiz Vys. Temp.* **11**, 1309 (1973) [*High Temp.* **11**, 1176 (1973)].
[430] R. C. Zeller and R. O. Pohl, *Phys. Rev. B* **4**, 2029 (1971).
[431] M. P. Zaitlin and A. C. Anderson, *Phys. Rev. B* **12**, 4475 (1975).

phonon is equal to its wavelength. Below 1°K the measured[432] K actually varies as T^2, but is a factor of about 50 larger than the minimum K. The nature of K in this region has been explained by Walton[433] and Zaitlin and Anderson[434] using two different models. It is clear, however, from Fig. 21 and from Walton's calculations that the plateau below 20°K is caused by a rapid increase in the mean-free-path above the minimum value.

The nature of the K of glass was pointed out by Kittel[435] and Kingery[436] as that of a material where the mean-free-path is of the order of a few interatomic spacings. Figure 21 appears to be the first calculation of the absolute K of a glass over a wide temperature range with no adjustable parameters. Clearly a glass exhibits the minimum possible lattice thermal conductivity. Further calculations could be made for glasses[430] of GeO_2 and Se, but the general conclusions are expected to be the same.

23. LIQUIDS

In Section 22 above it was seen that vitreous SiO_2 has a K very close to the minimum. Since the disorder in a glass and a liquid are similar, do all liquids have K values close to the minimum? The experimental evidence with regard to this question is meager, but the answer appears, in general, to be no. If two liquids are mixed and the K of the homogeneous mixture is less than the average K expected from the constituents, then the phonon mean-free-path has been further reduced by the additional bonding and density fluctuations on the atomic scale. Such thermal conductivity data exist[296] for a number of organic liquid pairs. The maximum reduction of K found at room temperature is about 15%, and many pairs exhibit only a 2–3% reduction. Thus these pure liquids appear to be already close to the minimum K.

For the molten alkali-halides the K_L of the pure liquid can be substantially reduced by adding more scattering centers into the liquid. For example, KCl melts have been doped with[437,438] $MgCl_2$ and $PbCl_2$.

---

[432] J. C. Lasjaunias, A. Ravev, M. Vandorpe, and S. Hunklinger, *Solid State Commun.* **17**, 1045 (1975).
[433] D. Walton, *Solid State Commun.* **14**, 335 (1974).
[434] M. P. Zaitlin and A. C. Anderson, *Phys. Status Solidi B* **71**, 323 (1975).
[435] C. Kittel, *Phys. Rev.* **75**, 972 (1949).
[436] W. D. Kingery, *J. Am. Ceram. Soc.* **44**, 302 (1961).
[437] P. V. Polyakov and E. M. Gildebrandt, *Teplofiz. Vys. Temp.* **12**, 892 (1974) [*High Temp.* **12**, 780 (1974)].
[438] P. V. Polyakov, E. M. Gildebrandt, G. A. Kotelnikova, and V. N. Mozhaeva, *Zh. Fiz. Khim.* **47**, 2155 (1973) [*Russ. J. Phys. Chem.* **47**, 1221 (1973)].

The maximum measured reduction in the K_L of pure KCl appears to lie between 20% and 40%. This magnitude of reduction might have been expected from the K_S and $K'_{SMIN\infty}$ for solid KCl at the melting point in Table XVII. Other alkali-halides in the liquid state behave similarly.[373,376,439,440] Thus it is reasonable to conclude that the K of most pure liquids is not quite as low as the minimum value, and some further reduction in K can be produced by a variety of means.

VII. Further Problems

There are several questions concerning K at high temperatures that have not been dealt with to any great extent in this paper. The most extensive of these is the question of anisotropy. In special cases the anisotropy in K can become very large, for example, as observed in graphite[441] and boron nitride.[442] Just how the magnitude of the anisotropy depends on the other crystal parameters is not clear. There have been several attempts made to calculate the anisotropy. One of the earliest was that of Wooster,[443] one of the more recent is that of Benin.[444] Further work in this area is needed.

Associated with the problem of anisotropy is the effect of uniaxial stress[276] on K. This was mentioned in Section 10. It seems that more experimental work on this problem is needed.

The question of the scattering of acoustic phonons by optic phonons has been treated in Section 12 with a very simple model for crystals with $n = 2$. A more sophisticated model is desirable, and in particular one which can be applied to crystals for which $n \geq 3$. The explanation of the results for $\alpha\text{-}Al_2O_3$ shown in Fig. 20 where at high temperatures $\epsilon \simeq 2.6$ would be particularly interesting.

The minimum value of K for a solid is of considerable technological importance in designing thermoelectric refrigerators and generators. It would be interesting to be able to produce semiconductor crystals whose K approaches this minimum at temperatures near room temperature. From Fig. 21 it is clear that this can be accomplished by turning the crystal into a glass. From Section 7 it would appear that a crystal with $n > 30,000$ would behave similarly. What other phonon scattering mecha-

---

[439] A. G. Turnbull, *Aust. J. Appl. Sci.* **12**, 30 (1961).
[440] C. T. Ewing, J. R. Spann, and R. R. Miller, *J. Chem. Eng. Data* **7**, 246 (1962).
[441] G. A. Slack, *Phys. Rev.* **127**, 694 (1962).
[442] A. Simpson and A. D. Stuckes, *J. Phys. C* **4**, 1710 (1971).
[443] W. A. Wooster, *Z. Kristallogr.* **95**, 138 (1936).
[444] D. Benin, *Phys. Rev. A* **7**, 334 (1973).

nisms might product this type of result? Any such mechanism would have to be quite strong over a large part of the phonon frequency spectrum. There are no obvious candidates at present.

Acknowledgments

I wish to thank C. T. Walker and R. Berman for many helpful discussions concerning the topic of this review and for encouraging me to undertake and complete the task of writing. Thanks also go to P. G. Klemens, T. F. McNelly, R. O. Pohl, and D. Walton for their insights into and comments on some of the topics presented.

Configurational Thermodynamics of Solid Solutions

D. de Fontaine

*School of Engineering and Applied Science, University of California, Los Angeles, California**

I.	Introduction	74
II.	State of Order	75
	1. Site Occupation	75
	2. Short-Range and Long-Range Order	77
	3. Concentration Waves	80
	4. Replacive Scattering	82
	5. Discussion of "State of Order"	86
III.	Internal Energy	88
	6. Pair Potentials	90
	7. Singularities of $V(h)$	99
	8. Symmetry of $V(h)$	104
	9. Continuum Gradient Expansion	112
	10. Discussion of Internal Energy	114
IV.	Elastic Interactions	116
	11. Microscopic Elastic Theory	117
	12. Continuum Elastic Theory	124
	13. Particle Interactions	130
	14. Displacive Scattering	138
	15. Discussion of Elastic Interactions	143
V.	Free Energy Models	146
	16. Configurational Free Energy	146
	17. Generalized Bragg–Williams Model	148
	18. Taylor and Fourier Expansions	154
	19. Landau–Lifshitz Theory	157
	20. Cluster Variation Method	164
	21. Discussion of Free Energy Models	175
VI.	Stable and Unstable States	179
	22. Ordered Ground States	180
	23. Coherent Phase Diagrams	194
	24. Stability Analysis	210

* Present address: Department of Materials Science and Mineral Engineering, University of California, Berkeley, California.

VII.	Fluctuations and Kinetics	231
	25. Local Atomic Arrangements	232
	26. Kinetic Effects	251
VIII.	Conclusion	272

I. Introduction

All solid phases are solutions since impurities will always be present in the form of either substitutional or interstitial solute. Even in "pure" crystals whose solute content is immeasurably small, vacancies, considered as solute impurities, must be present at all finite temperatures. The study of solid solutions, both experimental and theoretical, thus constitutes an essential and very broad field of solid state physics, and one which has developed significantly within the last decade or so.

The purpose of this article is to review current theoretical models applicable to concentrated solid solutions. The emphasis is on concentrated solutions because dilute ones can be considered as special cases and because the interesting effects of clustering and ordering are most apparent at high solute concentrations. Extended ranges of solubility are often found in metallic systems, and for this reason most examples selected will refer to metals and alloys, although the models presented could equally be applied to off-stoichiometric compounds, metallic or inorganic, and to amorphous solids. Wide solid solution fields are particularly common in crystals whose average lattice is either face-centered or body-centered cubic, hence most examples presented will refer to these lattice structures. To simplify the algebra, only substitutional solid solutions will be described in detail. This is not as restrictive as it may appear, since interstitial solutions can be considered as substitutional solutions of filled sites on otherwise empty sublattices. Furthermore, defects such as split interstitials, for example, can be represented as substitutional defects with tetragonal strain fields. More generally yet, certain displacive phase transformations can be modeled by classical order-disorder theories applied to "solid solutions" of displacive defects among undisplaced lattice sites. Likewise, magnetic transitions can be regarded as taking place in "spin solutions."

The present treatment is limited to those transformations and phase changes which conserve the basic lattice framework of the crystalline solid solution. We call these *coherent* phase transformations. Replacement of an atom or defect type by another on a lattice site is allowed, along with displacements from the average lattice positions. Discontinuous changes of lattice framework, hence dislocations, incoherent interfaces, free surfaces, and grain boundaries will not be treated here. Note, however, that coherent transformations often play an important role in phase transformations for which the end products are *incoherent* with the initial state. This is because the first products of transformations occurring far from equilibrium are generally coherent phases, and although coherency is lost at later stages, the final microstructure retains, as it were, the morphology of the initial coherent products.

In the study of solid solutions, one is concerned first of all with the description of the state of order of the system through suitable averaging procedures. Section II is devoted to this topic. Internal energy of solutions, from both electronic and elastic standpoints, is treated in Sections III and IV, respectively. In Section V, free energy models are presented: generalized Bragg–Williams model, Landau theory, cluster variation method. These models are applied to the study of phase equilibrium (ground states, phase diagrams) in Section VI, and to the study of fluctuations and transformation kinetics in Section VII. It will become apparent as each topic is brought up that two basic modes of description are available: the *wave method* and the *cluster method,* the advantages and disadvantages of which are further discussed in the Conclusion (Section VIII).

II. State of Order

The distribution of defects on lattice sites is generally not random, i.e., some correlation is usually present. In this section, we examine how correlations are defined, how various averaging techniques lead to the concepts of short-range and long-range order, and how correlations are related to scattering of radiation. The useful concept of concentration waves is also introduced.

1. Site Occupation

Let there be n types of "defects" which may be solute atoms, distortion centers, local displacements, magnetic spins, etc. An average lattice is defined in such a way that each defect or host atom may be associated uniquely with a lattice point. A lattice vector will be denoted by

$$\mathbf{x}(p) = p_\alpha \mathbf{a}_\alpha \quad (\alpha = 1, 2, 3; \text{ summation implied}) \quad (1.1)$$

where $\mathbf{a}_\alpha$ are lattice translation vectors and p_α are integers, or half-integers if the conventional rather than primitive Bravais lattice cell is used.

Site occupation operators are defined by

$$\sigma_i(p) = \begin{cases} 1 & \text{if defect of type } i \text{ is at site } (p), \\ 0 & \text{if defect of type } j \neq i \text{ is at site } (p), \end{cases} \quad (1.2)$$

with

$$i = 0, 1, 2, \ldots, n, \quad (1.3)$$

the subscript 0 being reserved for the "host." The crystal average of σ_i, or concentration of defect of type i is defined as

$$\bar{c}_i = \frac{1}{N} \sum_p \sigma_i(p), \quad (1.4)$$

where the sum extends over all N lattice points (p) of the crystal or portion of crystal considered. Deviations from the mean occupation are given by

$$\gamma_i(p) = \sigma_i(p) - \bar{c}_i = \begin{cases} 1 - \bar{c}_i \\ -\bar{c}_i \end{cases}. \quad (1.5)$$

We have the following conservation relations, or sum rules:

$$\sum_{i=0}^{n} \sigma_i(p) = \sum_{i=0}^{n} \bar{c}_i = 1, \quad (1.6)$$

$$\sum_{i=0}^{n} \gamma_i(p) = \sum_p \gamma_i(p) = 0. \quad (1.7)$$

In an $(n + 1)$-component solution, there are thus n independent concentrations.

In the case of a continuum or of an amorphous solid, a fine-meshed lattice is defined, with each small cell either occupied or empty. The same formalism can then be used with the subscript 0 now referring to the "empty" alternative. It is customary then to define a volume element ΔV containing C cells, and a shape function $s(p)$ equal to unity if cell (p) is in ΔV about $\mathbf{x}(p)$, zero otherwise. The local concentration or density function

$$c_i[\mathbf{x}(p)] = \frac{1}{C} \sum_{p'} s(p') \sigma_i(p - p') \quad (1.8)$$

follows with

$$0 \leq c_i(\mathbf{x}) \leq 1. \quad (1.9)$$

2. SHORT-RANGE AND LONG-RANGE ORDER

Consider an m-point cluster of lattice points $(p_1, p_2, \ldots, p_m)$ associated with given lattice site (p). The *state of order* of a solution can be characterized by specifying sets of configuration variables $\langle \ldots \rangle$ which denote the probability of finding, in an ensemble of $\mathcal{M}$ systems, a cluster with i-type defect at $(p + p_1)$, j-type at $(p + p_2), \ldots$, l-type at $(p + p_m)$. We have, for the expectation value of that particular cluster (or cluster frequency or probability):

$$\langle \sigma_i(p + p_1)\sigma_j(p + p_2) \cdots \sigma_l(p + p_m) \rangle$$
$$= \sum_{\text{states}} \sigma_i(p + p_1)\sigma_j(p + p_2) \cdots \sigma_l(p + p_m)\mathcal{P}(\text{state}) \quad (2.1)$$

where $\mathcal{P}$ is the probability of finding the system in the state considered. For example, if a canonical ensemble is used

$$\mathcal{P}(\text{state}) = \mathcal{Z}^{-1} e^{-E(\text{state})/k_B T} \quad (2.2)$$

where $\mathcal{Z}$ is the partition function, E denoting the internal energy of a system, T, the absolute temperature, and k_B, Boltzmann's constant. Canonical ensemble averaging implies that each system must have same defect concentrations $\bar{c}_i (i = 0, 1, \ldots, n)$, as given by Eq. (1.4).

a. Short-Range Order Parameter

For the point cluster, in a canonical ensemble, the trivial single-site average is obtained;

$$\langle \sigma_i(p) \rangle = \bar{c}_i, \quad (2.3)$$

all points of the system's Bravais lattice being assumed equivalent.

Of more interest is the pair probability $\langle \sigma_i(p)\sigma_j(p + r) \rangle$, or rather, the pair correlation function

$$q_{ij}(r) = \langle \gamma_i(p)\gamma_j(p + r) \rangle = \langle \sigma_i(p)\sigma_j(p + r) \rangle - \bar{c}_i \bar{c}_j. \quad (2.4)$$

We also have, by application of the mean value theorem,[1]

$$q_{ij} = \bar{c}_i[P_{ij}(r) - \bar{c}_j] = \bar{c}_j[P_{ji}(r) - \bar{c}_i], \quad (2.5)$$

where

$$P_{ij} = \langle \sigma_i(p)\sigma_j(p + r) \rangle / \langle \sigma_i(p) \rangle \quad (2.6)$$

is the probability of finding a j-type defect at $(p + r)$, given one of i-type at (p). By Eqs. (2.4) and (2.5), the pair correlation function at the origin, or autocorrelation, is

$$q_{ij}^0 \equiv q_{ij}(0) = \bar{c}_i(\delta_{ij} - \bar{c}_j) = \bar{c}_j(\delta_{ji} - \bar{c}_i), \quad (2.7)$$

where δ_{ij} vanishes unless i and j are equal. The normalized pair correlation function gives the Warren–Cowley *short-range order* (SRO) parameters [1,2]

$$\alpha_{ij}(r) = q_{ij}(r)/q_{ij}^0 \quad (2.8)$$

whose value at the origin ($r = 0$) must be unity. In an $(n + 1)$-component solution there are $n(n + 1)/2$ linearly independent pair correlation functions.

For random distribution of defects we have

$$\langle \sigma_i(p)\sigma_j(p')\rangle^{(R)} = \langle \sigma_i(p)\rangle\langle \sigma_j(p')\rangle = \bar{c}_i\bar{c}_j \quad (2.9)$$

for $(p) \neq (p')$, so that by Eqs. (2.4), (2.7), and (2.8),

$$\alpha_{ij}^{(R)}(r) = \delta(r) \quad (2.10)$$

where the discrete (or lattice) δ-function is unity at the origin, zero elsewhere. In the general case, correlation is expected to decrease in magnitude as the pair spacing $\mathbf{x}(r)$ increases.

b. Long-Range Order Parameter

Nontrivial single-site averages are often required. These can be obtained by sublattice averaging as follows[3]: Figure 1 shows an ensemble of systems, each system containing $N_1 \times N_2 \times N_3$ unit cells. Assume that initially all systems had identical distributions $\sigma_i(p)$. Then allow some defect exchanges to take place within each system, while conserving the value of the internal energy E_0. This procedure automatically insures long-range correlations between systems so that the microcanonical ensemble average gives

$$c_i(p) \equiv \langle \sigma_i(p)\rangle_0 = \frac{1}{\mathcal{M}} \sum_{\text{syst.}} \sigma_i(p). \quad (2.11)$$

A fictitious average crystal can now be defined, such as the one outlined in Fig. 1, each lattice point of which is a microsystem[4] characterized by

---

[1] D. de Fontaine, *J. Appl. Crystallogr.* **4**, 15 (1971).
[2] B.E. Warren, "X-Ray Diffraction," p. 229. Addison-Wesley, Reading, Massachusetts, 1969.
[3] D. de Fontaine, *J. Phys. Chem. Solids* **34**, 1285 (1973).
[4] R. Cadoret, *Phys. Status Solidi* B **46**, 291 (1971).

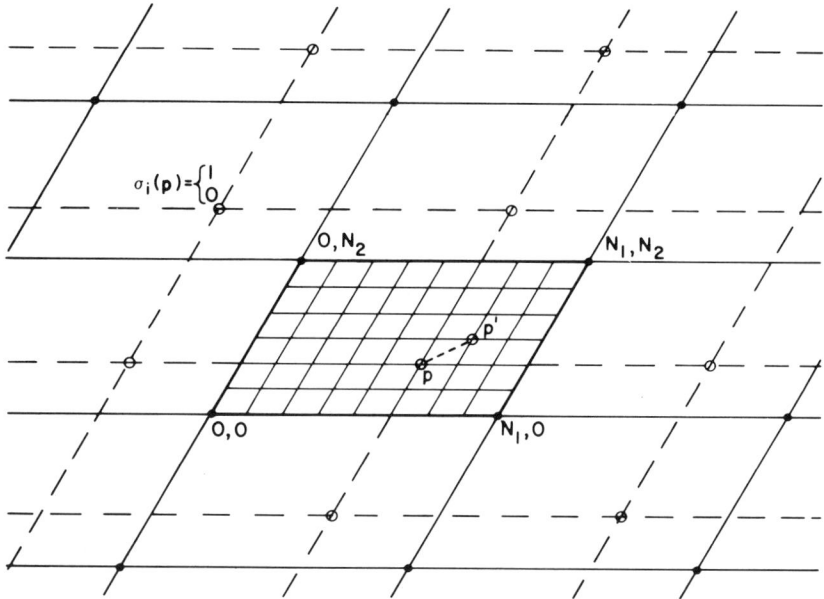

FIG. 1. Ensemble of supercells each containing $N_1 \times N_2 \times N_3$ unit cells. Superlattice through (p) shown dashed [D. de Fontaine, *J. Phys. Chem. Solids* **34**, 1285 (1973)].

continuously varying concentrations

$$0 \leq c_i(p) \leq 1. \qquad (2.12)$$

The resulting sample average crystal or supercell necessarily has periodic boundary conditions. The grids through the lattice points of the ensemble [one such grid through (p) is shown in Fig. 1 by dashed lines] may be interpreted as sublattices in the Bragg–Williams sense. For example, β-brass ordering can be described by an average system consisting of a single bcc unit cell, with one "average atom" at 000, and the other at $\tfrac{1}{2}\tfrac{1}{2}\tfrac{1}{2}$.

In general, *long-range order* (LRO) *parameters* are defined as linear combinations of the microcanonical average concentrations $c_i(p)$. There may be N such LRO parameters, $N = N_1 \times N_2 \times N_3$ being chosen large enough to display the essential features of the single-site distributions one wishes to describe. Hence, the concept of LRO is intimately related to the rather artificial procedure of defining sublattices and, as will be seen later, of using the zeroth approximation

$$\langle \sigma_i(p)\sigma_j(p')\rangle \to \langle \sigma_i(p)\rangle_0 \langle \sigma_j(p')\rangle_0 = c_i(p)c_j(p'). \qquad (2.13)$$

More specifically, LRO parameters are usually introduced in conjunction with order-disorder phase transformations. In that case, order parameters are defined as sublattice average that are zero above the transition temperature and nonzero below.

3. Concentration Waves

Since boundary effects are ignored, the crystalline system can be taken as a parallelepiped (supercell) containing N primitive unit cells on which periodic boundary conditions are imposed. The reciprocal unit cell and each Brillouin zone will thus also contain N points. Define the discrete Fourier transform operators

$$\mathcal{F} \equiv \frac{1}{N} \sum_p e^{-i\mathbf{k}(h)\cdot\mathbf{x}(p)} \tag{3.1}$$

and

$$\mathcal{F}^{-1} \equiv \sum_h e^{+i\mathbf{k}(h)\cdot\mathbf{x}(p)} \tag{3.2}$$

where the first sum is over the N lattice points of the periodic crystal, and the second is over the N points of the first Brillouin zone. The wave vector is defined by

$$\mathbf{k}(h) = 2\pi h_\alpha \mathbf{b}_\alpha \tag{3.3}$$

where

$$h_\alpha = m_\alpha / N_\alpha \quad (m_\alpha = 0, \pm 1, \pm 2, \cdots; \text{ no summation}) \tag{3.4}$$

and where $\mathbf{b}_\alpha$ are primitive translation vectors of the reciprocal lattice such that

$$\mathbf{a}_\alpha \cdot \mathbf{b}_\beta = \delta_{\alpha\beta}. \tag{3.5}$$

If a periodic continuum, or an infinite one, had been chosen to represent the system, the choice of Fourier operators would have been different, as indicated in Table I. In what follows, the context will indicate the form of the transform operator to be used. In general, uppercase symbols will be used to indicate the Fourier transform of the corresponding lowercase symbol. Thus

$$\Gamma_i(h) = \mathcal{F} \gamma_i(p), \tag{3.6}$$

the transform of the concentration deviations, defined by Eq. (1.5), represent concentration plane waves or *static concentration waves* (CW) in the terminology of Khachaturyan.[5] The expectation value of the CW

---

[5] A.G. Khachaturyan, *Phys. Status Solidi* B **60**, 9 (1973).

TABLE I. TYPES OF FOURIER OPERATORS (SUMMATION INTEGRATION) TO BE USED WITH INDICATED DIRECT AND RECIPROCAL SPACES

Reciprocal	$\mathscr{F}^{-1}$	$\mathscr{F}$	Direct
Discrete, periodic	$\sum$	$\dfrac{1}{N}\sum$	Discrete, periodic
Discrete, aperiodic	$\sum$	$\dfrac{1}{V}\int$	Continuous, periodic
Continuous, aperiodic	$\int$	$\dfrac{1}{(2\pi)^3}\int$	Continuous, aperiodic

amplitude squared is the Fourier transform of the pair correlation function, or concentration intensity for short:

$$Q_{ij}(h) = \langle \Gamma_i^*(h)\Gamma_j(h) \rangle = \mathscr{F} q_{ij}(p) \qquad (3.7)$$

where use has been made of the translational invariance of $\langle \gamma_i(r) \gamma_j(p+r) \rangle$. Note that the intensity Q_{ij} vanishes at the origin; its value at $h \neq 0$ for a random solution is

$$\bar{Q}_{ij} = q_{ij}^0/N. \qquad (3.8)$$

The integrated intensity is the value of the pair correlation function at the origin

$$\sum_h Q_{ij}(h) = q_{ij}^0 = \langle \gamma_i(p)\gamma_j(p) \rangle \qquad (3.9)$$

and, by Eq. (2.7), is thus a constant for fixed average concentrations. This important conservation rule is sometimes known as the "spherical constraint." Higher-order constraints such as[6]

$$\langle \gamma_i^m(p) \rangle = \bar{c}_i(1 - \bar{c}_i)[(1 - \bar{c}_i)^{m-1} - (-\bar{c}_i)^{m-1}] \qquad (m = 2, 3, \ldots, N) \qquad (3.10)$$

can also be derived by noting that γ_i^m can only be equal to $(1 - \bar{c}_i)^m$ or $(-\bar{c}_i)^m$, these values occurring with frequency $\bar{c}_i$ and $(1 - \bar{c}_i)$, respectively. These higher-order constraints can also be Fourier transformed to yield nonlinear integrated CW conservation rules, such as the ones used by Clapp and Moss[7] in their analysis of the ground states of ordered structures. In such derivations, one makes use of the discrete δ-function, used previously in Eq. (2.10), and regarded here as the Fourier back transform of unity. There exists a corresponding Fourier-space (or recip-

---

[6] D.W. Hoffman, private communication.
[7] P.C. Clapp and S.C. Moss, *Phys. Rev.* **171**, 754 (1968).

rocal- or k-space) discrete δ-function equal to unity at all reciprocal lattice points, zero elsewhere.

Note that integrated intensity conservation rules break down if SRO intensity is defined as the square of the Fourier coefficients of locally averaged concentration variables, following the small-volume averaging of Eq. (1.8), or the sublattice averaging of Eq. (2.11). In such cases, the average value of the product of occupation variables is replaced by the product of the averages, as in Eq. (2.13). For example, for completely disordered solutions, all averaged concentration deviation variables $\langle \gamma_i(p) \rangle_0$ vanish, so that the integrated intensity vanishes also. Actually, the integrated intensity of averaged CW measures the degree of clustering or ordering of a solution. The familiar application to small-angle scattering will be discussed in the next section.

The use of CW in ordering reactions is illustrated in Fig. 2. A ternary bcc solution of stoichiometric composition ABC_2 is first ordered to a CsCl structure by an $(h) = (100)$ wave of appropriate amplitude with "polarization" C–D in the concentration triangle. This wave places C atoms at the cell centers, and "average AB" atoms at the corners. Hence, the amplitude of this wave can be used as a long-range order parameter, the presence of "average atoms" indicating the implicit use of sublattice averaging. A $\frac{1}{2}\frac{1}{2}\frac{1}{2}$ wave of A–B polarization further orders the system to the Heusler structure ABC_2.

In general, concentration waves can be regarded as defined either by Eq. (3.6) or as the Fourier transform of averaged concentration variables. In the latter case, $\langle \Gamma_i(h) \rangle_0$ is a linear combination of average concentrations, which conforms to the general definition of LRO parameters and can thus be used as such, as shown in the above example. This procedure has been used primarily by the Russian school.[5]

4. Replacive Scattering

The relationship between concentration waves and scattering of radiation by a crystal is particularly simple when solute atoms or defects simply replace the host on lattice sites without causing any local relaxations, i.e., displacements, in the defect neighborhood. We call this *replacive scattering*; displacive scattering is discussed in Section 14.

a. Scattered Intensity

For generality, the finite size of the diffracting crystallite must be taken into account[8]: let $w(p)$ be the crystal shape function, equal to unity for

---

[8] A. Guinier, "Théorie et Technique de la Radiocristallographie," 3rd ed., p. 416. Dunod, Paris, 1964.

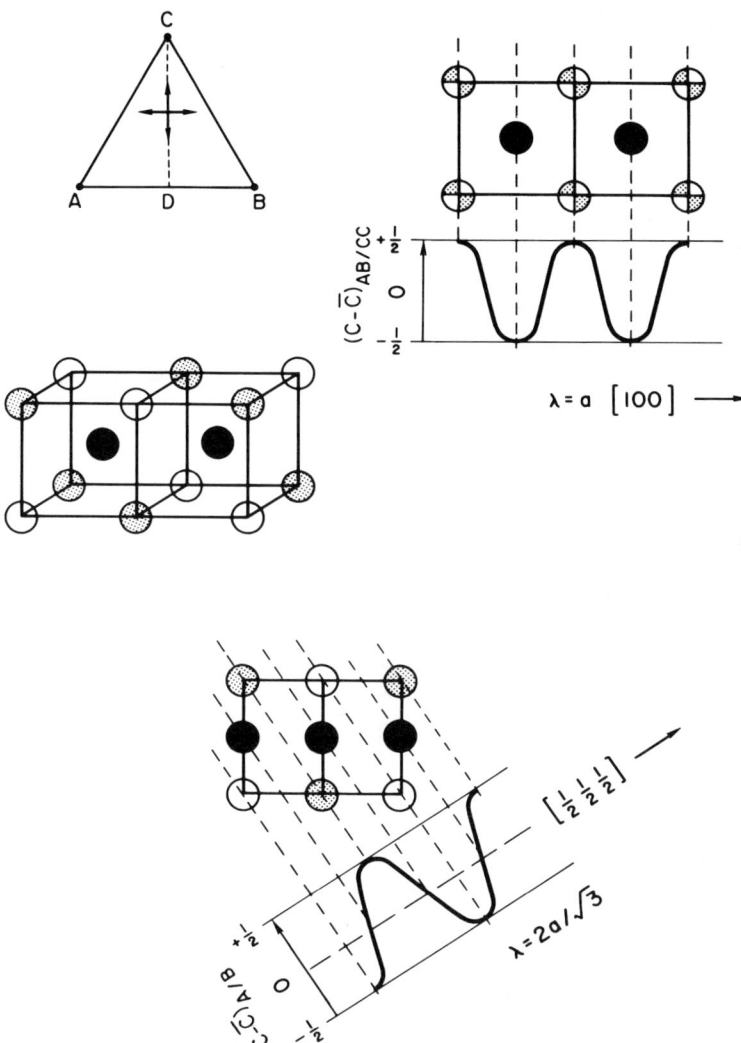

FIG. 2. [100] and [½½½] ordering waves in bcc lattice, together producing ABC$_2$ Heusler structure. Polarization of waves (CD and AB) shown in upper triangle.

(p) inside the crystallite, zero for (p) elsewhere in the perfect lattice (with periodic boundary conditions). The scattering geometry is as indicated in Fig. 3. The ith point scatterer has scattering factor, or cross section, $f_i(K)$, where K is the magnitude of the difference vector between outgoing and incoming monochromatic plane waves of wavelength λ. In

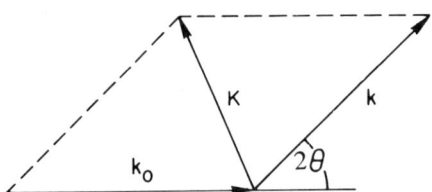

FIG. 3. Scattering geometry; k_0 is incident wave, k is scattered wave, K is scattering vector.

the language of neutron diffraction, this is the momentum transfer vector q. The scattering vector is continuously varying, and its magnitude is given by

$$K = (4\pi \sin \theta)/\lambda \qquad (4.1)$$

where θ is the Bragg angle.

In the kinematical approximation, it is assumed that the scattering is sufficiently weak so that the total scattering amplitude is the sum of single scattering events. The scattering amplitude from a lattice of point scatterers can then be expressed easily by making use of the occupation operator notation:

$$A(\mathbf{K}) = \sum_{j=0}^{n} f_j(K) \sum_p \sigma_j(p) w(p) e^{-i\mathbf{K}\cdot\mathbf{x}(p)}. \qquad (4.2)$$

The expected intensity is

$$I(\mathbf{K}) = \langle |A(\mathbf{K})|^2 \rangle$$

$$= \sum_{i,j=0}^{n} f_i(K) f_j(K)$$

$$\cdot \sum_{p,p'} \langle w(p) w(p') \sigma_i(p) \sigma_j(p') \rangle e^{-i\mathbf{K}\cdot[\mathbf{x}(p')-\mathbf{x}(p)]}. \qquad (4.3)$$

The autocorrelation of the shape function

$$\omega(r) = \sum_p w(p) w(p+r) \qquad (4.4)$$

is constant throughout the ensemble averaging process, and the pair correlation function is independent of (p) so that, by Eq. (2.4), the intensity simplifies to

$$I(\mathbf{K}) = \sum_{i,j=0}^{n} f_i(K) f_j(K) \sum_r \omega(r) [\bar{c}_i \bar{c}_j + q_{ij}(r)] e^{-i\mathbf{K}\cdot\mathbf{x}(r)}, \qquad (4.5)$$

where (r) is the difference $(p' - p)$. Define the real function

$$\Omega = N\mathcal{F}\omega \qquad (4.6)$$

which is sharply peaked about the origin of reciprocal space. The total diffracted intensity (4.5) can then be written as the sum of an average crystal intensity

$$I_{av}(\mathbf{K}) = f(K)^2 \Omega(\mathbf{K} - \mathbf{g}) \qquad (4.7)$$

with sharp peaks at the reciprocal lattice points $\mathbf{g}$ (Bragg reflections), with average scattering factor

$$\bar{f} = \sum_{i=0}^{n} \bar{c}_i f_i \qquad (4.8)$$

and a so-called *short-range order intensity*

$$I_{SRO}(\mathbf{K}) = N \sum_{i,j=0}^{n} f_i(K) f_j(K) \sum_{\mathbf{k}} Q_{ij}(\mathbf{k}) \Omega(\mathbf{K} - \mathbf{k} - \mathbf{g}), \qquad (4.9)$$

which is the convolution of the concentration intensity spectrum in the first BZ with the transform of the autocorrelation of the crystal shape function.

For a random distribution of defects we have, by Eqs. (3.8) and (2.7),

$$I_{SRO}^R = N(\overline{f^2} - \bar{f}^2) \qquad (4.10)$$

which is the so-called *Laue monotonic scattering*.

The SRO intensity, limited to the first BZ, can be simplified further by use of Eq. (1.7):

$$I_{SRO}(\mathbf{k})/\overline{f(k)}^2 = N^2 \sum_{i,j=1}^{n} \Delta f_i \Delta f_j Q_{ij}(\mathbf{k}), \qquad (4.11)$$

where, for simplicity, Ω has been replaced by N times the δ-function. Note that the summations are now over n defect types only ("host" excluded). If the scattering factor differences

$$\Delta f_i = [f_i(k) - f_0(k)]/\bar{f}(k) \qquad (i = 1, 2, \ldots, n) \qquad (4.12)$$

are assumed to be approximately independent of k, then Eq. (4.11) shows that all $n(n + 1)/2$ linearly independent pair correlation functions can be determined uniquely, in principle, by Fourier transforming the same number of normalized experimental (and suitably corrected) SRO intensities obtained with *distinct* radiations.[1] For binary ($n = 1$) solutions, a single radiation suffices, for ternaries ($n = 2$), three radiations are required optimally, for example, one X-ray and two neutron experiments, the latter two performed with different isotopes. In any case, the pair probability plays a central role in the determination of the state of order, since none of the other multiple correlations can be determined directly by scattering experiments. Indirect methods for higher cluster determinations will be examined in Section 25.

b. Integrated Intensity

For any defect rearrangement, the value of normalized I_{SRO} integrated over the BZ must be constant because of the conservation rule (3.9). If the concentration intensity spectrum $\langle Q \rangle_0$, say, is defined simply as the amplitude squared of concentration waves defined on the basis of either Eq. (1.8) or Eq. (2.11), then the intensity is no longer conserved. The "window-averaging" of Eq. (1.8) is used implicitly for the interpretation of small-angle scattering data. Let $S(\mathbf{k})$ be the Fourier transform of $s(\mathbf{x})$; then we must have

$$\langle Q \rangle_0 = |S|^2 Q \tag{4.13}$$

(a binary solution is considered, so that subscripts i and j may be deleted). Since $|S|^2$ attains its maximum value of unity at the origin, the spectrum $\langle Q \rangle_0$ remains close to the actual intensity spectrum Q for small k, i.e., for small diffraction angle θ. At higher angles, $|S|^2$ decreases rapidly, so that $\langle Q \rangle_0$ is then much smaller than Q, actually immeasurable beyond some effective cutoff angle. Hence

$$\sum_{BZ} \langle Q \rangle_0 \leq \bar{c}(1 - \bar{c}) \tag{4.14}$$

so that this "integrated intensity" merely measures the amount of inhomogeneities (defect clusters) of characteristic dimension greater than the reciprocal of the effective cutoff imposed in reciprocal space. According to this interpretation, there is no contradiction between conservation of SRO intensity and small-angle scattering practice. The conservation rule is particularly useful in the case of order–disorder studies characterized by diffuse intensity at BZ boundary points; the nonconservation is useful in clustering studies characterized by diffuse intensity in the BZ center.

5. Discussion on "State of Order"

In describing the state of order of a solution, it is neither feasible nor desirable to specify the occupation of all lattice sites. Instead, expectation values of cluster frequencies are defined by suitable averaging procedures. The point cluster yields simply the average concentration, the pair yields pair correlation functions, i.e., SRO parameters. Pairs constitute the watershed: with triplet, quadruplet, etc. probabilities, we enter the land of many-body problems where Fourier transforms are often more trouble than they are worth, and where therefore the simple correspondence between scattered intensity and cluster probabilities is lost.

Yet, much as one would like to remain within the two-body formulation, a satisfactory state-of-order description requires the use of multiplet probabilities whenever essentially "nonlinear" or many-body phenomena are considered, such as phase transitions (particularly first-order), nucleation, dislocation propagation, etc. As pointed out by Clapp,[9] simply visualizing the appearance of typical partially ordered states requires some knowledge of higher correlations.

Diffraction experiments cannot yield multiplet probabilities beyond the pair, so that indirect methods must be devised. A commonly used method is the Kirkwood superposition approximation which, for triplets, is

$$\langle \sigma_i(p)\sigma_j(p')\sigma_k(p'') \rangle$$
$$\rightarrow C \langle \sigma_i(p)\sigma_j(p') \rangle \langle \sigma_j(p')\sigma_k(p'') \rangle \langle \sigma_k(p'')\sigma_i(p) \rangle, \quad (5.1)$$

where the factor C is adjusted to make the sum of the triplet probabilities unity. However, this method can be seriously in error.[10,11]

A Monte Carlo simulation method was proposed by Cohen and co-workers[12,13]: atomic (or defect) interchanges are performed in such a way that computed pair correlations steadily approach those determined by a given diffraction experiment to given accuracy. Triplet and higher correlations are obtained by direct computation. The difficulty lies in interpreting the considerable computer output which this method generates. A third method, due to Clapp,[9-11] can be derived from Kikuchi's cluster variation method[14] (CVM, covered in Section 20), and its description is deferred until Section 25. A general method of deriving cluster probabilities from a set of independent (from the CVM viewpoint) multiplet correlation functions has recently been derived and applied to the ferromagnet Ising model.[15]

It is in principle possible to determine pair and multiplet probabilities

---

[9] P.C. Clapp, *Phys. Rev. B* **4**, 255 (1971).
[10] P.C. Clapp, *J. Phys. Chem. Solids* **30**, 2589 (1969).
[11] P.C. Clapp, in "Critical Phenomena in Alloys, Magnets and Superconductors" (R.E. Mills, E. Ascher, and R.I. Jaffee, eds.), p. 299. McGraw-Hill, New York, 1971.
[12] P.C. Gehlen and J.B. Cohen, *Phys. Rev.* **139**, A844 (1965).
[13] J.E. Gragg, Jr., P. Bardhan, and J.B. Cohen, in "Critical Phenomena in Alloys, Magnets and Superconductors" (R.E. Mills, E. Ascher, and R.I. Jaffee, eds.), p. 309. McGraw-Hill, New York, 1971.
[14] R. Kikuchi, *Phys. Rev.* **81**, 988 (1951).
[15] J.M. Sanchez and D. de Fontaine, *Phys. Rev. B* **17**, 2926 (1978).

directly by field ion microscopy.[16-18] Unfortunately, this technique is restricted to high melting-point materials and its use in state-of-order determination is presently limited by poor counting statistics and by ambiguities in the interpretation of the recorded patterns.

III. Internal Energy

The first step in obtaining a free energy functional suitable for concentrated solutions is the derivation of an internal energy function, or Hamiltonian. One may distinguish (a) the total *cohesive energy* of the solid, (b) the *energy of mixing* which is the difference between the energy of the actual solid solution and that of the sum of the energies in the pure unmixed state, and (c) the *configurational energy* which is the difference between the energy of a given atomic (or defect) configuration and the energy of, say, a random distribution of atomic species. Generally, the energy of mixing (b) is a small fraction of the total cohesive energy (a), and the configurational energy (c) is a small fraction of the energy of mixing.

The problem of calculating the internal energy of a solid is an extremely difficult one which formally necessitates the solution of the Schrödinger equation for arbitrary atomic configurations. Certain recently developed approximate techniques are quite promising, however. The central problem often consists in finding suitable perturbation expansions which converge sufficiently rapidly so that the expansion may be limited to second order. For pure *simple metals*, it is possible to construct a *pseudopotential*, so that the Schrödinger equation may be solved to second order in this small energy-dependent potential.[19-21] It then follows that the band structure energy may be written conveniently as a sum of pairwise effective interactions. Unfortunately, the list of metals for which pseudopotentials can be tabulated is not as extensive as one would like. For other metals, particularly the transition elements, the Fermi surfaces have complicated shapes deviating very significantly from the spherical one, and there is considerable overlap of the d-bands in the crystalline state. The pseudopotential formulation is therefore impractical, although exten-

---

[16] W. Dubroff and G.E. Machlin, *Acta Metall.* **16**, 1313 (1968).
[17] H.N. Southworth and B. Ralph, *Philos. Mag.* [8] **14**, 383 (1966); **21**, 23 (1970).
[18] H. Berg, Jr., T.T. Tsong, and J.B. Cohen, *Acta Metall.* **21**, 1589 (1973).
[19] W.A. Harrison, "Pseudopotentials in the Theory of Metals." Benjamin, New York, 1966.
[20] V. Heine, *Solid State Phys.* **24**, 1 (1970).
[21] V. Heine and D. Weaire, *Solid State Phys.* **24**, 249 (1970).

sions to d-band states have been given by Harrison[22] and by Waeber and Shively.[23] Actually, it appears that the cohesive energy of transition metals cannot be written as a sum of pair or even of multiplet (cluster) interactions.[24]

For solid solutions, the problem is compounded by the fact that ionic potentials differ from site to site, so that the crystal potential loses the translation symmetry of the lattice. The customary way out of that difficulty was to assume an average potential given by a weighted sum of the constituent ion potentials.[25,26] However, this *virtual crystal* method has been shown to give rather poor account of itself in the case of concentrated solutions. Particularly when the difference of the constituent ionic potentials is large with respect to the width of the d-band, a completely different approach must be used. Currently, the most satisfactory method is that of the *coherent potential approximation* (CPA): for the disordered case, one seeks an average single-site potential to be obtained self-consistently,[27] rather than the simple weighted sum. The calculation of cohesive energies can then proceed by standard techniques. It is impractical to obtain (small) ordering energies by taking the difference of two large cohesive energies in the ordered and disordered states. Rather, a way was recently discovered[28] of expanding the concentration fluctuation term, usually neglected in single-site CPA energy calculations, in a convergent generalized perturbation expansion. It is then possible to obtain LRO[29] and SRO[28] configurational energies, though of course not the total cohesive energy, as sums of effective pairwise (and multiplet) interactions, just as in the usual phenomenological theories. This is an important result, as it is clearly desirable to justify, and if possible to calculate, pair interaction parameters which play such a basic role in alloy theory.

As indicated by the title of this article, we are solely concerned here with the configurational portion of the internal energy, corresponding to isochoric atomic rearrangements on lattice sites. It therefore does not matter, for present purposes, how the cohesive energy is arrived at, and what types of potentials are actually used. Hence, for the sake of illus-

[22] W.A. Harrison, *Phys. Rev.* **181**, 1036 (1969).
[23] W.B. Waeber and J. E. Shively, *Physica (Utrecht)* **65**, 213 and 240 (1973); **68**, 409 (1973).
[24] J. Friedel, *in* "The Physics of Metals-I. Electrons" (J.M. Ziman, ed.), p. 340. Cambridge Univ. Press, London and New York, 1969.
[25] W.A. Harrison, "Pseudopotentials in the Theory of Metals," p. 142. Benjamin, New York, 1966.
[26] J.A. Inglesfield, *Acta Metall.* **17**, 1395 (1969).
[27] H. Ehrenreich and L.M. Schwartz, *Solid State Phys.* **31**, 149 (1976).
[28] F. Ducastelle and F. Gautier, *J. Phys. F* **6**, 2039 (1976).
[29] F. Gautier, F. Ducastelle, and J. Giner, *Philos. Mag.* [8] **31**, 1373 (1975).

tration only, the simplest method will be used: that of the pseudopotential in second-order perturbation scheme. This method is shown to lead a general internal energy expression in Fourier space, consisting of both unrelaxed lattice and relaxation (elastic) energies. The resulting *k-space potential* $V(\mathbf{k})$ is regarded as the fundamental configurational energy function; its Fourier back transform, a derived quantity, provides the effective two-body potential in real space. Important symmetry properties are completely model-independent (within the two-body assumption), and certain singular properties of $V(\mathbf{k})$ also appear to be fairly general. These properties are examined in Sections 8 and 7, respectively. Elastic energy is given a more detailed treatment in Section IV. The use of many-body interactions is briefly discussed in Section 23, for example.

6. Pair Potentials

The fundamental quantity to be calculated is the energy-wavenumber characteristic $W_{ij}(k)$ from which everything else can be made to follow, in principle. Harrison's treatment[19] will be applied to multicomponent solutions, for the sake of generality. For concentrated solid solutions, the advantages of performing all computations in k-space are considerable, as will presently be shown. For detailed numerical work pertaining to very dilute solutions or point defects, direct-space summation through the use of distorted-lattice Green's function is probably preferable; for simple metals the Green's function can be derived from pseudopotential theory[30,31]; for transition metals, a tight-binding approximation may be used.[32]

a. Pseudopotential Approach

One begins by an orthogonalized plane wave approach to the solution of Schrödinger's equation for a collection of N ions of $(n + 1)$ types in an electron sea, the whole assembly being contained in a box having periodic boundary conditions. By rearranging terms in Schrödinger's equation, one exhibits an operator $w(\mathbf{r})$ which acts as a weak effective potential seen by the electrons. It is possible to separate the total pseudopotential $w(\mathbf{r})$ in a sum of ionic pseudopotentials

$$w(\mathbf{r}) = \sum_{l} w_{l}(|\mathbf{r} - \mathbf{r}_{l}|) \tag{6.1}$$

centered at the ionic positions $\mathbf{r}_{l}$. The subscript l also indicates the nature of the ion (or defect) at $\mathbf{r}_{l}$, each ionic pseudopotential depending radially

---

[30] F. Benedek and P.S. Ho, *J. Phys. F* **3**, 1285 (1973; **4**, 183 (1974).
[31] P.S. Ho and R. Benedek, *IBM J. Res. Dev.* **18**, 386 (1974).
[32] K. Masuda, *J. Phys. F* **5**, 205 (1975).

on the position vector originating from the ion center. The weakness of the potential often allows one to solve the equation by second-order perturbation. There results zeroth and first-order terms which need not concern us here, since they are structure-insensitive (although volume-dependent and concentration-dependent). The important term is the second-order one which can be written

$$E_2(\mathbf{k}) = \frac{1}{N^2} \sum_{\mathbf{K}}{}' \sum_{l,m} e^{-i\mathbf{K}\cdot\mathbf{r}_{lm}} \frac{\langle \mathbf{k} + \mathbf{K} | w_l | \mathbf{k}\rangle \langle \mathbf{k} | w_m | \mathbf{K} + \mathbf{k}\rangle}{(\hbar^2/2m)(k^2 - |\mathbf{K} + \mathbf{k}|^2)}, \quad (6.2)$$

where $\hbar$ is Planck's constant divided by 2π, m is the electron mass, $\mathbf{k}$ is the free-electron wave vector so that, by analogy with scattering theory (Section 4), $\mathbf{k}$ denotes the incident wave, $\mathbf{k} + \mathbf{K}$ the scattered wave, and $\mathbf{K}$ the scattering vector. The interionic distance is denoted by

$$\mathbf{r}_{lm} = \mathbf{r}_m - \mathbf{r}_l. \quad (6.3)$$

The bra–ket notation in Eq. (6.2) is defined by

$$\langle \mathbf{k} + \mathbf{K} | w_l(r) | \mathbf{k}\rangle = \frac{1}{\Omega_0} \int_{\Omega_0} e^{-i(\mathbf{k}+\mathbf{K})\cdot\mathbf{r}} w_l(r) e^{i\mathbf{k}\cdot\mathbf{r}} d^3\mathbf{r}, \quad (6.4)$$

where Ω_0 is the average volume per ion. Since the pseudopotential w_l is an operator, the k dependence of the integral cannot be eliminated, so that the symbol $\langle \mathbf{k} + \mathbf{K} | w_l | \mathbf{k}\rangle$ denotes, in general, a *nonlocal pseudopotential form factor*. Both local and nonlocal form factors have been calculated for simple metals,[19,33,34] and can be computed in principle for other cases as well.[28] Hence, we assume in what follows that expressions such as (6.4) are known functions of $\mathbf{k}$ and $\mathbf{K}$.

The so-called *band structure energy* U_{bs} is obtained by summing $E_2(\mathbf{k})$ over all occupied states up to the Fermi wave vector $\mathbf{k}_F$. The electron–electron interactions are taken into account self-consistently in U_{bs}, but the ion–ion Coulomb repulsion U_c, also taken to be a sum of pairs, must still be added:

$$U_c = \frac{1}{2} \sum_{l \neq m} e^2 \frac{Z_l Z_m}{r_{lm}} = \frac{1}{2N} \sum_{\mathbf{K}}{}' \frac{4\pi e^2}{\Omega_0 K^2} \sum_{l \neq m} C(K) Z_l Z_m e^{-i\mathbf{K}\cdot\mathbf{r}_{lm}}, \quad (6.5)$$

where Z_l is the valence of the ion at $\mathbf{r}_l$. The second form of U_c in Eq. (6.5) is its Fourier representation with $C(K)$ an appropriate convergence factor which is implicitly corrected for in the structure-independent terms of the energy.[19] The summations in Eq. (6.5) are restricted to $l \neq m$, whereas no such restriction exists in Eq. (6.2). It is preferable to work

---

[33] A.E.O. Animalu and V. Heine, *Philos. Mag.* [8] **12**, 1249 (1965).
[34] M.L. Cohen and V. Heine, *Solid State Phys.* **24**, 37 (1970).

with strictly pairwise ($l \neq m$) terms, correcting for the $l = m$ terms of Eq. (6.2) in the structure-independent portion of the energy. The pair energy is thus

$$U = U_c + U'_{bs} = \frac{1}{2N} {\sum_{\mathbf{K}}}' \sum_{lm} W_{lm}(K) e^{-i\mathbf{K}\cdot\mathbf{r}_{lm}}, \qquad (6.6)$$

where U'_{bs} contains only $l \neq m$ contributions to the band structure energy. The *energy-wavenumber characteristic* is given by the real function

$$W_{lm}(K) = \frac{4\pi e^2 Z_l Z_m}{\Omega_0 K^2} C(K)$$

$$+ \frac{2}{N} \sum_{k<k_F} \frac{\langle \mathbf{k} + \mathbf{K} | w_l | \mathbf{k} \rangle \langle \mathbf{k} | w_m | \mathbf{K} + \mathbf{k} \rangle}{(\hbar^2/2m)(k^2 - |\mathbf{K} + \mathbf{k}|^2) \epsilon(K)} \qquad (6.7)$$

depending on the magnitude of the (practically continuous) scattering vector $\mathbf{K}$.

In Eq. (6.7), electron–electron interactions have been taken into account self-consistently by screening the potential by the dielectric function

$$\epsilon(K) = 1 + \frac{me^2}{2\pi k_F \hbar^2 \kappa^2} \left(\frac{1-\kappa^2}{2\kappa} \ln \left| \frac{1+\kappa}{1-\kappa} \right| + 1 \right) \qquad (6.8)$$

where

$$\kappa = K/2k_F, \qquad (6.9)$$

k_F being the radius of the Fermi sphere

$$k_F = (3\pi^2 \bar{Z}/\Omega_0)^{1/3} \qquad (6.10)$$

with

$$\bar{Z} = \sum_{i=0}^{n} \bar{c}_i Z_i. \qquad (6.11)$$

The simple form of screening used in Eq. (6.7) is convenient in discussing singularities in the energy-wavenumber characteristic as will be shown in Section 7. The nonlocal notation for the form factor is retained to show that, in general, W_{lm} cannot be written as a product of individual functions $W_l W_m$. This often-assumed factorization is not required in what follows. Equations (6.7)–(6.11) emphasize the fact that W_{lm} depends on the average concentrations, at least through the average valence $\bar{Z}$, and on volume (thus on many-body interactions), at least through the atomic volume Ω_0. Thus, for example, in a binary solution AB, there

CONFIGURATIONAL THERMODYNAMICS OF SOLID SOLUTIONS 93

are three concentration-dependent characteristics: W_{AA}, W_{BB}, and $W_{AB} = W_{BA}$.

b. Lattice Pair Interactions

The subscripts l, m play a dual role in Eq. (6.6): that of identifying the ion-type (or "nonion," such as a vacancy) and its location. These roles must be separated in order to exhibit the characteristic feature of the second-order perturbation method applied to solid solutions: the separation of the energy into a structure-independent potential and a structure factor. This will be accomplished here in the "lattice" framework by establishing a one-to-one correspondence between ions and lattice sites $(p, p', \ldots)$. The ion position vector associated with lattice site (p) may therefore be written

$$\mathbf{r}(p) = \mathbf{x}(p) + \mathbf{u}(p) \qquad (6.12)$$

$\mathbf{x}$ being a lattice vector, and $\mathbf{u}$ a displacement assumed to be small. The interionic vector $\mathbf{r}(p, p')$ may be split into a lattice vector $\mathbf{x}(p, p')$ and a displacement difference $\mathbf{u}(p, p')$, as shown in Fig. 4. The occupation operators $\sigma_i(p)$ defined in Eq. (2.2) are then used to transform Eq. (6.6) to the required form

$$U = \frac{1}{2N} \sum_{\mathbf{K}}{}' \sum_{i,j=0}^{n} W_{ij}(\mathbf{K}) \sum_{p \neq p'} \sigma_i(p) \sigma_j(p') e^{-i\mathbf{K} \cdot [\mathbf{x}(p,p') + \mathbf{u}(p,p')]}. \qquad (6.13)$$

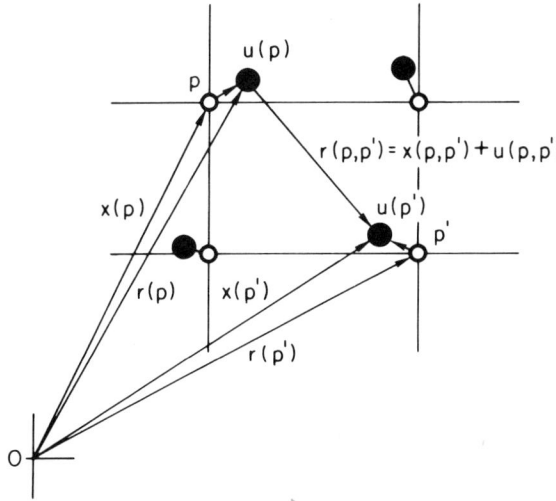

FIG. 4. Lattice positions (open circles) and actual atomic positions (filled circles). $\mathbf{x}(p)$ is lattice vector, $\mathbf{r}(p)$ is atomic position vector, $\mathbf{u}(p)$ is displacement vector.

Under the assumption of small displacements **u**, the exponential may be expanded to second order:

$$e^{-i\mathbf{K}\cdot\mathbf{u}(p,p')} \cong 1 - ik_\alpha u_\alpha(p,p') - \tfrac{1}{2}K_\alpha K_\beta u_\alpha(p,p')u_\beta(p,p') \quad (6.14)$$

with implied summation over Greek Cartesian subscripts. By Eq. (2.5) we have

$$\sigma_i(p)\sigma_j(p') = \bar{c}_i\bar{c}_j + [\bar{c}_i\gamma_j(p) + \bar{c}_j\gamma_i(p')] + \gamma_i(p)\gamma_j(p'). \quad (6.15)$$

Each of the three types of terms of Eq. (6.15) must now be taken with each of those of (6.14) to yield nine terms: a zeroth-order term, two linear terms in the variables γ_i, u_α which vanish identically, three quadratic terms which are the important ones, and higher-order terms. The derivation parallels that given in real space by Shirley.[35] After some algebra, one finds

$$U = \sum_{i=0}^{n} E_i + E + \text{higher-order terms}, \quad (6.16)$$

with

$$E_i = \frac{N}{2}\bar{c}_i \sum_p{}' \omega_{ij}(p) \quad (6.17)$$

and

$$E = \frac{1}{2}\sum_{p,p'}\left[\sum_{i,j=0}^{n}\omega_{ij}(p'-p)\gamma_i(p)\gamma_j(p') \right.$$
$$+ 2\sum_{i=0}^{n}\psi_{i\alpha}(p'-p)u_\alpha(p)\gamma_i(p')$$
$$\left. + \varphi_{\alpha\beta}(p'-p)u_\alpha(p)u_\beta(p')\right] \quad (6.18)$$

with

$$\omega_{ij}(p) = \frac{1}{N}\sum_{\mathbf{K}}{}' W_{ij}(\mathbf{K})e^{-i\mathbf{K}\cdot\mathbf{x}(p)} \quad (6.19)$$

$$\psi_{i\alpha}(p) = \frac{1}{N}\sum_{j=0}^{n}\bar{c}_j \sum_{\mathbf{K}}{}' K_\alpha W_{ij}(\mathbf{K})e^{-i\mathbf{K}\cdot\mathbf{x}(p)} \quad (6.20)$$

$$\varphi_{\alpha\beta}(p) = \frac{1}{N}\sum_{i,j=0}^{n}\bar{c}_i\bar{c}_j \sum_{\mathbf{K}}{}' K_\alpha K_\beta W_{ij}(\mathbf{K})e^{-i\mathbf{K}\cdot\mathbf{x}(p)}. \quad (6.21)$$

---

[35] G.G. Shirley, *Phys. Rev. B* **10**, 1149 (1974).

We must also have

$$\omega_{ij}(0) = - \sum_{p}{}' \omega_{ij}(p) = \omega_{ij}^0 \qquad (6.22)$$

$$\psi_{i\alpha}(0) = 0 \qquad (6.23)$$

$$\varphi_{\alpha\beta}(0) = - \sum_{p}{}' \varphi_{\alpha\beta}(p) = \varphi_{\alpha\beta}^0. \qquad (6.24)$$

In Eq. (6.16), E_i represents the energy of a hypothetical crystal containing $N_i = \bar{c}_i N$ ions of type i occupying the positions of the average lattice. The quadratic term E is the pair energy to be used in further developments of the theory. As noted by Shirley,[35] it is really not justified to discard the higher-order terms in Eq. (6.16). If these terms are retained, the algebra becomes nearly untractable however, so that terms higher than quadratic will be blithely ignored in the following treatment for the sake of simplicity. The only justification that can be offered rests on the questionable "zeroth approximation": that of replacing $\langle u_\alpha \gamma_i \gamma_j \rangle \ldots$ correlations by products of sublattice averages $\langle u_\alpha \rangle_0 \langle \gamma_i \rangle_0 \langle \gamma_j \rangle_0 \ldots$ These latter products (and the fourth-order ones) can then be regarded as negligible compared to quadratic terms in the case of solutions deviating not too much from the disordered state. Also, the simple form of Eq. (6.18) has yielded useful results previously.[36,37]

The summation over lattice sites in Eq. (6.18) contains the self-terms $(p = p')$, for the sake of making the notation more compact. The expectation value of the total energy of a solution with random distribution of defects is given precisely by these self-terms, since, for $p \neq p'$, all correlation must vanish:

$$\langle U(\text{random}) \rangle = \sum_{i=0}^{n} E_i + \frac{N}{2} \left[\sum_{i,j=0}^{n} \omega_{ij}^0 q_{ij}^0 + \varphi^0 \langle u^2 \rangle \right] \qquad (6.25)$$

where

$$\varphi^0 = \varphi_{11}^0 + \varphi_{22}^0 + \varphi_{33}^0 \qquad (6.26)$$

is the trace of the φ matrix, or *force-constant matrix*. In the absence of displacements, the mean square static displacement $\langle u^2 \rangle$ vanishes, and then the meaning of Eq. (6.25) is the familiar one of the sum of the energies of the pure states plus the energy of random mixing.

c. Relaxation Energy

The first term of Eq. (6.18) represents the pair energy of a solution with all ions exactly occupying the lattice sites: it is thus the *unrelaxed*

---

[36] H.E. Cook and D. de Fontaine, *Acta Metall.* **17**, 915 (1969).
[37] H.E. Cook and D. de Fontaine, *Acta Metall.* **18**, 189 (1970).

lattice energy. The other two terms together must then represent the *relaxation energy*, which at equilibrium must decrease the total energy. Hence, the force in the α direction acting on ion at lattice site (p) in the unrelaxed state and tending to move it toward its equilibrium position is

$$f_\alpha(p) = -\sum_{p'}{}' \sum_{i=0}^{n} \psi_{i\alpha}(p' - p)\gamma_i(p), \qquad (6.27)$$

with $\psi_{i\alpha}$ an appropriate "solute-lattice" coupling parameter.[36]

To calculate the static equilibrium position of the ions under this set of forces, we assume "mechanical" (or "elastic") equilibrium at all times

$$\partial E/\partial u_\alpha = 0. \qquad (6.28)$$

First, the pair energy E must be expressed in Fourier space:

$$E = \frac{N}{2} \sum_h \left[\sum_{i,j=0}^{n} \Omega_{ij}(h)\Gamma_i^*(h)\Gamma_j(h) \right.$$
$$\left. + 2\sum_{i=0}^{n} \Psi_{i\alpha}(h)U_\alpha^*(h)\Gamma_i(h) + \Phi_{\alpha\beta}(h)U_\alpha^*(h)U_\beta(h) \right] \qquad (6.29)$$

which transforms the double sums over lattice sites in Eq. (6.18) to a single sum over N discrete points in the first Brillouin zone. As before, uppercase symbols denote the Fourier transforms of the lowercase ones. Since we are now performing the discrete (or lattice) transform, according to the first line of Table I, the coefficients of the variables in the quadratic form (6.29) must be given by:

$$\Omega_{ij}(h) = \sum_p \omega_{ij}(p)e^{i\mathbf{k}(h)\cdot\mathbf{x}(p)}$$

$$= \frac{1}{N}\sum_{\mathbf{k}}{}' W_{ij}(K)\sum_p e^{-i(\mathbf{K}-\mathbf{k})\cdot\mathbf{x}(p)}$$

$$= \sum_{\mathbf{g}} W_{ij}(|\mathbf{k} + \mathbf{g}|) \qquad (6.30)$$

$$\Psi_{j\alpha}(h) = i\sum_{l=0}^{n} \bar{c}_l \sum_{\mathbf{g}} (k_\alpha + g_\alpha)W_{lj}(|\mathbf{k} + \mathbf{g}|) \qquad (6.31)$$

$$\Phi_{\alpha\beta}(h) = \sum_{i,j=0}^{n} \bar{c}_i\bar{c}_j \sum_{\mathbf{g}} (k_\alpha + g_\alpha)(k_\beta + g_\beta)W_{ij}(|\mathbf{k} + \mathbf{g}|), \qquad (6.32)$$

obtained from Eqs. (6.19), (6.20), and (6.21), respectively. Functions Ω_{ij} and $\Phi_{\alpha\beta}$ are real and symmetrical in the exchange of subscripts, while $\Psi_{j\alpha}$ is pure imaginary. The sum over reciprocal lattice vectors $\mathbf{g}$ in Eqs.

(6.30), (6.31), and (6.32) gives to the corresponding functions the translational symmetry of the reciprocal lattice. Furthermore, since the original energy-wavenumber characteristic W_{ij} had spherical symmetry, the new functions must also possess the point group symmetry of the reciprocal lattice, which is the same as that of the direct lattice.

The equilibrium condition (6.28) now yields, in k-space notation,

$$U_\alpha = -\sum_{i=0}^{n} \Phi_{\alpha\beta}^{-1} \Psi_{i\beta} \Gamma_i, \qquad (6.33)$$

where $\Phi_{\alpha\beta}^{-1}$ are the elements of the matrix inverse to Φ; in other words, Φ^{-1} is the Fourier transform of the Green's tensor.

The displacements can now be eliminated from (6.29) by means of (6.33) to yield the very simple expression

$$E = \frac{N}{2} \sum_h{}' \sum_{i,j=0}^{n} \mathcal{V}_{ij}(h) \Gamma_i^*(h) \Gamma_j(h) \qquad (6.34)$$

with effective k-space pair potential

$$\mathcal{V}_{ij} = \Omega_{ij} - \Psi_{i\alpha} \Phi_{\alpha\beta}^{-1} \Psi_{j\beta}^*. \qquad (6.35)$$

The first term of Eq. (6.35) contributes to the unrelaxed lattice energy, the second one to the relaxation energy. The fundamental k-space potential $\mathcal{V}_{ij}(h)$ has some of the symmetry properties of Ω_{ij}, as will be discussed in detail later (Section 8). Another Fourier back transform yields the pair energy in direct space

$$E = \frac{1}{2} \sum_{p,p'} \sum_{i,j=0}^{n} \nu_{ij}(p' - p) \gamma_i(p) \gamma_j(p'). \qquad (6.36)$$

Equations (6.34) or (6.36) are the basic expressions for the energy to be used in the remainder of this article. The form of E, which is our Hamiltonian function, is very simple, but the pair parameters

$$\nu_{ij}(p) = N\mathcal{F}^{-1} \mathcal{V}_{ij}(h) \qquad (6.37)$$

are extremely difficult to calculate from first principles: one must in principle obtain pseudopotential form factors (6.4) for all (i, j) pairs, construct the energy-wavenumber characteristics (6.7), convolute these with the reciprocal lattice according to the formulas (6.30), (6.31), (6.32), take a 3×3 matrix inverse at every point in the first BZ, and finally back-transform (6.35). Small wonder that the pair interaction parameters $\nu_{ij}(r)$ are generally regarded as empirical constants given *a priori* or determined from bulk experimental measurements. Fortunately, many interesting properties of solutions can be discussed on the basis of just two or three pair parameters, at least in the case of small elastic distortions.

d. Independent Variables

If ions or defects cannot be destroyed or created, the conservation rules (1.6) and (1.7) must be obeyed. In that case, dependent concentration deviations may be eliminated, resulting in energy formulas similar to (6.34) and (6.36), but having pair functions given by one of the following expressions:

$$V_{ij} = \mathcal{V}_{ij} - \mathcal{V}_{i0} - \mathcal{V}_{0j} + \mathcal{V}_{00} \qquad (i, j \neq 0) \tag{6.38}$$

or

$$V_{ij} = 2\mathcal{V}_{ij} - \mathcal{V}_{ii} - \mathcal{V}_{jj} \qquad (i \neq j). \tag{6.39}$$

These formulas effectively reduce the $\mathcal{V}$ symmetric matrix as indicated

$$\begin{pmatrix} 0,0 & \cdot & \cdot & \cdot & \cdot & \cdot \\ 1,0 & 1,1 & \cdot & \cdot & \cdot & \cdot \\ 2,0 & 2,1 & 2,2 & \cdot & \cdot & \cdot \\ \cdot & \cdot & \cdot & & & \\ \cdot & \cdot & \cdot & & & \\ n,0 & n,1 & \cdot & \cdot & & n,n \end{pmatrix} \tag{6.40}$$

The full triangle contains the pair functions remaining under scheme (6.38), the dashed triangle those remaining under scheme (6.39). Scheme (6.38) was in fact used to derive Eq. (4.11).

With either reduction method, the expectation value of the pair energy can be written with the help of Eq. (3.7) as

$$\langle E \rangle = \frac{N}{2} \sum_{h}{}' \sum_{i,j} V_{ij}(h) Q_{ij}(h) \tag{6.41}$$

or, with the help of Eq. (2.4) as

$$\langle E \rangle = \frac{N}{2} \sum_{p} \sum_{i,j} v_{ij}(p) q_{ij}(p), \tag{6.42}$$

where the sum over i, j is the appropriate one for the reduction scheme selected. The $n(n + 1)/2$ pair correlation functions used in these equations are linearly independent.[1] Upon comparing Eq. (6.41) to Eq. (4.11), one sees that $\langle E \rangle$ can be regarded as an integrated SRO intensity, the scattering factors in this case being just the k-space potentials.

In binary systems, the only independent pair potential is V_{11}, so that subscripts may be dropped altogether in scheme (6.38):

$$E = \frac{N}{2} \sum_h{}' V(h) |\Gamma(h)|^2. \tag{6.43}$$

The energy E of multicomponent systems can likewise be brought into the form of a sum of amplitude squared by diagonalizing the matrix (6.40). A system of n independent systems is therefore obtained, each one characterized by a particular direction in composition space given by the corresponding eigenvector. Examples of diagonalizations will be given in later sections. In the meantime, we investigate the properties of a given $V(h)$ potential, subscripts being deleted as the properties in question do not depend on the components chosen.

7. Singularities of $V(h)$

We shall examine later under what conditions the state of order is governed primarily by the properties of the k-space potential $V(h)$. We may anticipate by stating that for cases in which the configurational entropy takes a particularly simple form, minimization of the free energy F will be accomplished when E reaches its minimum allowable value. This will naturally occur for states of order characterized by maxima in the $Q(h)$ pair correlation spectrum located in the regions of the absolute minima of $V(h)$. Consequently much can be predicted about the types of ordering reactions to expect from a study of the shape of the $V(h)$ function, in particular from a search of its absolute minima. This study is complicated by the existence of singularities of the k-space potential. These singularities are of two types: (a) singularity at $V(0)$ due to the fact that the relaxation energy is undefined at the origin of the reciprocal lattice, and (b) singularities at $\mathbf{K} = 2\mathbf{k}_F$ in the energy-wavenumber characteristic $W(K)$ itself. These two topics are taken up in turn in this section.

a. Singularity at the Origin

By Eqs. (6.31) and (6.32), both $\Phi_{\alpha\beta}$ and Ψ_α vanish at the origin $h = 0$, so that the value of the relaxation energy coefficient, given by the second term of Eq. (6.35), is undefined. Actually,

$$R = \Psi_\alpha \Phi_{\alpha\beta}^{-1} \Psi_\beta^* \tag{7.1}$$

is multivalued at the origin. The long wavelength limit of R for given fixed $\mathbf{k}/k$ does exist and is given by continuum elasticity (as will be

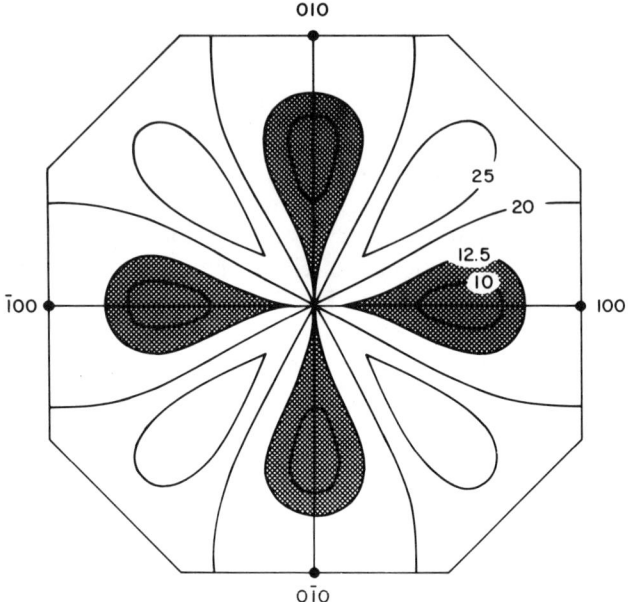

FIG. 5. Elastic energy $V(h)$ contours for AuNi in a (001) section of the first Brillouin zone. Low-energy regions shaded [D. de Fontaine and H. E. Cook, *in* "Critical Phenomena in Alloys, Magnets and Superconductors" (R. E. Mills, E. Ascher, and R. I. Jaffee, eds.), p. 257. McGraw-Hill, New York, 1971].

shown in Section IV), but the value of $R(0)$ depends on the direction of approach to the origin.

The fact that $R(0)$ is undefined causes no problem in the energy expressions, since the origin is excluded from the summation for two reasons: it was already excluded in Eq. (6.2), and concentration wave amplitudes vanish there as well. What is more troublesome is that $R(h)$, and hence $V(h)$, has no Taylor's expansion at the origin of k-space. Consequently, whenever elastic energy contributions are significant, the vicinity of the origin cannot be treated in the same way as the other points in the Brillouin zone. As an example, Fig. 5 shows $V(h)$ energy contours for AuNi calculated on the basis of purely elastic interactions.[38] Cubic symmetry is respected, but the energy contours go right through the origin, instead of being practically spherical about the BZ center, as would be the case if $V(h)$ could be expanded in a Taylor's series. One may there-

[38] D. de Fontaine and H.E. Cook, *in* "Critical Phenomena in Alloys, Magnets and Superconductors" (R.E. Mills, E. Ascher, and R.I. Jaffee, eds.), p.257. McGraw-Hill, New York, 1971.

fore expect, in general, anisotropic streaking of SRO intensity about reciprocal lattice points, characteristic of anisotropic clustering in crystal space.

Another consequence of this singularity is that the Fourier transform (6.37), performed as in Eq. (6.19) with origin excluded, will yield a large number of significant pair interaction parameters $v(p)$. In fact, only an infinite set of pair parameters can reproduce such contours at the origin as are shown in Fig. 5, for example. In other words, elastic distortions induce extremely long-range interactions, so that the description of $V(h)$ by means of two or three pair interaction parameters is then clearly unsatisfactory.

b. Singularities at the Fermi Surface

The origin of this singularity lies in the energy-wavenumber characteristic itself, hence we need only examine the real space interaction ω defined by Eq. (6.19). The summation may be replaced by an integral[19]

$$\omega(r) = \frac{\Omega_0}{\pi^2} \int_0^\infty W(K) \frac{\sin Kr}{Kr} K^2 \, dK. \tag{7.2}$$

For a nearly spherical Fermi surface, the integral can be evaluated for large r, giving the asymptotic formula

$$\omega(r)_{\text{asy}} = A \frac{\cos(2k_F r + \varphi)}{(2k_F r)^3}, \tag{7.3}$$

where A is a constant amplitude and φ is a phase shift required by the presence of different type ions in the solid solution. Actually, Eq. (7.3) describes the oscillations of the potential quite well even at the first few neighboring lattice positions. The wavelength of these *Friedel oscillations* is 2π times the reciprocal of the Fermi sphere diameter, the oscillations themselves arising from the logarithmic singularity in the derivative of the dielectric function $\epsilon(K)$ at $K = 2k_F$, i.e., at $\kappa = 1$ in Eq. (6.8).

The dielectric function in Eq. (6.8) was calculated according to the general formula,

$$\epsilon(K) = 1 - \frac{4\pi e^2}{K^2} \sum_{\mathbf{k}} \frac{f(k) - f(K+k)}{(\hbar^2/2m)(k^2 - |\mathbf{K}+\mathbf{k}|^2)}, \tag{7.4}$$

the second term of which is similar to that of Eq. (6.7), particularly in the case of local pseudopotential form factors which may then be taken outside the summation in Eq. (6.7). The Fermi–Dirac distribution functions $f(k)$ have been introduced in the numerator of Eq. (7.4) in order to

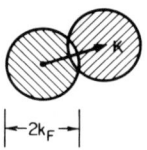

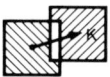

FIG. 6. Loss of contact of Fermi surfaces causes abrupt change in dielectric function $\epsilon(K)$ at $K = 2k_F$. Singularity is pronounced when Fermi surfaces have flat portions.

convert the sum over $\mathbf{k} < \mathbf{k}_F$ in Eq. (6.7) to an unrestricted sum.[39] The only terms that contribute to the sum in Eq. (7.4) are those for which the state with wave vector $\mathbf{k}$ is empty and that with $\mathbf{K} + \mathbf{k}$ filled, or vice versa. Hence, the functional form of the sum changes rather abruptly when the Fermi surfaces centered at the origin and at $\mathbf{K}$ just separate, i.e., at $\mathbf{K} = 2\mathbf{k}_F$, as shown in Fig. 6. There results a singularity in $\epsilon(K)$, which is a mild one when the Fermi surface is spherical. In that case, Eq. (6.8) follows from (7.4), and the Friedel oscillations fall off as $1/r^3$.

When the Fermi surface has flat portions, the oscillations fall off as $1/r$[40] and the dielectric function itself (rather than its derivative) has a singularity at $\mathbf{K} = 2\mathbf{k}_F$. Such is the origin of the Kohn anomaly,[41] which may result in so-called *charge density waves*,[42,43] a spectacular example of which is to be found in TaS_2.[44] Of more direct interest here is the fact that in the case of Fermi surfaces with "flats," the singularity of $\epsilon(K)$ creates a minimum in $W(K)$ and hence in $V(h)$ at $\mathbf{K}(h) = 2\mathbf{k}_F$.[45] The analog of the charge density waves in these cases are concentration waves of wave vector $2\mathbf{k}_F$. These waves scatter incident radiation (X rays, electrons), so that SRO diffuse intensity maxima will be found to follow the loci of $\epsilon(K)$ singularities in reciprocal space. Under favorable conditions, portions of the Fermi surface are actually mapped out by the

---

[39] J.M. Ziman, "Principles of the Theory of Solids," p. 126. Cambridge Univ. Press, London and New York, 1964.
[40] S.C. Moss and R.H. Walker, *J. Appl. Crystallogr.* **8**, 96 (1974).
[41] W. Kohn, *Phys. Rev. Lett.* **2**, 393 (1959).
[42] A.W. Overhauser, *Phys. Rev. B* **3**, 1888 (1971).
[43] W.L. McMillan, *Phys. Rev. B* **12**, 1187 and 1197 (1975).
[44] J.A. Wilson, F.J. DiSalvo, and S. Mahajan, *Adv. Phys.* **24**, 117 (1975).
[45] S.C. Moss, *Phys. Rev. Lett.* **22**, 1108 (1969).

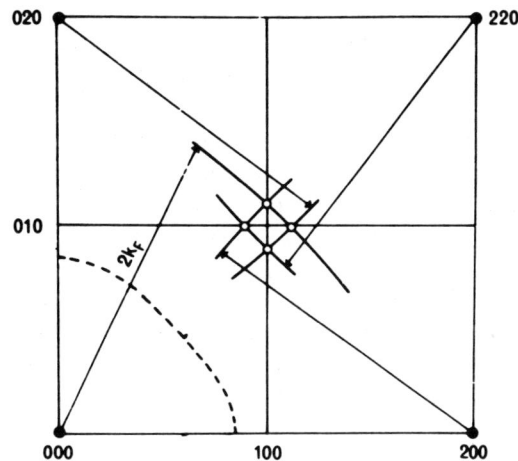

FIG. 7. Loci of singularities at $2k_F$ convoluted with reciprocal lattice. Fermi surfaces for Cu_3Au shown as dashed curve; note flatness around $\langle 110 \rangle$ [S. C. Moss, *Phys. Rev. Lett.* **22**, 1108, (1969)].

diffraction pattern as shown for example by Castles *et al.*[46] for the case of long-period ordering of vacancies in off-stoichiometric titanium oxide, and by Moss[45] in Cu_3Au for the case of long-period modulations of Cu–Au ordering waves. The analysis given by Moss of the splitting of the 110 superlattice reflection into four satellites is shown in Fig. 7. Note that the arc of diffuse intensity at $\mathbf{K} = 2\mathbf{k}_F$ must be convoluted with the reciprocal lattice, according to formula (6.30) for $\Omega(h)$.

The same sort of explanation may be advanced for the origin of other long-period superlattices. A good review of this topic is given by Sato and Toth[47] who initially proposed an explanation based on the "rigid band" approach. It is now known, however, that the Fermi sphere's touching a BZ boundary gives no appreciable lowering of the band-structure energy.[48] If the Fermi surface has extended flats, then the problem may be regarded approximately as one-dimensional. In that case, an exact calculation does demonstrate the existence of a deep minimum of U_{bs} at $\mathbf{K} = 2\mathbf{k}_F$,[48] lending support to the Sato and Toth explanation. In any case, the periodicity of the concentration wave is directly related to the caliper dimension of the Fermi surface in that wave direction, so that the wavelength varies continuously with such parameters as the electron-

[46] J.R. Castles, J.M. Cowley, and A.E.C. Spargo, *Acta Crystallogr., Sect. A* **27**, 376 (1971).
[47] H. Sato and R.S. Toth, *in* "Alloying Behavior in Concentrated Solid Solutions" (T. B. Massalski, ed.), p. 295. Gordon & Breach, New York, 1965.
[48] V. Heine and D. Weaire, *Solid State Phys.* **24**, 289 (1970).

atom ratio. Hence, the static CW wavelength, which is generally incommensurate with the lattice parameters, can vary monotonically with alloying element concentration, an experimentally well-documented effect.[47]

In all such explanations, the assumption is tacitly made that structure can be related directly to the k-space internal energy coefficient $V(h)$. This is only correct if the free energy can be expressed adequately as a Taylor's expansion to second order in the concentration wave amplitudes (see Section 18). Many of the discrepancies between observed diffraction patterns and theoretical models based on the "singularity" idea, such as the absence of diffuse streaks connecting the long-period superlattice satellites, may very well be related to the neglect of higher-order terms coming primarily from the configurational entropy.

8. Symmetry of $V(h)$

Despite present theoretical ambiguities, it does appear that certain states of order can be related to singularities in $V(h)$. More importantly, certain types of ordered structures can be related directly to the minima of $V(h)$. In this section, we examine the regular points in the BZ at which $V(h)$ has vanishing gradient. The treatment follows that given in an earlier paper.[49] We shall neglect here complications due to the singularity at the origin of k-space with the understanding that if elastic distortions are important, the following considerations apply (at the origin) only to the unrelaxed potential $\Omega(h)$, rather than to the complete $V(h)$ itself.

a. Special Points

The necessary condition for point $(h°)$ to be at a minimum of $V(h)$ is given by the set of equations

$$\frac{\partial V}{\partial h_\alpha°} = 0 \qquad (\alpha = 1, 2, 3). \tag{8.1}$$

The wave vectors satisfying Eq. (8.1) also yield the locations of the maxima and saddle points of $V(h)$. The nature of the extrema can be ascertained by expanding $V(h)$ to second order about $(h°)$:

$$V(h) = V_0 + \frac{1}{2} \sum_{\alpha=1}^{3} \lambda_\alpha(h°) H_\alpha^2. \tag{8.2}$$

In Eq. (8.2), the linear term vanishes because of Eq. (8.1), V_0 is the value of V at the extremum, λ_α are the eigenvalues of the real symmetrical

---

[49] D. de Fontaine, *Acta Metall.* **23**, 553 (1975).

matrix of second derivatives of $V(h)$ with respect to h_α, h_β, and the coordinates H_α are measured along the three principal axes of $V(h°)$ from the position of the extremum.

Maxima, minima, and saddle points must always be present. In fact, because the function $V(h)$ has the translational symmetry of the reciprocal lattice, a topological theorem due to Morse[50] is applicable, from which it follows that the number of various types of extrema must verify certain inequalities.[51,52] In particular, the minimum number of extrema is one maximum, one minimum, and six saddle points.

The function $V(h)$ possesses not only the translational but also the point group symmetry of the reciprocal lattice. Thus, if a solution of Eq. (8.1) is found at some $\mathbf{k}(h°)$, other extrema of the same type must exist at all other points belonging to the star of the vector $\mathbf{k}$, the star of a wave vector consisting of all those vectors which transform into one another by the operations of symmetry of the space group of the reciprocal lattice. If a symmetry element (rotation, rotation–inversion, or mirror plane) of the space group in k-space is located at point (h), the vector representing the gradient $\nabla_h V(h)$ of an arbitrary potential energy function $V(h)$ at that point must lie along or within the symmetry element. If two or more symmetry elements intersect at (h), one must necessarily have

$$|\nabla_h V(h)| = 0 \qquad (8.3)$$

since a finite-magnitude vector cannot lie simultaneously in two intersecting straight lines (or on a line and a plane) having only a point in common. At these so-called *special points,* Eq. (8.1) is thus satisfied by symmetry requirements alone, so that $V(h)$ must present an extremum regardless of the choice of pairwise interaction energies. This universal character of the special points was pointed out by Lifshitz[53] in his study of phase transitions of the second kind and by Khachaturyan.[5,54]

b. Cubic Crystals

The special points in k-space for the fcc and bcc lattices will be found listed in the International Tables for X-Ray Crystallography[55] under the space groups of their reciprocals: group Im3m (bcc) and Fm3m (fcc),

---

[50] M. Morse, "Functional Topology and Abstract Variational Algebra," Mém. Sci. Math., Fasc. 92. Gauthier-Villars, Paris, 1935.
[51] L. Van Hove, *Phys. Rev.* **89,** 1189 (1953).
[52] J.C. Phillips, *Phys. Rev.* **104,** 1263 (1956).
[53] E.M. Lifshitz, *J. Phys. (Moscow)* **7,** 61 and 251 (1942).
[54] A.G. Khachaturyan, *Sov. Phys.—Solid State (Engl. Transl.)* **5,** 16 and 548 (1963).
[55] N.F.M. Henry and K. Lonsdale, eds., "International Tables for X-ray Crystallography," Vol. I. Kynoch Press, Birmingham, 1952.

TABLE II. SPECIAL POINTS IN THE fcc BRILLOUIN ZONE[a]

Wave vector			Point group symmetry		
Miller	BSW[b]	Wyckoff[c]	Internat.	Schönflies	Class
$\langle 000 \rangle$	Γ	a	m3m	O_h	Cubic
$\langle 100 \rangle$	X	b	4/mmm	D_{4h}	Tetragonal
$\langle \frac{1}{2}\frac{1}{2}\frac{1}{2} \rangle$	L	c	$\bar{3}$m	D_{3d}	Trigonal
$\langle 1\frac{1}{2}0 \rangle$	W	d	$\bar{4}$2m	D_{2d}	Tetragonal

[a] D. de Fontaine, *Acta Metall.* **23**, 553 (1975).
[b] L.P. Bouckaert, R. Smoluchowski, and E. Wigner, *Phys. Rev.* **50**, 58 (1936).
[c] R.W.G. Wyckoff, "The Analytical Expression of the Results of the Theory of Space Groups." Carnegie Inst., Washington, D.C., 1922.
[d] N.F.M. Henry and K. Lonsdale, eds., "International Tables for X-Ray Crystallography," Vol. I. Kynoch Press, Birmingham, 1952.

respectively. The four special points for each of these lattices are listed in Tables II and III. The usual reciprocal space notation, the standard wave vector group notation,[56] and the Wyckoff symbols are given.[57] The point symmetry at the special point, known as the wave vector group, is also given in Tables II and III, along with the symmetry class.

If the function $V(h)$ is everywhere regular, as we are assuming here, then it may be expanded in a Fourier series limited to the first few neighbor pair parameters. These parameters $v(p)$, formally given by Eq. (6.37), should be regarded as empirical constants introduced for the sole purpose of conveniently representing the k-space potential $V(h)$. They should not necessarily be regarded as actual pair interaction parameters. For fcc and bcc lattices, Eq. (6.37) gives

$$V(h) = \sum_s v^{(s)} Y^{(s)}(h), \qquad (8.4)$$

where the function $Y^{(s)}$, for arbitrary coordination shell s, is given by a modified version of a formula given by Squires[58]:

$$Y^{(s)}(h) = \frac{z^{(s)}}{2} \sum_{\alpha=1}^{3} \cos(2\pi h_1 p_\alpha^{(s)}) [\cos(2\pi h_2 p_{\alpha+1}^{(s)}) \cos(2\pi h_3 p_{\alpha+1}^{(s)})$$
$$+ \cos(2\pi h_3 p_{\alpha+1}^{(s)}) \cos(2\pi h_2 p_{\alpha+2}^{(s)})], \qquad (8.5)$$

with

$$\alpha \to (\alpha - 1) \text{ modulo } (3) + 1, \qquad (8.6)$$

[56] L.P. Bouchaert, R. Smoluchowski, and E. Wigner, *Phys. Rev.* **50**, 58 (1936).
[57] R.W.G. Wyckoff, "The Analytical Expression of the Results of the Theory of Space Groups." Carnegie Inst., Washington, D.C., 1922.
[58] G.L. Squires, *Ark. Fys.* **25**, 21 (1963).

TABLE III. SPECIAL POINTS IN THE bcc BRILLOUIN ZONE[a]

Wave vector			Point group symmetry		
Miller	BSW[b]	Wyckoff[c]	Internat.[d]	Schönflies	Class
⟨000⟩	Γ	a	m3m	O_h	Cubic
⟨100⟩	H	b	m3m	O_h	Cubic
⟨½½½⟩	P	c	$\bar{4}$3m	T_d	Cubic
⟨1½0⟩	N	d	mmm	D_{2h}	Orthorhombic

[a] D. de Fontaine, *Acta Metall.* **23**, 553 (1975).
[b] L.P. Bouckaert, R. Smoluchowski, and E. Wigner, *Phys. Rev.* **50**, 58 (1936).
[c] R.W.G. Wyckoff, "The Analytical Expression of the Results of the Theory of Space Groups." Carnegie Inst., Washington, D.C., 1922.
[d] N.F.M. Henry and K. Lonsdale, eds., "International Tables for X-Ray Crystallography," Vol. I. Kynoch Press, Birmingham, 1952.

and where $z^{(s)}$ is the number of lattice points in the coordination shell s, $p_\alpha^{(s)}$ are integers and half-integers denoting the Cartesian coordinates of a point in the first octant of the shell s, and h_α denote Cartesian coordinates in the first Brillouin zone.

Matrices of second derivatives of the $Y^{(s)}$ shell functions for the first five coordination shells have been evaluated at all special points of the fcc and bcc lattices.[49] Their eigenvalues $\lambda_\alpha^{(s)}$ can generally be determined by inspection and are recorded in Tables 3 and 4 of de Fontaine[49] along with the values of the shell function itself at the special points. The corresponding principal directions (eigenvectors) are given here in Table IV. All eigenvalues must be equal at points of cubic symmetry, at most one eigenvalue may differ from the others at points of tetragonal and trigonal symmetry, while all three may differ at points of orthorhombic symmetry. The signs of the eigenvalues determine the nature of the extrema of the shell function.

The usefulness of these eigenvalue tables follows from a very remarkable property of the special points: for a given special point, the diagonalization operators of the matrices of the various $Y^{(s)}$ are the same for all coordination shells, since the eigenvectors (listed in Table IV) are deter-

TABLE IV. PRINCIPAL (EIGENVECTOR) DIRECTIONS CORRESPONDING TO SPECIAL POINTS[a]

Symmetry of special point	Principal directions
Cubic and tetragonal	[100], [010], [001]
Trigonal	[111], [1$\bar{1}$0], [11$\bar{2}$]
Orthorhombic	[110], [1$\bar{1}$0], [001]

[a] D. de Fontaine, *Acta Metall.* **23**, 553 (1975).

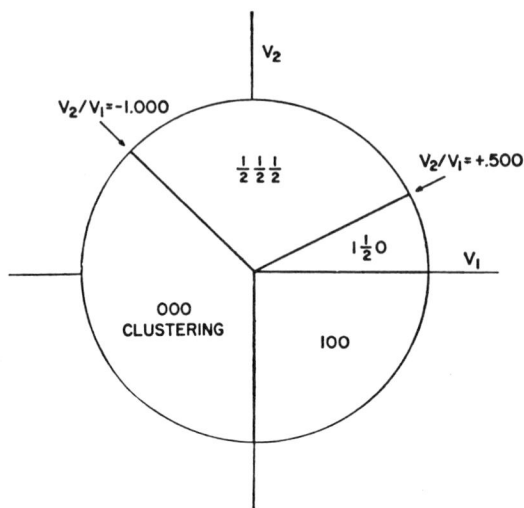

FIG. 8. Locations of the minima of $V(h)$ as a function of the ratio V_2/V_1 for the fcc lattice [P. C. Clapp and S. C. Moss, *Phys. Rev.* **171**, 754 (1968)].

mined solely by symmetry considerations. It follows that the matrix of second derivatives of an arbitrary $V(h)$ function can be diagonalized at the special points by the very same orthogonal operator which diagonalizes its component shell functions $Y^{(s)}$ and therefore the "special-point" eigenvalues are additive:

$$\lambda_\alpha = \sum_s v^{(s)} \lambda_\alpha^{(s)} . \qquad (8.7)$$

Thus, the nature of the special extrema for an arbitrary $V(h)$ function can be determined by performing a weighted sum of standard eigenvalues.

Pictorial representations are useful for classifying $V(h)$ functions according to the special point at which the lowest minimum is located. If only first- and second-neighbor pair parameters are retained, only the ratio V_2/V_1 (simplified notation for $v^{(2)}/v^{(1)}$) need be investigated. Figures 8 and 9 give the location of the $V(h)$ minima as a function of this ratio for fcc and bcc lattices, respectively. Such plots were first obtained by Villain[59] for magnetic structure and by Clapp and Moss[7] for binary solid solutions. When only V_1 and V_2 are used, absolute minima of $V(h)$ are always located at one of the special points. Note however that bcc special point $\langle \frac{1}{2}\frac{1}{2}0 \rangle$ is not represented in Fig. 9; actually, one has to go to

[59] J. Villain, *J. Phys. Chem. Solids* **11**, 303 (1959).

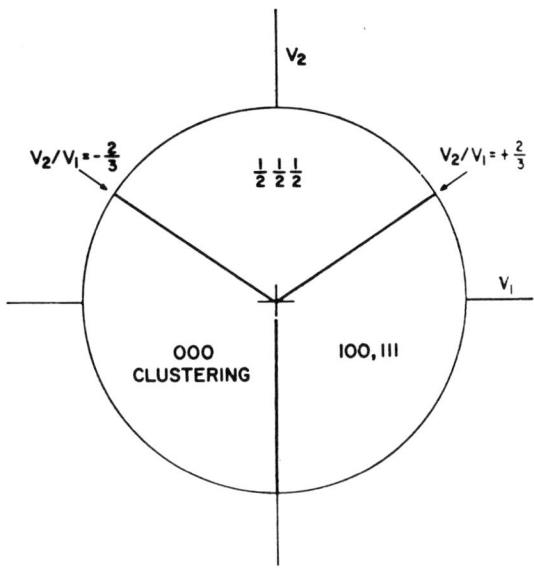

FIG. 9. Location of the minima of $V(h)$ as a function of the ratio V_2/V_1 for the bcc lattice [P. C. Clapp and S. C. Moss, *Phys. Rev.* **171**, 754 (1968)].

third-neighbor parameter V_3 in order to produce an absolute minimum at that point.[60]

If third-neighbor parameters are retained as well, the location of lowest special point can be represented graphically as a function of the two ratios V_2/V_1 and V_3/V_1. Figure 10 illustrates this for the fcc lattice. In this case, absolute minima can also be found at $\langle\frac{2}{3}00\rangle$ for certain ranges of the pair parameter ratios, and was so indicated on the original plots given by Clapp and Moss.[7,60] However, the $\langle\frac{2}{3}00\rangle$ regions are not included here, since the precise location of the minimum will actually vary continuously with the parameter ratios. This is because $\langle\frac{2}{3}00\rangle$ not being a special point, the position of the corresponding minimum is not "anchored" there by symmetry.[49] In fact, no definite ordered structure can be found based on this minimum.[60] Morinaga[61] has pointed out that, for large V_3, the lowest special points $\langle 100\rangle$ and $\langle\frac{1}{2}\frac{1}{2}\frac{1}{2}\rangle$ may in fact be saddle points; absolute minima occur then at nonspecial points. However, as will be discussed in Section VI, only the special points are of interest, as these (and only these) can be related to fundamental instabilities of solid solutions. Hence, in Figs. 10a and b, the diagrams of Clapp and Moss[7]

[60] P.C. Clapp, private communication.
[61] M. Morinaga, *Acta Metall.* **25**, 957 (1977).

and of Morinaga[61] have been modified so as to indicate only special-point regions. Regions where the three eigenvalues λ_α ($\alpha = 1, 2, 3$) fail to be all positive are shaded. The boundaries of these (vanishing eigenvalues) are easily calculated by use of Eq. (8.7) and of the eigenvalue tables given elsewhere.[49] It appears, as discussed in Section VI, that known ordered structures can be grouped into "families," each member of a

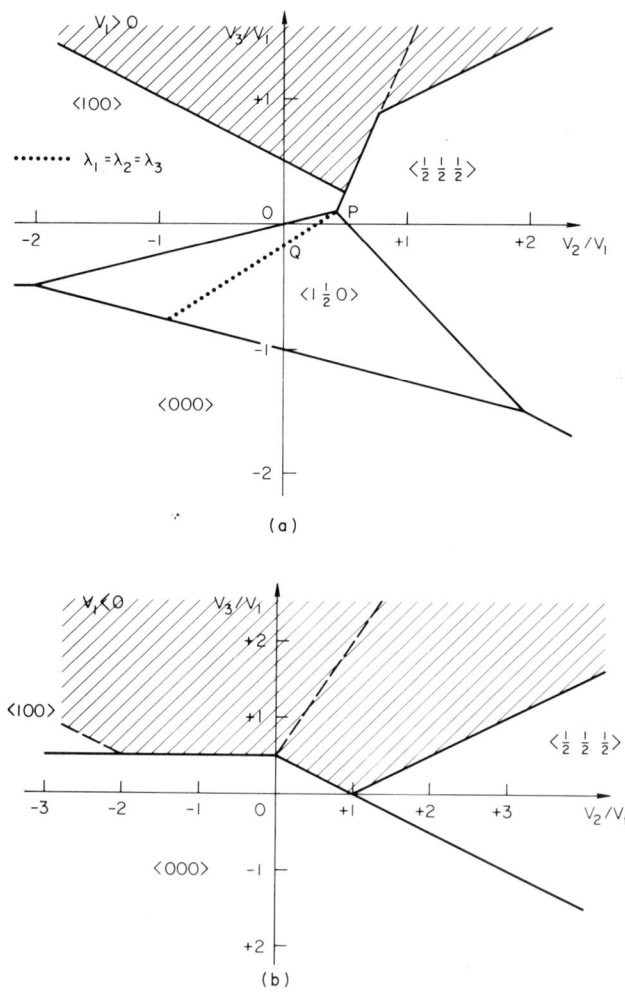

FIG. 10. Location of special-point minima of $V(h)$ as a function of the ratios V_2/V_1 and V_3/V_1 for the fcc lattice. Regions where eigenvalues λ_α fail to be all three positive are shown shaded. (a) $V_1 > 0$. Dotted line is locus of equality of all three $\langle 1\frac{1}{2}0 \rangle$ eigenvalues. (b) $V_1 < 0$.

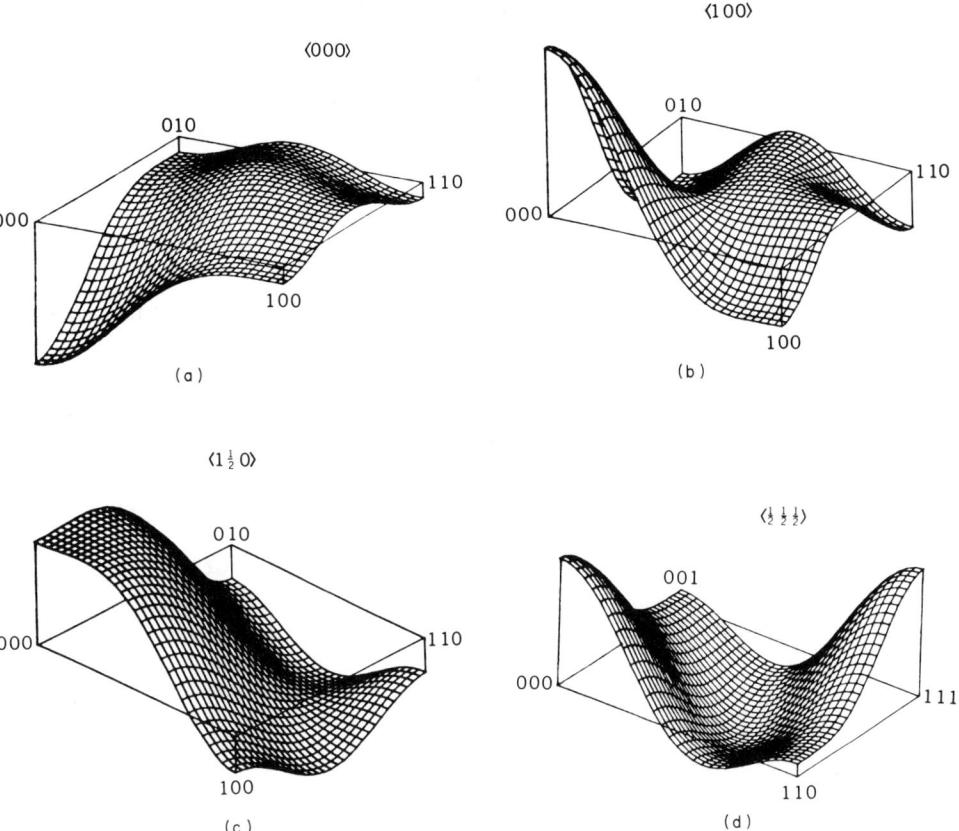

FIG. 11. Computer-generated perspective plots of $V(h)$ surfaces [P. Talley, UCLA Report (1977) (unpublished)]. (a) (001) section; $V_1 = -1$, $V_2 = -1$, $V_3 = 0$; $\langle 000 \rangle$-type instability. (b) (001) section; $V_1 = +1$, $V_2 = 0$, $V_3 = 0.75$; $\langle 100 \rangle$-type instability. (c) (001) section; $V_1 = +1$, $V_2 = 0$, $V_3 = -0.2$; $\langle 1\frac{1}{2}0 \rangle$-type instability (coordinates of point Q, Fig. 10a). (d) ($\bar{1}10$) section; $V_1 = +1$, $V_2 = +1$, $V_3 = 0$; $\langle \frac{1}{2}\frac{1}{2}\frac{1}{2} \rangle$-type instability.

given family having the same parent special-point instability. Hence, there are only four fcc-based and four bcc-based families, despite the large variety of known or conjectured ordered structures.[62,63]

Typical $V(h)$ surfaces for the four fcc families are shown in perspective in Fig. 11, either in (001) or ($\bar{1}10$) sections.[64] The parameters used in constructing the surface in Fig. 11b were chosen so as to place the ratios

[62] J. Kanamori, *J. Japan Inst. Metals* **15**, 35 (1976) (in Japanese).
[63] J. Kanamori and Y. Kakehashi, *J. Phys. (Paris)* **38**, C7-274 (1977).
[64] P. Talley, UCLA Report (1977) (unpublished).

of V_2/V_1 and V_3/V_1 in the shaded portion of Fig. 10a. Thus, point 100 in Fig. 11b is really a saddle point, the true minimum being located closer to 000. We may thus expect a long-period modulation of the $\langle 100 \rangle$ ordering wave, but clearly, the resulting structure will belong to the $\langle 100 \rangle$ family. The pair parameter ratios chosen for Fig. 11c were those of point Q in Fig. 10a, also chosen for the amplification rate plot of Fig. 67 (see Section 26).

9. Continuum Gradient Expansion

It is often convenient to consider the solid solution as a continuum, and to replace summations by integrations. This is most easily accomplished for the internal energy by starting from the expansion about a $V(h)$ minimum given by Eq. (8.2). Since the origin of k-space is singular, care must be taken, in what follows, to replace $V(0)$ by $\Omega(0)$ whenever the expansion is about the origin and elastic distortions are important. The continuum correction for the relaxation energy $R(h)$ will be given in Section 12.

Let the component of the wave vector $\mathbf{K}$ from the point of the minimum in the α principal direction be (see Table IV for principal directions in fcc and bcc crystals)

$$K_\alpha = (2\pi/a_\alpha)H_\alpha \tag{9.1}$$

in which a_α is the magnitude of the lattice translation vector in the α direction, and where H_α was defined in Eq. (8.2). This latter equation, by Eq. (9.1) becomes

$$V(\mathbf{K}) = V_0 + 2 \sum_{\alpha=1}^{3} \Lambda_\alpha K_\alpha^2 \tag{9.2}$$

where

$$V_0 \equiv V(h^\circ) = \sum_p v(p) e^{i\mathbf{k}(h^\circ) \cdot \mathbf{x}(p)} \tag{9.3}$$

and

$$\Lambda_\alpha = (a_\alpha/4\pi)\lambda_\alpha(h^\circ) \quad \text{(no summation over } \alpha\text{).} \tag{9.4}$$

The internal energy (6.43) may then be written

$$E = \frac{N}{2} V_0 \sum_\mathbf{k} |\Gamma(\mathbf{k})|^2 + N \sum_{\alpha=1}^{3} \Lambda_\alpha \sum_\mathbf{K} |iK_\alpha \Gamma(K)|^2 \tag{9.5}$$

which by Parseval's theorem (continuous-periodic version, second line of Table I) becomes

$$E = N_V \int_{V_N} \left\{ \frac{1}{2} V_0 [c(\mathbf{x}) - \bar{c}]^2 + \sum_{\alpha=1}^{3} \Lambda_\alpha \left(\frac{\partial c}{\partial x_\alpha} \right)^2 \right\} d^3\mathbf{x}, \qquad (9.6)$$

where N_V is the number of atoms per unit volume, i.e., the inverse of Ω_0 defined in Eq. (6.4). The integration is over the volume V_N of the periodic region. Since an averaging method such as that of Eq. (1.8) has been used implicitly in deriving (9.5), the amplitudes $\Gamma(k)$ vanish for a completely disordered solution. Hence the internal energy E is in fact a *configurational energy* in this case (which could be denoted ΔE), the difference between the energy of a particular (continuum) configuration and that of the fully disordered state (random configuration).

This is the familiar *square-gradient* continuum approximation,[65] the concentration gradient taking care of the nonlocal character of the energy in real space. Note that the expansion in Eq. (9.2) was performed about any $V(h)$ minima, including BZ boundary points, so that the continuum gradient formula (9.6) may be used in ordering as well as clustering cases. This is not generally recognized because the expansion in earlier treatments was performed only about the origin. It is merely required that point (h°) be a true minimum so that all three eigenvalues $\lambda_\alpha(h^\circ)$ be positive, insuring that the *gradient energy coefficients* be positive. The concentration wave $c(\mathbf{x}) - \bar{c}$ [obtained by some form of sublattice averaging or window averaging, as in Eqs. (2.13) or (1.8)], must therefore be regarded as a long-period modulation of the ordering wave. An example is given in Fig. 12: the concentration profile $c(\mathbf{x})$ is seen to define an *ordered domain*.

The origin of reciprocal space for cubic crystals has cubic symmetry (see Tables II and III), so that all three eigenvalues have common value λ. Hence, for clustering cubic systems

$$E = N_V \int_{V_N} [\Omega(c) + \Lambda(\nabla c)^2] d^3\mathbf{x} + E_e \qquad (9.7)$$

where E_e is the continuum elastic energy, to be evaluated in Section 12, and where we have

$$\Omega(c) = \tfrac{1}{2} \Omega^0 [c(\mathbf{x}) - \bar{c}]^2 \qquad (9.8)$$

with

$$\Omega^0 = \sum_p \omega(p). \qquad (9.9)$$

[65] J.W. Cahn and J.E. Hilliard, *J. Chem. Phys.* **28**, 258 (1958).

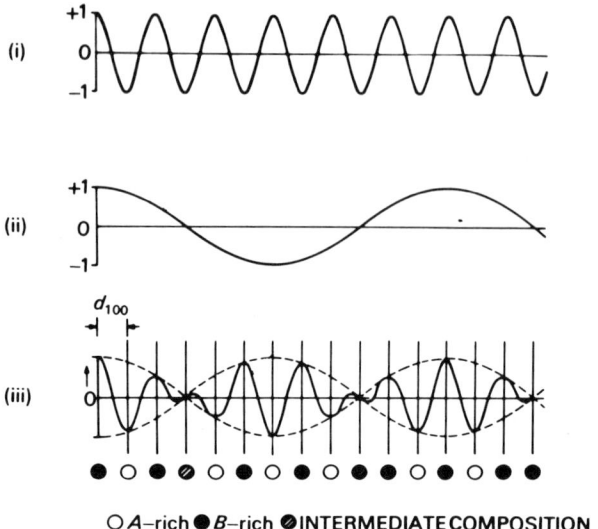

FIG. 12. Long-period modulation of order. (i) Ordering wave $\mathbf{k}(h°)$. (ii) Modulating wave $c(\mathbf{x}) - \bar{c}$. (iii) Resulting ordered domains.

The sum $\Omega°$ of pair parameters is proportional to the second derivative with respect to average concentration squared of the bulk internal energy of a solid solution of uniform concentration equal to the average $\bar{c}$. The first term of Eq. (9.7) is the internal energy counterpart of the Cahn and Hilliard[65] free energy expression,

$$F = \int_{V_N} [f(c) + \kappa(\nabla c)^2] \, d^3\mathbf{x} \quad (9.10)$$

which may be obtained simply from Eq. (9.7) by adding an appropriate entropy contribution. In Eq. (9.10), $f(c)$ is the bulk Helmholtz free energy per unit volume. The gradient energy coefficient

$$\kappa = N_V(a/4\pi)^2 \lambda_\alpha(0) \quad (9.11)$$

is positive and has units of reciprocal length.

10. Discussion of Internal Energy

It was shown, in Sections 6–9, how a quadratic form in the concentration deviations γ_i and displacement components u_α could be derived for the pairwise internal energy E, all parameters of which could be obtained uniquely, in principle, from appropriate energy-wavenumber characteristics $W_{ij}(K)$. Emphasis was placed on the k-space formulation because of its great formal simplicity: in Fourier space, sums of two-body interactions are replaced by sums of single-body terms. Also, the terms of the sum factor very conveniently in products of k-space potentials and structure factors. Hence, the potentials may be computed once and for all, the energy of various defect configurations being computed by merely altering the structure factor.

A number of parameters and functions in direct and reciprocal space were introduced in Section III: the unrelaxed lattice energy coefficient Ω_{ij}, the force function Ψ_i, the transformed force-constant matrix Φ, the k-space potential V_{ij}, the gradient energy vector Λ. Since each of these can be derived from the energy-wavenumber characteristic W_{ij}, it would be of course of primary interest to have values of these k-space functions. At best, we may count on known pseudopotential form factors for pure simple metals. Generalizations to concentrated alloys of arbitrary metals appears to be a long way off. The defining equations (6.30)–(6.32) for Ω_{ij}, $\Psi_{i\alpha}$, $\Phi_{\alpha\beta}$ indicate, particularly for the latter, that $W_{ij}(K)$ is required to converge rapidly beyond the first Brillouin zone. For simple metals, pseudopotential theory indicates that this is usually the case.

One may attempt empirical determinations of the energy-wavenumber characteristic, or its Fourier transform (7.2), or at least the asymptotic form (7.3). Various authors[66-69] have used experimental SRO X-ray intensity data in conjunction with fluctuation theory to determine the constants A and φ in Eq. (7.3), but it is not yet known, through independent checks, how reliable these experimental determinations are. Another alternative is to use semiempirical pair potentials such as those derived by Machlin.[70] Certain bulk properties appear to be predicted quite well by these potentials, but a determination of a $V(h)$ potential from these has not been attempted. Finally, one may simply postulate a quadratic form for the energy E, and then estimate the coefficients by comparing to experimental data. For example, the knowledge of the ordered ground-state structure will yield ranges of values for the pair parameters through the use of the maps of Figs. 8, 9, and 10, or from the more complete ground-state diagrams.[63,71-73] Short-range order fluctuations yield maximum information however, as will be shown in Section 25.

[66] S.C. Moss and P.C. Clapp, *Phys. Rev.* **171**, 764 (1968).
[67] S.W. Wilkins, *Phys. Rev. B* **2**, 3935 (1970).
[68] S.W. Wilkins and C.G. Shirley, *J. Appl. Crystallogr.* **8**, 197 (1975).
[69] P. Bardhan and J.B. Cohen, *Acta Crystallallogr., Sect. A* **32**, 597 (1976).
[70] E.S. Machlin, *Acta Metall.* **22**, 95, 109, 367, and 1433 (1974); **24**, 543 (1976).
[71] M.J. Richards and J.W. Cahn, *Acta Metall.* **19**, 1263 (1971).
[71a] S.M. Allen and J.W. Cahn, *Scripta Metall.* **7**, 1261 (1973).
[72] S.M. Allen and J.W. Cahn, *Acta Metall.* **20**, 423 (1972).
[73] T. Kudō and S. Katsura, *Prog. Theor. Phys.* **56**, 435 (1976).

Diffuse intensity measurements can even furnish such basic information as the shape of the Fermi surface, as mentioned in Section 7.[60] The basic theory does not appear to be completely worked out yet, although the role of the singularities at $2\mathbf{k}_F$ is qualitatively well understood. To complicate matters, diffuse intensity in the shape of Fermi surfaces has also been interpreted by Sauvage and Parthé[74] as arising from scattering by vacancy clusters in off-stoichiometric carbides, and by De Ridder et al.[75,76] as arising from scattering by Ni–Mo clusters in Ni_4Mo, for example. The connection between these viewpoints has not been established.

What is often overlooked in these experimental comparisons is that the internal energy functions, all the "physics" of which resides in $V(h)$, tells only part of the story. One should actually compare experimental data with the free energy, bearing in mind that the configurational entropy is essentially many-body. Such comparisons are more difficult and thus far less advanced, as will be apparent when free energy models are discussed in Section V. Before that, the more specialized topic of elastic interactions is briefly presented in Section IV.

IV. Elastic Interactions

The subject of elastic interactions of crystalline defects has been studied extensively these past few years. Each of the topics briefly treated below: microscopic elastic theory or lattice statics (Section 11); continuum elastic theory (Section 12); cluster interactions (Section 13); and displacive scattering (Section 14) could in fact each be the object of a separate review paper. The interested reader will find pertinent reviews in published conference proceedings such as the ones on "Interatomic Potentials and Simulation of Lattice Defects" (Battelle Inst. 1971),[77] "Point Defects and Their Aggregates in Metals" (Univ. of Sussex, 1972),[78] and on "Lattice Distortions and Local Atomic Arrangements"

---

[74] M. Sauvage and E. Parthé, *Acta Crystallogr., Sect. A* **28**, 607 (1972).

[75] R. De Ridder, G. Van Tendeloo, and S. Amelinckx, *Acta Crystallogr., Sect. A* **32**, 216 (1976).

[76] R. De Ridder, G. Van Tendeloo, D. Van Dijck, and S. Amelinckx, *J. Phys. (Paris)* **38**, C7-178 (1977).

[77] P.C. Gehlen, J.R. Beeler, Jr. and R.I. Jaffee, eds., "Interatomic Potentials and Simulation of Lattice Defects." Plenum, New York, 1972.

[78] "International Conference on Point Defects and Their Aggregates in Metals" (R. Bullough, ed.), *J. Phys. F* **2**, 233–496 (1973).

(KFA Jülich, 1974),[79] for example. Here we touch only briefly on these topics in so far as they relate to the general subject of this article.

11. Microscopic Elastic Theory

The basic ideas of microscopic elasticity (or lattice statics) were introduced in Section 6. The great difficulty of evaluating parameters from first principles, either through Eqs. (6.20, 6.21 or 6.31, 6.32), was alluded to. Consequently, the approach generally taken is a phenomenological one: values of parameters are assumed and varied systematically until a good fit is obtained to selected macroscopic measurements. In particular, it is customary to assume fictitious forces $f_i(p)$ acting on site (p), and to modify the force model and force magnitudes until satisfactory agreement is achieved between calculated and measured X-ray or neutron diffuse intensity from defect solid solutions. These so-called *Kanzaki forces* were actually introduced by Matsubara[80] and the methods of calculation were subsequently extended by Kanzaki[81] and by Krivoglaz and Tikhonova.[82]

a. Transformation Defects

Instead of forces, one may alternately define phenomenological coupling parameters $\varphi_{i\alpha}$ which are related to the Kanzaki forces by[36]

$$f_\alpha(p) = -\sum_{i=1}^{n} \sum_{r}{}' \varphi_{i\alpha}(r)\gamma_i(p + r). \qquad (11.1)$$

This equation differs from (6.27) in the following respects: first, the parameters $\varphi_{i\alpha}$ are given *a priori* and are not necessarily related to any pseudopotential formulation; second, the summation is over defects ($i = 1, \ldots, n$) only, the host having implicitly been removed by the full triangle scheme of (6.40). The advantage of this procedure was pointed out by Cook[83] who introduced *transformation defects* defined by their effects on neighboring lattice sites, the effect being chosen so as to simulate a structural transformation which is known to take place in the system studied. The transformation in question can be fairly complex;

---

[79] "International Discussion Meeting on Studies of Lattice Distortions and Local Atomic Arrangements by X-ray, Neutron and Electron Diffraction" (W. Schmetz, ed.), *J. Appl. Crystallogr.* **8**, 79–229 (1975).
[80] T.J. Matsubara, *J. Phys. Soc. Jpn.* **7**, 270 (1952).
[81] H. Kanzaki, *J. Phys. Chem. Solids* **2**, 24 and 107 (1957).
[82] M.A. Krivoglaz and E.A. Tikhonova, *Ukr. Fiz. Zh.* **3**, 297 (1958).
[83] H.E. Cook, *Acta Metall.* **21**, 1431 (1973).

such as a Martensitic transformation giving rise to incoherent (dislocation) interfaces.[84] Cook considers two basic types of fictitious defects: a Bain type and a Burgers type.[83] The former creates a local distortion of arbitrary symmetry in the neighborhood of the defect, the latter creates a displacement or shift of certain neighboring atoms. For example, Cu apparently precipitates in an Al-Cu solid solution initially in the form of platelets, one Cu layer thick. Hence cubic symmetry is broken as the Cu atoms cluster, which must necessarily mean that the values of the pair interaction parameters (6.37) depend on the local surrounding atomic configuration. This in turn introduces in the energy (6.36) terms of order higher than the second in concentration, in other words, *anharmonic* contributions. In a Cu platelet, Cu-Cu bonding within the Cu plane must differ from Cu-Al bonding across the plane, so that tetragonal rather than cubic distortion is expected at a Cu site. Hence, the Cu solute in this state of clustering should be modeled by an "ellipsoidal atom," i.e., by a local distortion tensor of the form given later by Eq. (11.10). This is further discussed in Section 13,a. These ellipsoidal or Bain-type defects can then be introduced in a lattice statics energy formula, such as Eq. (6.34), as shown presently. On the other hand, Burgers defects can be used to model the omega phase transition[85] in Group IVB alloys for example, since this transformation is characterized by the collapse of 111 bcc lattice planes,[86] hence by atomic displacements or shuffles. Dislocations can also be described through the use of displacive defects, the atomic displacements being given by the Burgers vector of the dislocation, hence the term "Burgers defects." As an example, atomic displacements associated with dislocations, including the core region, have been calculated by Boyer and Hardy[87] by lattice statics in the harmonic approximation, i.e., by use of Eq. (6.33). Other treatments are reviewed by Bullough and Tewary.[88]

In this article, we are primarily concerned with *replacive* effects, hence only Bain-type defects will be considered explicitly. In contradistinction to displacive (Burgers) defects, replacive (Bain) defects are conserved in most applications, hence the parameters $\bar{c}_i$ defined by Eq. (1.4) are constant, and it is preferable in that case to use the deviation operators $\gamma_i(p)$ rather than the $\sigma_i(p)$ themselves. The formalism of previous sec-

---

[84] H.E. Cook, *Acta Metall.* **23**, 1027 (1975).
[85] H.E. Cook, *Acta Metall.* **23**, 1041 (1975).
[86] D. de Fontaine, *Acta Metall.* **18**, 275 (1970).
[87] L.L. Boyer and J.R. Hardy, *Philos. Mag.* [8] **24**, 647 (1971).
[88] R. Bullough and V.K. Tewary, "Lattice Theories of Dislocations," in "Dislocation Theory—A Collective Treatise" (F.R.N. Nabarro, ed.), ch. 5. M. Dekker, New York, 1975.

tions can then be taken over without essential modifications. In this way, even anharmonic effects, such as those mentioned above in the case of Al–Cu clustering, can be treated approximately by a harmonic theory resembling that of Section 6. The pertinent formulas to be used are

$$E_e = \frac{N}{2} \sum_h \sum_{i,j} \mathscr{E}_{ij}(h) \Gamma_i^*(h) \Gamma_j(h) \tag{11.2}$$

for the elastic energy E_e, with k-space elastic potential

$$\mathscr{E}_{ij} = \Phi_{ij} - \Phi_{i\alpha} \Phi_{\alpha\beta}^{-1} \Phi_{j\beta}^*, \tag{11.3}$$

the double sum over (i, j) in (11.2) being the appropriate one for reduction scheme (6.38). These equations correspond to (6.34) and (6.35) but differ from the latter in that $\Phi_{i\alpha}$ is the Fourier transform of phenomenological parameters related to fictitious Kanzaki forces, and Φ_{ij} is the Fourier transform of φ_{ij} parameters, similar to the ω_{ij} of Section 6, but derived by suitable models from the $\varphi_{i\alpha}$ parameters.[36,83] The $\varphi_{\alpha\beta}$ are here assumed to include noncentral interactions. Equations (6.23) and (6.24) are taken over without modification, but the sign in (6.22) is customarily changed to $(+)$,[36] thus

$$\varphi_{ij}(0) = + \sum_p{}' \varphi_{ij}(p). \tag{11.4}$$

This amounts to a change in reference state: in Section 6 the reference state was that consisting of $n + 1$ "pure solids" each containing $\bar{c}_i N$ ($i = 0, 1, \ldots, n$) atoms (ions) exactly occupying the sites of the average lattice of the solid solution of composition $\bar{c}_i$. According to the definition (11.4) however, the reference state is that of $n + 1$ stress-free pure solids with defects (atoms, ions) occupying exactly the sites of the pure solid lattices, the lattice parameters and symmetry of which may differ from those of the solid solution lattice.

The various phenomenological parameters are related to the elastic constants of the material $c_{\alpha\beta\gamma\delta}$ and to an appropriate distortion tensor $\eta_{i\alpha\beta}$ by the so-called *long-wave relations*[36]

$$-\sum_p \varphi_{\alpha\beta}(p) x_\beta(p) x_\delta(p) = \Omega_0 (c_{\alpha\beta\gamma\delta} + c_{\alpha\delta\gamma\beta}) \tag{11.5}$$

$$\sum_p \varphi_{i\alpha}(p) x_\beta(p) = \Omega_0 \eta_{i\gamma\delta} c_{\alpha\beta\gamma\delta} \tag{11.6}$$

$$\sum_p \varphi_{ij}(p) = \Omega_0 \eta_{i\alpha\beta} \eta_{j\gamma\delta} c_{\alpha\beta\gamma\delta}, \tag{11.7}$$

the first of which is borrowed from lattice dynamics. In these equations, Ω_0 is the atomic volume, $\varphi_{i\alpha}$ and φ_{ij} refer to the Bain component of the transformation defect, and $\eta_{i\alpha\beta}$ is the Bain strain for unit concentration change of defect of type i. A purely dilatational strain in crystals of cubic symmetry is characterized by the tensor (the subscript denoting defect type being temporarily discarded for simplicity)

$$\eta_{\alpha\beta} = \eta \delta_{\alpha\beta} \qquad (11.8)$$

where the scalar

$$\eta = \frac{1}{a_0} \frac{da_0}{d\bar{c}} \qquad (11.9)$$

is the one introduced by Cahn in his theory of spinodal decomposition.[89] In Eq. (11.8), a_0 is the lattice parameter and $\bar{c}$ the average concentration of solute atoms.

If the solute atoms or defects have tetragonal instead of cubic strain fields, as in the example of Al–Cu platelets mentioned above, then we have

$$\{\eta_{\alpha\beta}\} = \begin{pmatrix} \eta_{11} & 0 & 0 \\ 0 & \eta_{11} & 0 \\ 0 & 0 & \eta_{33} \end{pmatrix} \qquad (11.10)$$

with tetragonality ratio

$$t = \eta_{11}/\eta_{33} . \qquad (11.11)$$

Other symmetries are possible. Another useful concept is that of the *dipole (or double-force) tensor* $P_{\alpha\beta}$ used by the German school.[90] It is related to the distortion tensor by a formula derived from Eq. (11.6)[91]:

$$P_{\alpha\beta} = \Omega_0 \eta_{\gamma\delta} c_{\alpha\beta\gamma\delta} . \qquad (11.12)$$

b. k-Space Elastic Energy

The great advantage of expressing energy as a quadratic form in configuration variables, i.e., as in the "harmonic" formulas (6.34) or (11.2), is that all the physics of the problem is contained in the k-space functions $\mathcal{V}_{ij}(h)$ or $\mathcal{E}_{ij}(h)$ which can be calculated quite independently of the ion or defect configurations. This has practical implications for computer calculations: the k-space elastic energy coefficient $\mathcal{E}_{ij}$ can be computed at each of the N points of the primitive reciprocal unit cell (or BZ), once force constant and Kanzaki force models have been adopted, and the

---

[89] J.W. Cahn, *Acta Metall.* **10**, 179 (1962).
[90] R. Siems, *Phys. Status Solidi* **42**, 105 (1970).
[91] A.S. Nowick and B.S. Berry, "Anelastic Relaxation in Crystalline Solids," p. 186. Academic Press, New York, 1972.

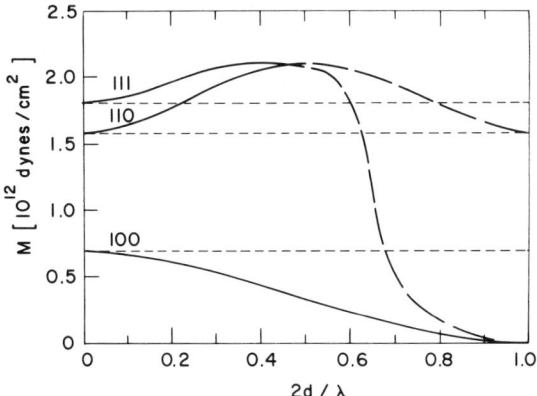

FIG. 13. Curves of effective modulus $\mathscr{E}(h)/(2\Omega_0\eta^2)$ for a bcc solution in quasi-nearest-neighbor force model. The modulus goes to zero at $h = (2d/\lambda) = 1$ for the $\langle 100 \rangle$ and $\langle 111 \rangle$ directions, resulting in no elastic energy for the CsCl-type ordered structure [H. E. Cook and D. de Fontaine, *Acta Metall.* **17**, 915 (1969)].

result stored on magnetic tape. Elastic energies for arbitrary defect configurations can then be performed very efficiently by first obtaining the Γ_i concentration wave spectra by a fast Fourier transform algorithm,[92] and then by performing the single sum over the N points in the reciprocal unit cell, as indicated in Eq. (11.2). A large number of different defect configurations, of arbitrary complexity, can thus be run very inexpensively for N even as large as $32^3 = 32,768$.[93]

It is instructive to plot values of the k-space elastic energy coefficient along high-symmetry directions such as $\langle 100 \rangle$, $\langle 110 \rangle$, and $\langle 111 \rangle$ in lattices of cubic symmetry. Figure 13 shows curves for $\mathscr{E}(h)/(2\Omega_0\eta^2)$ for a binary substitutional solution in a bcc lattice in a quasi nearest-neighbor Kanzaki force model,[36] the solute atoms being associated with a local cubic distortion as given by Eq. (11.9). Each curve represents the elastic energy of a concentration wave of unit amplitude in the indicated crystallographic directions. In the long-wavelength limit ($h = 0$), the $\langle 100 \rangle$ wave has lowest energy, the $\langle 111 \rangle$ highest. This is in accordance with continuum elasticity[89] for cubic crystals having positive value of the anisotropy parameter

$$A = 2C_{44} - C_{11} + C_{12}, \qquad (11.13)$$

---

[92] J.W. Cooley and J.W. Tuckey, *Math Comput.* **19**, 296 (1965).

[93] E. Seitz, D. de Fontaine, and F. Plesset, UCLA-ENG-7689 (unpublished report). School of Engineering and Applied Science, University of California, Los Angeles, 1976.

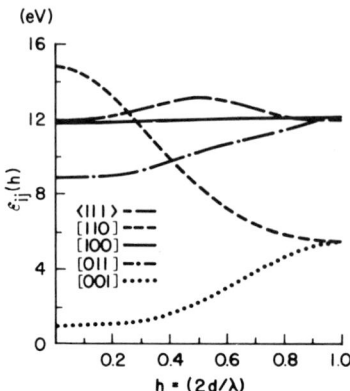

FIG. 14. Curves of $\mathscr{E}_{ij}$ for Ni along various directions calculated for the case of distortion tensors of tetragonal symmetry, with parallel tetragonality axes [H. Yamauchi and D. de Fontaine, UCLA-ENG-7398 (unpublished report), School of Engineering and Applied Science, University of California, Los Angeles, 1973].

where C_{11}, C_{12}, and C_{44} are the only nonvanishing elastic constants for cubic materials (in the Voigt notation).

Continuum elastic theory (Section 12) predicts no wavelength dependence of $\mathscr{E}(h)$. In that respect at least, the theory is defective since the $\langle h00 \rangle$ and $\langle hhh \rangle$ curves must coincide at $h = 1$ in bcc lattices for symmetry reasons.[37] The microscopic theory correctly predicts this behavior, as seen in Fig. 13, with vanishing elastic energy for the (100) wave at the BZ boundary. This is a reasonable result for the adopted force model since, according to Fig. 2, the (100) wave converts a bcc structure to a CsCl ($L1_0$) structure which easily accommodates a large atom in the cubic cell center and small ones at the corners (or vice versa). A (001) reciprocal lattice plane section with $\mathscr{E}(h)$ contours was shown previously in Fig. 5.

The introduction of polar or "ellipsoidal" defects with distortion tensor given by (11.10) lifts the degeneracy of the $\langle 100 \rangle$, $\langle 110 \rangle$, and $\langle 111 \rangle$ directions. This is illustrated in Fig. 14 calculated by means of a many-neighbor force model using the $\varphi_{\alpha\beta}$ force constants of Cu,[94] for which the lattice is fcc. The tetragonality ratio, defined by Eq. (11.11), was $t = 0.2$, and the defects were assumed to have tetragonality axis along the [001] direction.

A special case is of interest, that of two defects one (i) located at the origin of a periodic supercell containing N lattice sites, the other (j)

---

[94] H. Yamauchi and D. de Fontaine, UCLA-Eng-7389 (unpublished report), School of Engineering and Applied Science, University of California, Los Angeles, 1973.

located at arbitrary site (r). It is then easy to show that the elastic interaction energy $\epsilon_{ij}(r)$ between two defects separated by $\mathbf{x}(r)$ is given by the back Fourier transform of $\mathscr{E}_{ij}(h)$, with $\mathscr{E}_{ij}(0)$ set equal to zero. It also follows that $\mathscr{E}_{ii}(0)$ is twice the self-energy of defect of type i. Hence the total elastic energy can be given a real space representation

$$E_e = \frac{1}{2} \sum_{p,p'} \sum_{ij} \epsilon_{ij}(p' - p)\gamma_i(p)\gamma_j(p') \qquad (11.14)$$

which is analogous to Eq. (6.36). Interestingly, a plot of the real-space $\epsilon(r)$ as a function of $\mathbf{x}(r)$ exhibits positive and negative (repulsive and attractive) interactions looking for all the world like Friedel oscillations. One may see this, for example, in the results of calculations of the interaction energies of vacancy pairs in Al and Cu performed by Hardy and Bullough.[95,96]

c. Average Elastic Energy

Formulas such as (11.2) and (11.14) may be used for the case of a few interacting point defects, where the exact location of defects are known or assumed. For concentrated solid solutions, the primary subject of this article, only expectation values of site occupation are available. The average value of the elastic energy is thus, by Eqs. (3.7) and (11.2)

$$\langle E_e \rangle = \frac{N}{2} \sum_{ij} \sum_h \mathscr{E}_{ij}(h) Q_{ij}(h). \qquad (11.15)$$

For a random (completely disordered) solution we therefore have, by Eqs. (3.8) and (2.7),

$$\langle E_e \rangle^R = \frac{N}{2} \sum_{ij} \bar{c}_i (\delta_{ij} - \bar{c}_j) \bar{\mathscr{E}}_{ij} \qquad (11.16)$$

where $\bar{\mathscr{E}}_{ij}$ is the Brillouin zone average of the k-space elastic potential:

$$\bar{\mathscr{E}}_{ij} = \frac{1}{N} \sum_h \mathscr{E}_{ij}(h). \qquad (11.17)$$

For simplicity, let us now consider binary solutions. From the foregoing, the expectation value of the elastic energy may be written

$$\langle E_e \rangle = \frac{N}{2} \bar{\mathscr{E}} \bar{c}(1 - \bar{c}) + \frac{1}{2} \sum_{p \neq p'} \epsilon(p' - p) \langle \gamma(p)\gamma(p') \rangle. \qquad (11.18)$$

---

[95] J.R. Hardy and R. Bullough, *Philos. Mag.* [8] **15**, 237 (1967).
[96] R. Bullough and J.R. Hardy, *Philos. Mag.* [8] **17**, 833 (1968).

The first term of this equation represents the energy of a disordered solution and is equal to N times the elastic self-energy of a single defect times a parabolic dependence of the average concentration. This result was derived originally by Eshelby[97] and by Friedel.[98] The second term of Eq. (11.18) represents a sum of defect-pair interactions which vanishes in the case of a random solution. Hence this term is the configurational elastic energy. This proves the result that the interaction energy of a random distribution of defects vanishes (although the total energy does not). The Fourier space counterpart of Eq. (11.18) is

$$\langle E_e \rangle = \frac{N}{2}\bar{\mathscr{E}}\bar{c}(1-\bar{c}) + \frac{N}{2}\sum_h [\mathscr{E}(h) - \bar{\mathscr{E}}]\langle|\Gamma(h)|^2\rangle \qquad (11.19)$$

in which the k-space configurational energy coefficient is just the deviation of $\mathscr{E}(h)$ from its BZ average.

It is often convenient to express the average elastic energy as a quadratic form in CW amplitudes. To that effect, microcanonical averaging must be performed according to the general formula

$$\langle \gamma_i(p) \rangle_0 = \frac{1}{N}\sum_{p'} s(p'-p)\gamma_i(p) \qquad (11.20)$$

which contains both (1.8) and (2.11) as special cases, depending upon whether s is a small "window" about (p), or a "grid" through (p), over an ensemble of systems, respectively. Equation (4.13) follows, rewritten here as

$$\langle \Gamma_i^* \rangle_0 \langle \Gamma_j \rangle_0 = |S|^2 \langle \Gamma_i^* \Gamma_j \rangle \qquad (11.21)$$

where S, the Fourier transform of the averaging function s, must have value unity at the origin because of the normalization condition on s. Consequently, in order to write the energy as

$$\langle E_e \rangle_0 = \frac{N}{2}\sum_{i,j}\sum_h \mathscr{E}_{ij}(h)\langle \Gamma_i^*(h)\rangle_0 \langle \Gamma_j(h)\rangle_0 \qquad (11.22)$$

instead of the correct (11.15), it is necessary to neglect implicitly a portion $(1 - |S|^2)$ of the Q_{ij} spectrum. This has important consequences in comparing microscopic and continuum elastic theories, as shown in the next section.

12. Continuum Elastic Theory

If the CW spectrum of a solution is dominated by long wavelengths (more than about 10 lattice parameters), then the discrete nature of the

---
[97] J.D. Eshelby, *Solid State Phys.* **3**, 79 (1956).
[98] J. Friedel, *Adv. Phys.* **3**, 466 (1954).

underlying lattice becomes unimportant. In that case, and also in the case of amorphous solid solutions, it is more convenient to use continuum rather than microscopic elasticity. The results are then independent of the force models adopted. Short-wavelength CW's are always present, however, since any amount of compositional disorder must introduce a high-frequency "noise" in the Q_{ij} spectrum. Since the elastic energy of these short waves cannot be treated properly in the continuum approximation, it is best to attenuate the short-wavelength portion of the Fourier spectrum by a "window" function which takes the value unity in the vicinity of the origin of the BZ, and decays rather rapidly beyond a selected large-k cutoff, say 0.1 times the reciprocal lattice parameter. These requirements define an $S(h)$ function for Eq. (11.21) whose Fourier transform is a window function 10 lattice parameters wide, say. Hence, any compositional fluctuations on a scale less than about 10 lattice parameters will not be "seen" in this description.

The window averaging introduces continuously varying local concentration deviations in a small neighborhood of point $\mathbf{x}(p)$, for which the discrete → continuum transformation can be symbolized as follows

$$\gamma_i(p) \to \langle \gamma_i(p) \rangle_0 \equiv \hat{\gamma}_i(\mathbf{x}), \tag{12.1}$$

the caret over a symbol designating, here and in what follows, parameters pertaining to the continuum model. In Fourier space we have analogously

$$\Gamma_i(h) \to \langle \Gamma_i(h) \rangle_0 \equiv \hat{\Gamma}_i(\mathbf{k}). \tag{12.2}$$

a. Basic Formulas

The elastic energy $\hat{E}_e$ of a continuum whose elastic strain $e_{\alpha\beta}(\mathbf{x})$ in volume element $d^3\mathbf{x}$ varies continuously with $\mathbf{x}$ is given by

$$\hat{E}_e = \frac{1}{2} \int_{V_N} c_{\alpha\beta\gamma\delta} e_{\alpha\beta}(\mathbf{x}) e_{\gamma\delta}(\mathbf{x}) \, d^3\mathbf{x}, \tag{12.3}$$

where the integration is performed over the volume V_N of the periodic solid containing lattice sites. The volume element $d^3\mathbf{x}$ is the averaging "window" defined above; its dimensions will be allowed to go to zero for mathematical convenience. The symmetric elastic strain tensor is given by

$$e_{\alpha\beta} = \tfrac{1}{2}(\epsilon_{\alpha\beta} + \epsilon_{\beta\alpha}) \tag{12.4}$$

where $\epsilon_{\alpha\beta}$ are elastic distortions. The total local distortions are given by a sum of elastic and stress-free distortions[99]

$$u_{\alpha,\beta} \equiv \partial u_\alpha / \partial x_\beta = \epsilon_{\alpha\beta} + \epsilon_{\alpha\beta}^0 . \tag{12.5}$$

[99] T. Mura, *Proc. R. Soc. London, Ser. A* **280**, 528 (1964).

Let us now assume that the stress-free distortions are due to local defect concentration changes in $d^3\mathbf{x}$:

$$\epsilon_{\alpha\beta}^0(\mathbf{x}) = \sum_{i=1}^{n} \eta_{i\alpha\beta}\hat{\gamma}_i(\mathbf{x}) \tag{12.6}$$

where $\eta_{i\alpha\beta}$ has the same meaning as in Section 11.

By putting (12.4), (12.5), and (12.6) into (12.3) we obtain

$$\hat{E}_e = \frac{1}{2}\int_{V_N} c_{\alpha\beta\gamma\delta}\left[u_{\alpha,\beta}u_{\gamma,\delta} - 2u_{\alpha,\beta}\sum_{j=1}^{n}\eta_{j\gamma\delta}\hat{\gamma}_j \right. \\ \left. + \sum_{i,j}\eta_{i\alpha\beta}\eta_{j\gamma\delta}\hat{\gamma}_i\hat{\gamma}_j \right] d^3\mathbf{x}. \tag{12.7}$$

The total distortions $u_{\alpha,\beta}$ may be found by imposing the condition of minimum elastic energy (6.28). As shown in Section 6, this is best done in Fourier space, yielding the continuum counterpart of Eqs. (11.2) and (11.3):

$$\hat{E}_e = \frac{V_N}{2}\sum_{\mathbf{k}}\sum_{ij}\hat{\mathscr{E}}_{ij}(\mathbf{k})\hat{\Gamma}_i^*(\mathbf{k})\hat{\Gamma}_j(\mathbf{k}) \tag{12.8}$$

with

$$\hat{\mathscr{E}}_{ij} = \hat{\Phi}_{ij} - \hat{\Phi}_{i\alpha}\hat{\Phi}_{\alpha\beta}^{-1}\hat{\Phi}_{j\beta}^* \tag{12.9}$$

and

$$\hat{\Phi}_{ij} = c_{\alpha\beta\gamma\delta}\eta_{i\alpha\beta}\eta_{j\gamma\delta} \tag{12.10}$$

$$\hat{\Phi}_{j\alpha} = -ic_{\alpha\beta\gamma\delta}k_\beta\eta_{j\gamma\delta} \tag{12.11}$$

$$\hat{\Phi}_{\alpha\beta} = c_{\alpha\gamma\beta\delta}k_\gamma k_\delta. \tag{12.12}$$

The latter equality defines the so-called Green–Christoffel tensor,[100] the inverse $\hat{\Phi}_{\alpha\beta}^{-1}$ of which is the k-space Green's tensor for the problem. Equation (12.8) has the same form as (11.22), which was anticipated since the "window averaging" was used from the start in setting up the continuum formulation.

In principle then, the problem of determining the elastic energy of a continuum containing arbitrary distribution of defects of known distorting characteristics $\eta_{\alpha\beta}$ is completely solved by the set of Eqs. (12.8)–(12.12). There are two practical difficulties however: the inversion of the Green–Christoffel matrix and performing the summation in (12.8). In practice, this summation is generally replaced by an integral over the continuous wave vector $\mathbf{k}$, to which a short-wavelength cutoff must be applied, generally at the Debye frequency.[100,101] In the continuum approximation, the k-space elastic energy coefficient $\tilde{\mathscr{E}}$ depends only on the directional cosines

$$\boldsymbol{\kappa} = \mathbf{k}/k \qquad (12.13)$$

of the wave vector $\mathbf{k}$, not on its magnitude k.

b. Cubic Crystals

The Green's tensor for cubic crystals is well known,[102] so that it is a relatively straightforward matter to obtain the energy coefficient in that case. The formula for $\tilde{\mathscr{E}}$ is particularly simple for the case of defects with spherically symmetric dilatational strain fields, i.e., for $\eta_{\alpha\beta}$ given by Eq. (11.8). The formula has been given by a number of authors[103]; we give here Hilliard's version[104]:

$$Y(\boldsymbol{\kappa}) \equiv \tfrac{1}{2}\tilde{\mathscr{E}}(\boldsymbol{\kappa})/\eta^2 = \tfrac{1}{3}(C_{11} + 2C_{12}) \left\{ 3 - \frac{C_{11} + 2C_{12}}{C_{12} + C_{44}} \right.$$
$$\left. \times \left[\frac{B^2 + 2B(C-1)I_4 + 3(C-1)^2 I_6}{B^2(B+C) + B(C^2-1)I_4 + (C-1)^2(C+2)I_6} \right] \right\} \qquad (12.14)$$

with

$$B = C_{44}/(C_{12} + C_{44}); \qquad C = (C_{11} - C_{44})/(C_{12} + C_{44}) \qquad (12.15)$$

$$I_4 = \kappa_2^2\kappa_3^2 + \kappa_3^2\kappa_1^2 + \kappa_1^2\kappa_2^2; \qquad I_6 = \kappa_1^2\kappa_2^2\kappa_3^2, \qquad (12.16)$$

$\kappa_\alpha(\alpha = 1, 2, 3)$ being the directional cosines of the wave vector $\mathbf{k}$. Corresponding formulas can be written down for crystals of the tetragonal class.[105] General formulas have also been obtained for cubic crystals with $\eta_{\alpha\beta}$ given by the diagonal matrix (11.10).[101]

A useful approximate formula for Y in cubic crystals was given by Cahn[89]

$$Y(\boldsymbol{\kappa}) = \tfrac{1}{3}(C_{11} + 2C_{12}) \left[3 - \frac{C_{11} + 2C_{12}}{C_{11} + 2AI_4} \right], \qquad (12.17)$$

where A is the anisotropy parameter defined by Eq. (11.13). Equation

---

[100] V.K. Tewary, *Adv. Phys.* **22**, 757 (1973).
[101] H. Yamauchi and D. de Fontaine, *Acta Metall.* (in press).
[102] W.H. Zachariasen, "X-ray Diffraction in Crystals." Wiley, New York, 1945.
[103] For example, P.H. Dederichs and J. Pollman, *Z. Phys.* **255**, 315 (1972).
[104] J.E. Hilliard, *in* "Phase Transformations," p. 557. Am. Soc. Metals, Metals Park, Ohio, 1970.
[105] J.E. Hilliard, private communication.

(12.17), which is exact for the three directions $\langle 100 \rangle$, $\langle 110 \rangle$, and $\langle 111 \rangle$, shows that a concentration wave along $\langle 100 \rangle$ will have minimum energy, and one along $\langle 111 \rangle$ maximum energy for solids with positive anisotropy ($A > 0$), the usual case. For $A < 0$ (Nb, V, Mo), the opposite holds. In isotropic solids ($A = 0$), a further simplification of Eqs. (12.14) or (12.17) occurs:

$$Y_0 = C_{11} + C_{12} - 2(C_{12}^2/C_{11}) = E/(1 - \nu) \qquad (12.18)$$

where in this formula E is Young's modulus and ν is Poisson's ratio. In that case, even the orientation dependence of the concentration wave disappears, as expected, hence the notation Y_0. The elastic energy of an isotropic solution with isotropic defects is thus expressed simply by back-transforming Eq. (12.8):

$$\hat{E}_e = \eta^2 Y_0 \int_{V_N} [\hat{\gamma}(\mathbf{x})]^2 \, d^3\mathbf{x}, \qquad (12.19)$$

so that the elastic energy depends only on the integrated concentration deviations squared.

c. Elastic Energy Paradox

Consider a clustering solution whose internal energy is primarily of elastic origin, i.e., such that the component atoms of a binary solution, say, are of essentially different "sizes," creating large misfit distortions. Starting from the completely disordered state, how does the internal energy change as solute clusters grow in amplitude and size according to the theoretical models developed above?

(i) According to the microscopic theory, Eqs. (11.18) and (11.19), the random solute distribution has energy $(N/2)\bar{\mathscr{E}}\bar{c}(1 - \bar{c})$. The energy must decrease from this value upon clustering because shifting solute intensity $Q(h)$ toward the origin of k-space will result in lower integrated values of the product $\mathscr{E}(h)Q(h)$ since, for a clustering solution, $\mathscr{E}(h)$ [more generally $V(h)$] must attain its lowest value at the BZ center. In other words, the configurational energy coefficient $[\mathscr{E}(h) - \bar{\mathscr{E}}]$ must be negative for small $\mathbf{k}(h)$ in the elastically "soft" directions, $\langle 100 \rangle$ in the usual case of cubic crystals. Diffuse intensity streaks radiating along $\langle 100 \rangle$ from the reciprocal lattice positions are thus expected in diffraction patterns.

(ii) In the continuum formulation, the disordered state is characterized by constant local concentrations $c_i(\mathbf{x}) = \bar{c}_i$ (uniform solution), and hence concentration deviations $\hat{\gamma}_i(\mathbf{x})$ which vanish for all $\mathbf{x}$. The CW $\hat{\Gamma}_i(\mathbf{k})$ spectrum also vanishes everywhere, so that the elastic energy, by Eq. (12.8), vanishes identically. Upon clustering, CW intensity is continually created, hence $\hat{E}_e$ must steadily increase since $\hat{\mathscr{E}}$ is positive for all **k** (otherwise the crystal would be mechanically unstable).

(iii) Imagine a solid solution for which the k-space elastic energy coefficient $\mathscr{E}(h)$ is everywhere in the BZ equal to its average value $\bar{\mathscr{E}}$, as in the case of an isotropic continuum. Then, according to Eq. (11.19), $\hat{E}_e$ remains constant regardless of the way solute atoms are redistributed. This is the model of Friedel[98] and Eshelby.[97]

We therefore arrive at the paradoxical situation that according to (i) the elastic energy decreases upon clustering, according to (ii) it increases, while according to (iii) it remains constant. Model (iii) is clearly invalid since, even if the crystal is macroscopically isotropic [$A = 0$ in Eq. (12.17)], it must be microscopically anisotropic because of the discrete nature of the lattice; that is to say, the $\mathscr{E}$ surface depicted in Fig. 5 by contour lines can never be perfectly flat. Having discarded (iii), we are still left with the apparently mutually contradictory models (i) and (ii). This paradox is resolved when it is recalled that in setting up the continuum model, the short-wavelength portion of the concentration spectrum was discarded. For consistency, the $[1 - |S|^2]Q$ spectrum must be implicitly contained in a "chemical" energy which must be added to E_e in order to obtain the total internal energy. This "chemical" contribution may be added as a large negative constant $\mathscr{E}^0$ to $\hat{\mathscr{E}}(\mathbf{k})$, so that $[\mathscr{E}(\mathbf{k}) - \mathscr{E}^0]$ will be negative along the soft **k** directions, allowing the total internal energy to decrease in the course of clustering, in agreement with the correct result of the microscopic theory.

Of course, a complete formulation of the internal energy was already given in Section 6, where both "chemical" (electronic interactions) and "elastic" (relaxation) energies were considered. Why then construct an approximate continuum theory with chemical and elastic contributions treated from completely different viewpoints? This is simply for convenience in comparing coherent clustering and incoherent precipitation. The thermodynamics of the latter process are available for certain systems, and the elastic constants (or at least the Y_0 modulus) can often be measured separately. The two contributions can then be combined in an approximate model which conveniently bypasses interatomic potentials, force constants, Kanzaki forces, etc. Examples will be given in Section 23 where the difference between incoherent (equilibrium) and coherent (metastable) phase diagrams is discussed.

The result stated under (iii), above, is known as *Crum's theorem*,[106] the validity of which rests on the following assumptions: (1) the k-space elastic coefficient $\mathscr{E}_{ij}$ is a constant throughout k-space, which can only

---

[106] Cited by F.R.N. Nabarro, *Proc. R. Soc. London, Ser. A* **175**, 519 (1940).

be true for an isotropic continuum; (2) either periodic boundary conditions must be assumed or the crystal boundaries must be at infinity, where the concentration deviations and their spatial derivatives must vanish; (3) the individual crystalline defects considered must cause only isotropic expansion or contraction of neighboring elastic medium, otherwise $\mathscr{E}_{ij}$ cannot be isotropic; (4) linear elasticity must be assumed (harmonic approximation). Crum's theorem can therfore be stated as follows: *in an isotropic (linear) infinite elastic continuum, centers of dilatation do not interact through elastic strain fields.*

This nontrivial result has far-ranging consequences. In particular, it is instructive to investigate the conditions under which the theorem does *not* hold:

(1) The medium is not a continuum, i.e., the discrete nature of the lattice must be explicitly taken into account. It was shown[96] that in Copper various near-neighbor defect pairs have different elastic interactions energies, some repulsive, some attractive. Even in near isotropic cases (aluminum, for example), defect pairs interact strongly (positively and negatively) in the microscopic formulation. Likewise, if the medium can be approximated by a continuum, but is anisotropic, dilated (or contracted) defect clusters can interact elastically.

(2) The elastic medium is finite in extent. Then, as shown by Eshelby, dilatational defects can interact through image fields.[97]

(3) The defects (or defect clusters) cause local anisotropic distortions. This is the case for solid-solution precipitates of tetragonal symmetry in a cubic (or isotropic) matrix. Point defects in certain interstitial sites also create anisotropic distortions, and resulting elastic interactions can lead to interstitial ordering, which is a first step toward interstitial compound formation.

(4) If defect-rich regions have elastic properties that differ from relatively defect-free regions, elastic interactions can take place between defect clusters.[107] A complete theory of concentration-dependent elastic moduli would require anharmonic terms in the energy expression, and the simplicity of the present model would be lost.

The next section deals with some of the points listed above for the case of interactions of pairs of defect clusters or (coherent) "particles."

13. Particle Interactions

Formulas for defect cluster pair interactions take a particularly simple form in the Fourier representation of the preceding sections, both in the discrete (Section 13,a) and in the continuum version (Section 13,b).

---

[107] J.D. Eshelby, appendix to the paper by A.J. Ardell and R.B. Nicholson, *Acta Metall.* **14**, 1295 (1966).

Consider two nonoverlapping centrosymmetric particles a and b, each containing only a single type of defect (in addition to, possibly, the "host"). For example, the defects could be characterized by tetragonal strain fields for which the $\eta_{\alpha\beta}$ distortions are given by Eq. (11.10) and its two other possible permutations. Symbols 1, 2, and 3 will be used to indicate that the tetragonality axis for such defects points in the [100], [010], and [001] directions, respectively. In crystals of cubic symmetry, only two cases need be considered for these tetragonal defects: the case of parallel ($\|$) and that of perpendicular ($\perp$) tetragonality axes:

$$\| : \quad (a, b) = (3, 3), \qquad (13.1)$$

$$\perp : \quad (a, b) = (1, 2), \qquad (13.2)$$

a notation introduced by Khachaturyan and Shatalov.[108] Isotropic defects, with η given by Eq. (11.8), will be denoted by the index 0, so that combinations (0, 0) and (3, 0) should also be added to (13.1) and (13.2). Such are the cases usually considered, but the general formulas derived below are quite general.

a. Microscopic Formulation

Let the shape functions of the two particles be $s_a(p)$ and $s_b(p)$ such that

$$\gamma_a(p) = s_a(p) \qquad (13.3)$$

$$\gamma_b(p) = s_b(p - r) \qquad (13.4)$$

for particle a centered at the origin and b centered at arbitrary position $\mathbf{x}(r)$. The respective concentration amplitudes are

$$\Gamma_a(h) = S_a(h) \qquad (13.5)$$

$$\Gamma_b(h) = S_b(h) e^{-i\mathbf{k}(h)\cdot\mathbf{x}(r)}, \qquad (13.6)$$

S designating the Fourier transform of s. Introducing

$$\Gamma = \Gamma_a + \Gamma_b \qquad (13.7)$$

into Eq. (11.2) yields, by Eqs. (13.5) and (13.6),

$$E_e = E_{aa} + E_{bb} + E_{ab}(r) \qquad (13.8)$$

with

$$E_{aa} = \frac{N}{2} \sum_h{}' \mathscr{E}_{aa}(h) |S_a(h)|^2 \qquad (13.9)$$

---

[108] A.G. Khachaturyan and G.A. Shatalov, *Sov. Phys—Solid State (Engl. Transl.)* **11**, 118 (1969).

(and likewise for *b*) and with

$$E_{ab}(r) = N \sum_h{}' \mathscr{E}_{ab}(h) S_a^*(h) S_b(h) e^{-i\mathbf{k}(h)\cdot\mathbf{x}(r)}. \qquad (13.10)$$

The latter equation shows that the particle pair interaction energy $E_{ab}(r)$ for all (r) is given directly by the Fourier back transform of the product $\mathscr{E}_{ab} S_a^* S_b$. Equation (13.9) shows that a particle's self-energy is given by one-half the value of the back transform at the origin when the particles are identical in all respects. These facts make numerical computations very attractive with the use of the fast Fourier transform algorithm.[93,109] Some examples will now be given.

Isolated point defects can be treated in the above framework by making the shape functions *s* enclose a single lattice site. Split interstitials (dumbbells) consisting of two atoms placed symmetrically about a vacant site in $\langle 100 \rangle$ directions can be modeled approximately by a single substitutional tetragonal defect. Likewise, a Guinier–Preston (G.P.) zone in, say, Al-4 at. % Cu, can be modeled by a one-layer thick disk of defects, each defect being characterized by the distortion tensor given by (11.10), with only η_{33} different from zero, i.e., with vanishing tetragonality ratio t[110,111] (the magnitude of η_{33}, unimportant for relative energy calculations, was taken here to be unity). As mentioned above, the "ellipsoidal" nature of the Cu defect atoms takes care of the different interatomic bonding within and across the Cu disk. Figure 15[109] gives the interaction energy between two identical disks (G.P. zones) in the parallel configuration. One disk is located at the origin (solid outline). Values of the interaction energy, calculated according to Eq. (13.10), are given in electron volts at two positions of the second disk's center: -0.4 eV at the "stacking" position, and -2.9 eV at the rim position. The fact that the "stacking" position was found to be most favorable (minimum E) in the Al host (but not in the more anisotropic Cu host)[109] explains the (relative) stability of the θ'' structure, consisting of two Cu layers two lattice parameters apart, as observed in a further decomposition stage of supersaturated Al–Cu solid solutions.[112] If the second disk is taken to be orthogonal to the first one, the most favorable positions of the second disk are roughly along the $\langle 100 \rangle$ directions, at an optimal distance of the order of the disk diameter, as shown in Fig. 16.[109] Although the orthogonal configurations are less stable than the parallel, one expects all three $\langle 100 \rangle$ G.P. zone variants to be present in order to maximize the configurational entropy. Hence, from elastic energy considerations alone, a periodic three-dimensional super-

---

[109] E. Seitz and D. de Fontaine, *Acta Metall.* **26**, 1671 (1978).
[110] G. Sines and R. Kikuchi, *Acta Metall.* **6**, 500 (1958).
[111] D.J. Millar, G. Sines, and J.W. Goodman, *Acta Metall.* **23**, 245 (1975).
[112] A. Kelly and R.B. Nicholson, *Prog. Mater. Sci.* **10**, 151 (1963).

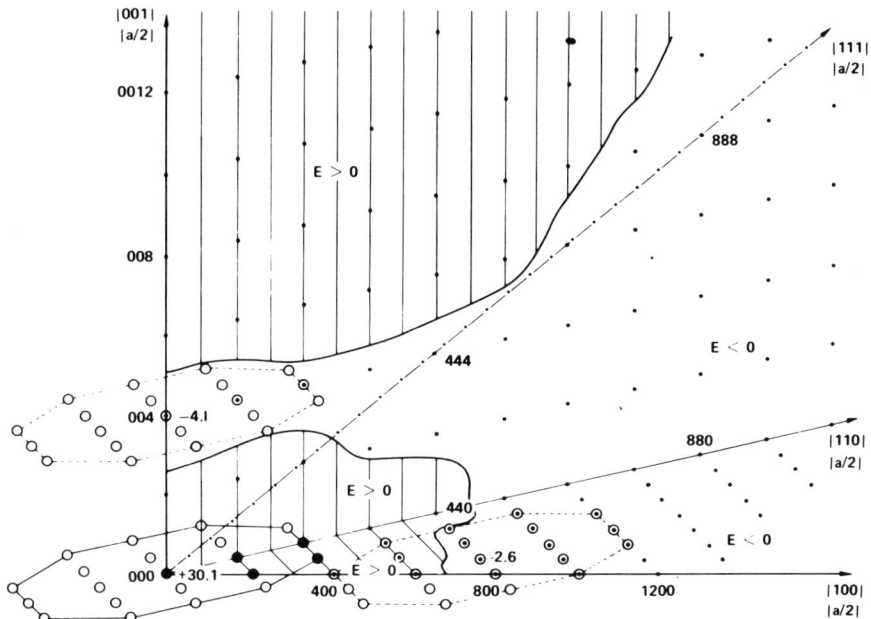

FIG. 15. Elastic interaction energies (eV) in Al for two parallel disks of tetragonal defects, repulsive regions shown shaded. First disk at origin, second disk (broken line) is shown in both most stable stacking position and most stable disk growth position [E. Seitz and D. de Fontaine, *Acta Metall.* **26**, 1671 (1978)].

array of mutually orthogonal ⟨100⟩ disks is the expected morphology in the case of either G.P. zones or coherent ordered precipitates with tetragonal distortion, as observed experimentally.[113,114] Continuum calculations lead to similar conclusions as shown next.

b. Continuum Formulation

The continuum version of Eq. (13.10) is obtained formally by placing carets ($\wedge$) over appropriate symbols, and by using the elasticity formulas developed in Section 12. Here, a few additional modifications will be made in order to recover the basic formulas derived by Khachaturyan and co-workers, who pioneered these calculations in a remarkable series of papers.[108,115,116]

[113] L.E. Tanner and J.J. Leamy, in "Order-Disorder Transformation in Alloys" (H. Warlimont, ed.), p. 180, and references cited therein. Springer-Verlag, Berlin and New York, 1974.
[114] P. Eurin, J.M. Penisson, and A. Bourret, *Acta Metall.* **21**, 559 (1973).
[115] A.G. Khachaturyan, *Phys. Status Solidi* **35**, 119 (1969).
[116] A.G. Khachaturyan and V.M. Airapetyan, *Phys. Status Solidi A* **26**, 61 (1974).

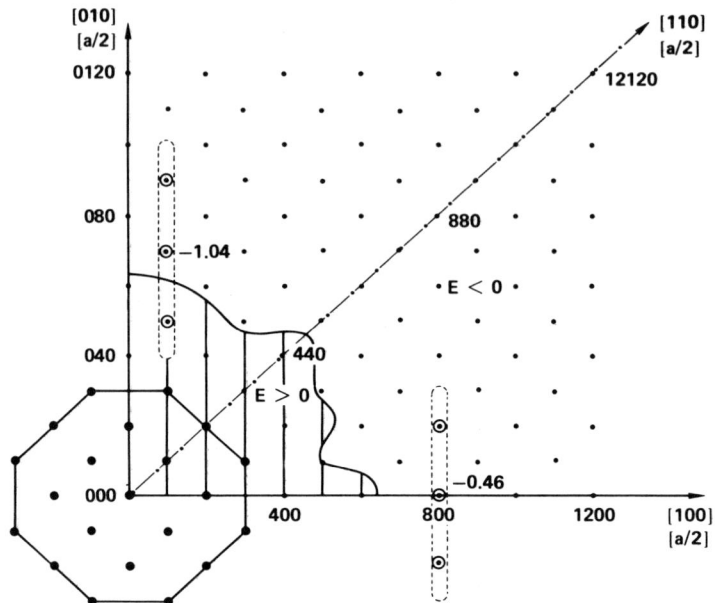

FIG. 16. Elastic interaction energies (eV) in Al for two disks in orthogonal configuration. Same tetragonal defects as in Fig. 15. Most favorable positions are shown for second disk (on edge, broken line) [E. Seitz and D. de Fontaine, *Acta Metall.* **26**, 1671 (1978)].

First we define a particle shape function $\xi(\mathbf{x})$ equal to unity if $\mathbf{x}$ is inside the cluster centered at the origin, 0 otherwise. Its Fourier transform Ξ is related to S defined in (13.5) by[101]

$$S(\mathbf{k}) = \Xi(\mathbf{k})[1 - \delta(\mathbf{k})]. \qquad (13.11)$$

Introducing this formula for s_a and s_b in Eq. (13.10) (continuum version) yields

$$\hat{E}_{ab}(\mathbf{x}) = \frac{1}{2} \hat{\Phi}_{ab} \left[\frac{V_a * V_b}{V_N} - \frac{V_a V_b}{V_N} \right]$$

$$- \frac{1}{2V_N} \sum_{\mathbf{k}} \hat{R}_{ab}\left(\frac{\mathbf{k}}{k}\right) \Xi_a^*(\mathbf{k}) \Xi_b(\mathbf{k}) e^{-i\mathbf{k}\cdot\mathbf{x}} \qquad (13.12)$$

where V_a and V_b are the particle volumes, and where the convolution product is given by

$$\frac{V_a * V_b}{V_N} = \int_{V_N} \xi_a(-\mathbf{y})\xi_b(\mathbf{x} - \mathbf{y}) \, d^3\mathbf{y}. \qquad (13.13)$$

Equations (13.12) and (13.13) are obtained by noting that the first term of Eq. (12.9), $\hat{\Phi}_{ab}$, is a constant according to Eq. (12.10), and that the second term $\hat{R}_{ab}$ denoted "relaxation energy coefficient" by analogy with its discrete counterpart (7.1), depends only on the direction κ of the wave vector **k**.

Khachaturyan and Shatalov[108] (KS) consider finite particles in an infinite continuum ($V_N \to \infty$), and replace the summation in Eq. (13.12) by an integral. For nonoverlapping particles the convolution in (13.12) vanishes, so that one arrives directly at the KS formula (in a different notation):

$$\hat{E}_{ab}(\mathbf{x}) = -\frac{1}{16\pi^3} \int \hat{R}_{ab}\left(\frac{\mathbf{k}}{k}\right) \Xi_a^*(\mathbf{k}) \Xi_b(\mathbf{k}) e^{-i\mathbf{k}\cdot\mathbf{x}} d^3\mathbf{k}. \quad (13.14)$$

The integration is over infinite k-space but is restricted in practice to a finite Debye cutoff.

Let us first investigate the asymptotic form of Eq. (13.14) valid for large interparticle separation x. Formally, this is accomplished by replacing the shape functions ξ by δ-functions weighted by the volume of the particle. Then, spherical coordinate systems are defined: one of which (r, θ_r, φ_r), referred to the cubic axes, fixes the position of second particle center with respect to the first (located at the origin), the other (k, θ_k, φ_k), referred to a new coordinate system rotated with respect to the first in such a way that the new z axis passes through the particle centers, fixes the wave vector **k**. It is then possible to remove the r dependence and the transforms of the shape functions from the integral in Eq. (13.14):

$$\lim_{r\to\infty} \hat{E}_{ab}(r, \theta_r, \varphi_r) = \frac{V_a V_b}{16\pi^2 r^3} I_{ab}(\theta_r, \varphi_r), \quad (13.15)$$

where I_{ab}, a triple integral over (rotated) k-space, contains the angular dependence of the interaction energy. For cubic crystals, the triple integral can be further reduced by using a method first introduced by Lifshitz and Rozentsweig[117] to determine the back transform of the Green's tensor $\Phi_{\alpha\beta}^{-1}$. This procedure is mentioned in an appendix to a paper by Shatalov and Khachaturyan,[118] and was used for the case of displacements around a dilatational strain center in a cubic continuum by Kanzaki[81] and by Flocken and Hardy.[119] The most complete calculation is that of Yamauchi[101] who derived the following formula for the

---

[117] I.M. Lifshitz and L.N. Rozentsweig, *Zh. Eksp. Teor. Fiz.* **17**, 783 (1947).
[118] G.A. Shatalov and A.G. Kharchaturyan, *Fiz. Met. Metalloved.* **25**, 56 (1968).
[119] J.W. Flocken and J.R. Hardy, *Phys. Rev. B* **1**, 2447 (1970).

orientation dependence:

$$I_{ab}(\theta_r, \varphi_r) = -\int_0^{2\pi} \left[\frac{\partial^2}{\partial \theta_k^2} \hat{R}_{ab}(\theta_k, \varphi_k; \theta_r, \varphi_r)\right]_{\theta_k=\pi/2} d\varphi_k, \quad (13.16)$$

where the integrand is obtained by straightforward (but tedious) application of the formulas of Section 12 specialized to cubic crystals. It is a fairly simple matter to perform the $(0, 2\pi)$ integration numerically, whereas a triple integration in k-space poses difficulties.

It is seen from Eq. (13.15) that the interparticle elastic potential decreases as $1/r^3$ for large r, a well-known result. The angular dependence tends to produce particle alignment during coarsening, Fig. 17 typifying the resulting morphology: as noted by KS, if two particles are oriented in such a way that I_{ab} is negative (attractive direction), then, under the coarsening condition

$$V_a + V_b = \text{const.}, \quad (13.17)$$

$\hat{E}_{ab}$ will be minimum for $V_a = V_b$, thus tending to equalize and stabilize

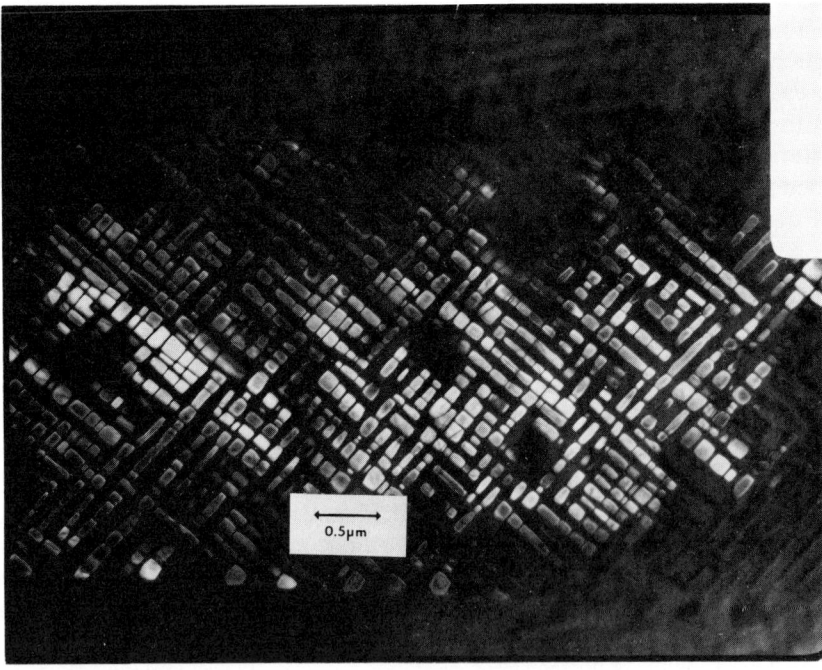

FIG. 17. Cuboid ordered Ni_3Al coherent precipitates in Ni-rich matrix. [From D. Chellman and A. J. Ardell, unpublished work at UCLA.]

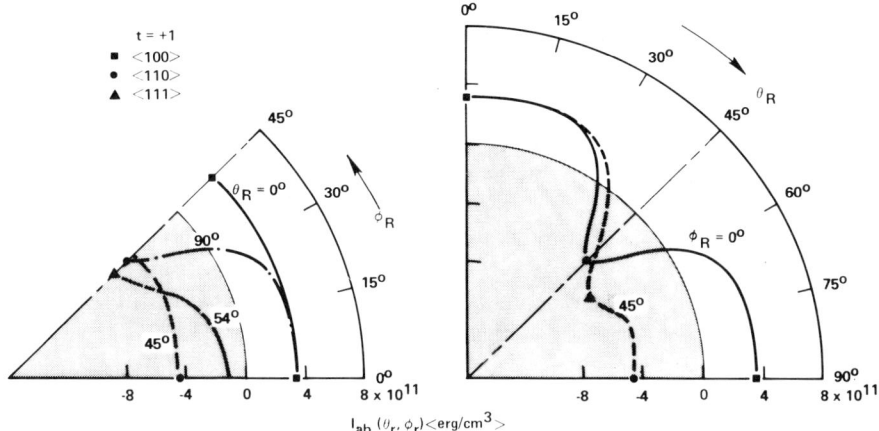

FIG. 18. Angular dependence $I_{ab}(\theta_r, \phi_r)$ (in erg/cm³) for asymptotic form (13.16) of interaction energy, for distortion tensor of cubic symmetry ($t = 1$) [H. Yamauchi and D. de Fontaine, *Acta Metall.* (in press)].

particle dimensions. If I_{ab} is positive (repulsive direction), then $\hat{E}_{ab}$ will decrease if the larger of the two particles grows at the expense of the smaller. Alignment of roughly equal-sized particles is thus expected along favorable directions, with significant slowing down of the coarsening rate below the one predicted by the Wagner–Lifshitz–Slyozov theory.[120,121] In fact, a periodic superarray of particles can only coarsen by the slow process of "superarray dislocation climb."

Figure 18 shows the angular dependence I_{ab} for the case of clusters of defects with purely dilatational strain fields ($t = 1$) in a Cu matrix ($A > 0$).[101] Maximum interparticle attraction occurs along $\langle 100 \rangle$ ($\theta_r = 0°$; $\theta_r = 90°$, $\varphi_r = 0°$), and maximum repulsion (shaded) occurs along $\langle 111 \rangle$ ($\theta_r = 54°$, $\varphi_r = 45°$). Similar plots have been constructed for tetragonality ratios $t = +2$ and -2, for both $\parallel$ and $\perp$ cases.[101]

The equilibrium interparticle spacing can only be determined by numerical integration of Eq. (13.14). This is conveniently done for pairs of spherical particles of equal radius r_s, for which the transformed shape function is[101,108]

$$\Xi_a(k) = \Xi_b(k) = 3V_a[\sin(kr_s) - kr_s \cos(kr_s)]/(kr_s)^3. \quad (13.18)$$

The calculation was done by KS for defect clusters with arbitrary tetragonality ratio t ($\parallel$ and $\perp$) in an isotropic continuum, and by Yamauchi for

---

[120] C. Wagner, *Z. Elektrochem.* **65**, 581 (1961).
[121] I.M. Lifshitz and V.V. Slyozov, *J. Phys. Chem. Solids* **19**, 35 (1961).

clusters with $t = 1, +2, -2$ ($\|$ and $\perp$), in Cu.[101] Figure 19, taken from the latter work, compares $\hat{E}_{ab}$ calculated for $A > 0$ (Cu), with $t = 1$, along $\langle 100 \rangle$ and $\langle 110 \rangle$ by the asymptotic formula (13.15) combined with (13.16) (solid curve), and by the complete formula (13.14) combined with (13.18) (dotted curve). The abscissa is in units of particle diameters. It is seen that the asymptotic formula is really quite good along the attractive direction for a particle separation slightly greater than the touching distance, and fairly good along a repulsive direction beyond two or three particle diameters. In these types of calculations, the minimum energy position is along the attractive $\langle 100 \rangle$ direction just beyond the touching distance, in accordance with the microstructure of Fig. 17. This micrograph is from a coarsened Ni–Al alloy, but Ni has roughly the same elastic constants as Cu, and Ni_3Al has cubic symmetry, thereby straining the matrix isotropically. The fact that the three-dimensional structure of coherent precipitates must be a simple cubic superarray under the condition $A > 0$, $t = 1$ was recently demonstrated theoretically by Khachaturyan and Airapetyan.[116]

14. Displacive Scattering

Scattered intensity formulas were derived in Section 4 under the assumption that all defects and ions (scattering centers in general) occupied exactly the average lattice sites. This artificial restriction will now be lifted by allowing displacements from the lattice sites. The general task of determining the scattered intensity from arbitrary configurations of replacive and displacive defects is extremely complex. Hence, we treat only a relatively simple problem: that of correcting the diffuse intensity for the purpose of recovering the SRO intensity, under the assumption of small displacements $\mathbf{u}(p)$. Rather than merely correcting for displacements by eliminating that component of the scattered intensity, it may be of interest to analyze the diffuse intensity directly to obtain information about the displacement fields themselves. This topic is only touched upon here, and the interested reader is referred to relevant review articles.[122–124]

a. Krivoglaz Correction

Because of the analogy between the kinematical diffraction theory (Section 4) and the second-order perturbation solution of the Schrödinger equation (Section 6), we may simply combine the formulas of the latter

[122] P.H. Dederichs, *J. Phys. F* **3**, 471 (1973).
[123] P. Ehrhart, H.G. Haubold, and W. Schilling, *Adv. Solid State Phys.* **14**, 87 (1974).
[124] G.S. Bauer, E. Seitz, and W. Just, *J. Appl. Crystallogr.* **8**, 162, (1975).

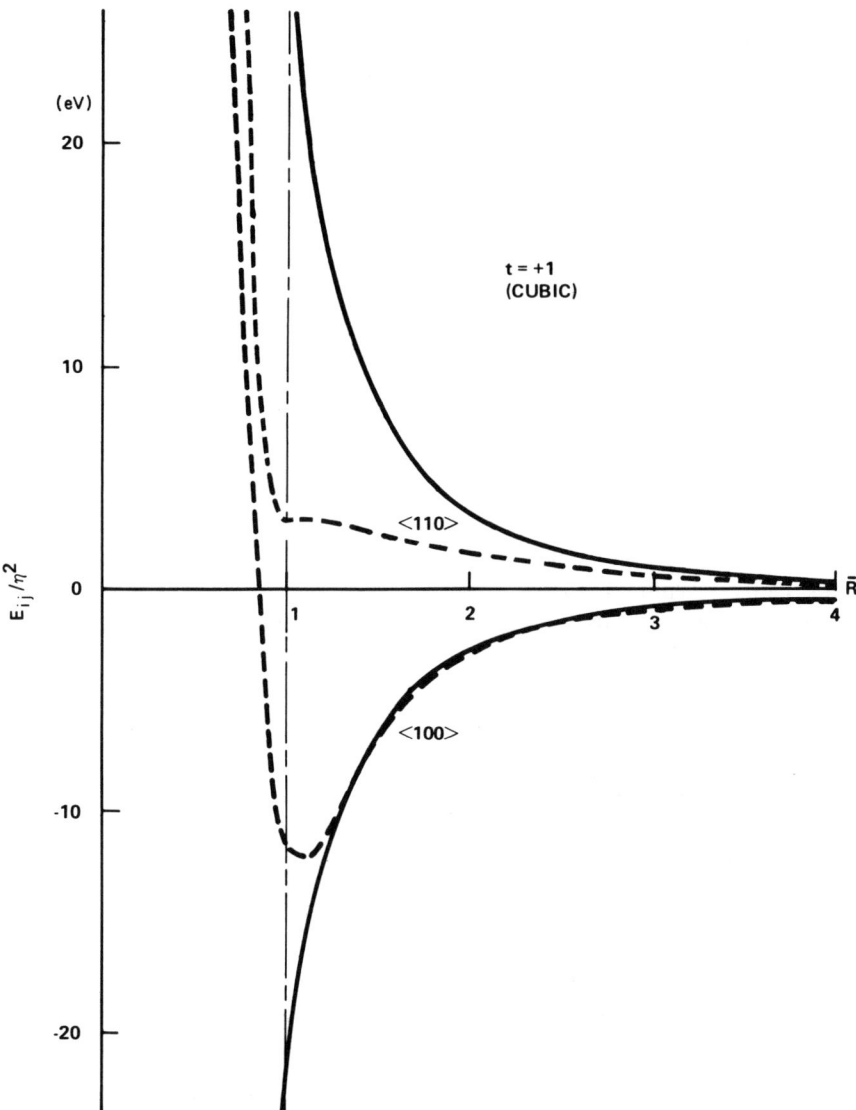

FIG. 19. Normalized interaction energy calculated by Eqs. (13.14) and (13.18) (broken curve) for two identical spherical particles with distortion tensor of cubic symmetry. Solid line is for asymptotic formula [H. Yamauchi and D. de Fontaine, *Acta Metall.* (in press)].

section with those of the former to obtain approximate formulas for diffracted intensity under displacive scattering; it is merely required to refrain from summing over the scattering vector **K** in Eqs. (6.19)–(6.21) or (6.29)–(6.32), and to replace the EWNC $W_{ij}(K)$ by the product of atomic scattering factors $f_i(K)f_j(K)$. Under the small-displacement assumption (6.14), one may thus extend the SRO intensity equation (4.11) for

$$\mathbf{k} = \mathbf{K} - \mathbf{g} \qquad (14.1)$$

not too far from the nearest Bragg peak **g**, as follows:

$$|A(\mathbf{K})|_D^2 = N\bar{f}^2(K) \sum_{\mathbf{k}} \Omega(\mathbf{K} - \mathbf{k}) \sum_{i,j=1}^{n} \Delta f_i(K)\Delta f_j(K)[\Gamma_i^*(\mathbf{k})\Gamma_j(\mathbf{k})$$
$$- 2i\bar{c}_j K_\alpha U_\alpha^*(\mathbf{k})\Gamma_i(\mathbf{k}) + \bar{c}_i\bar{c}_j K_\alpha K_\beta U_\alpha^*(\mathbf{k})U_\beta(\mathbf{k})], \qquad (14.2)$$

where Ω is the transform of the shape function of the crystal. Equation (14.2) represents the *diffuse intensity* (subscript D), combining both SRO intensity in the strict sense of Eq. (4.11), and the intensity due to displacements. The Bragg peak intensity is that given by Eq. (4.7), corrected by the Debye–Waller factor.[2]

As was done previously in Sections 6 and 11, the displacement amplitudes may be eliminated by Eq. (6.33), rewritten here as

$$U_\alpha = -i\sum_{j=1}^{n} \Theta_{j\alpha}\Gamma_j \qquad (14.3)$$

with

$$\Theta_{j\alpha} = \Phi_{\alpha\beta}^{-1} \operatorname{Im}(\Phi_{j\beta}), \qquad (14.4)$$

the imaginary part of the pure imaginary $\Phi_{j\beta}$ (or $\Psi_{j\beta}$, from Section 6) being taken so that $\Theta_{j\alpha}$ may be defined as a real function. Note that this definition differs somewhat from that given elsewhere.[36,125,126]

From Eq. (14.2), the expectation value of the diffuse intensity is then (with Ω replaced by N times the discrete δ-function)

$$I_D(\mathbf{K}) = N^2\bar{f}^2(K) \sum_{i,j=1}^{n} [\Delta f_i - K_\alpha\Theta_{i\alpha}(\mathbf{k})][\Delta f_j - K_\beta\Theta_{j\beta}(\mathbf{k})]Q_{ij}(\mathbf{k}) \qquad (14.5)$$

with Δf_i regarded as practically independent of the scattering vector K within a given reciprocal unit cell. Three types of terms appear in Eq. (14.5): (a) one independent of the vector Θ which is just I_{SRO} of Eq. (4.11), (b) one linear in Θ which is the so-called *size effect* term,[127] and

---

[125] H.E. Cook, *J. Phys. Chem. Solids* **30**, 1097 (1969).
[126] J.E. Gragg, Jr., *J. Phys. Chem. Solids* **32**, 1195 (1971).
[127] B.E. Warren, B.L. Averbach, and B.W. Roberts, *J. Appl. Phys.* **22**, 1493 (1951).

(c) one quadratic in Θ which, close to a Bragg peak, contributes to the so-called *Huang peak*.[128] The coefficients of Q_{ij} for these three terms are, respectively, (a) independent of **K**, (b) centrosymmetric about the origin and asymmetric about any other Bragg peak, (c) centrosymmetric about any Bragg peak.

For binary systems, Eq. (14.5) simplifies to

$$I_D(\mathbf{K}) = N^2 G(\mathbf{K}) Q(\mathbf{K} - \mathbf{g}) \qquad (14.6)$$

with

$$G(\mathbf{K}) = \bar{f}^2(K)[\Delta f - \mathbf{K} \cdot \Theta(\mathbf{K} - \mathbf{g})]^2. \qquad (14.7)$$

The recipe for obtaining the pair correlation function $q(r)$, or the SRO parameters $\alpha(r)$ from experimental diffuse intensity is in principle quite simple: compute $G(\mathbf{K})$, divide I_D by G to obtain the corrected Q spectrum, then take the back Fourier transform. Such is the method proposed by Krivoglaz.[82]

Even if the assumptions leading to Eqs. (14.6)–(14.7) are valid, the Krivoglaz correction is in fact not simple to use in practice because of the difficulty of determining the (vector) function Θ. Krivoglaz and Tikhonova[82] have given approximate formulas and Cook[125] has proposed a useful model. Gragg[126] suggested that Θ be obtained directly from experimental measurements, by making use of symmetry arguments first put forward by Borie and Sparks.[129] For cubic crystals (of lattice parameter a_0), Gragg obtains the formula

$$\Theta_1(lh) = \frac{-a_0}{2\pi(l_1 + l_2 + l_3)\bar{f}}$$
$$\cdot \frac{(1-h)I_D[l(2+h)] + 2hI_D[lh] - (1+h)I_D[l(2-h)]}{-h(2-h)I_D[l(2+h)] + 2(4-h^2)I_D[lh] + h(2+h)I_D[l(2-h)]}$$

$$(14.8)$$

with $l_\alpha = 1$ or 0, valid for the three symmetry directions:

$(l) \equiv (100):\quad \Theta_1(lh) \equiv \Theta_1(h00),\ \Theta_2(h00) = \Theta_3(h00) = 0, \qquad (14.9)$

$(l) \equiv (110):\quad \Theta_1(lh) \equiv \Theta_1(hh0) = \Theta_2(hh0),\ \Theta_3(hh0) = 0, \qquad (14.10)$

$(l) \equiv (111):\quad \Theta_1(lh) \equiv \Theta_1(hhh) = \Theta_2(hhh) = \Theta_3(hhh). \qquad (14.11)$

It was hoped to determine force constants $\varphi_{\alpha\beta}$ and coupling parameters $\varphi_{i\alpha}$, φ_{ij} directly by this method, and hence to calculate Θ for all other

---

[128] K. Huang, *Proc. R. Soc. London, Ser. A* **190**, 102 (1947).
[129] B. Borie and C.J. Sparks, *Acta Crystallogr., Sect. A* **27**, 198 (1971).

directions in k-space. This subsequently proved to be impractical.[130] Nevertheless, the possibility of obtaining Θ values from Eq. (14.8) by making measurements at various intervals along the $\langle 100 \rangle$, $\langle 110 \rangle$, and $\langle 111 \rangle$ directions only is an attractive one. Values of Θ along other directions could then be estimated by Houston's method,[131] thereby possibly providing a convenient alternative to the method of Borie and Sparks[129] which requires taking extensive data in complicated polyhedral regions in reciprocal space.[132]

b. Scattering from Isolated Defects

In studying concentrated substitutional solutions, one is mostly concerned with the state of order, hence with the pair correlation function which must be somehow extracted from diffuse intensity measurements. Therefore, scattering due to displacements is considered a nuisance and must be eliminated, for example, by one of the approximate methods mentioned above. In another context however, displacive scattering can be analyzed in its own right to yield valuable information about crystalline displacement fields. Such studies are generally performed on very dilute solid solutions in which defect-defect correlations may be neglected so that the defects may be regarded as diffracting independently.

Under this assumption, the Q spectrum in Eq. (14.6) is, by Eq. (3.8), the constant $Q = q^0/N$, so that the Krivoglaz $G(\mathbf{K})$ function itself should account for the experimental diffuse intensity. What one does in practice is to postulate a set of Kanzaki forces $\mathbf{f}(p)$, calculate the resulting $G(\mathbf{K})$, insert this into Eq. (14.7), compare with the experimental intensity, and repeat the procedure with modified Kanzaki forces until reasonably good agreement with experimental diffraction data is attained.[123,124] In this way the structure of the individual defect may be determined, i.e., its distortion tensor $\eta_{\alpha\beta}$ or the corresponding double-force tensor $P_{\alpha\beta}$ defined in Eq. (11.12). Close to a Bragg peak (origin excepted), the diffuse intensity is dominated by the Huang scattering, i.e., by the terms in Θ^2 in Eq. (14.7). For an isolated defect, this scattering pattern is characterized by a nodal plane of zero intensity passing through the reciprocal lattice point, the orientation of this plane depending on the symmetry of the strain field of the defect (i.e., its dipole tensor) and on the indices of the reciprocal lattice point.[81,122,133–135] One expects all variants of anisotropic

[130] J.E. Gragg, Jr., *J. Phys. Chem. Solids* **35**, 717 (1974).
[131] W.V. Houston, *Rev. Mod. Phys.* **20**, 161 (1948).
[132] M. Hayakawa, P. Bardhan, and J.B. Cohen, *J. Appl. Crystallogr.* **8**, 87 (1975).
[133] J.W. Flocken and J.R. Hardy, *Phys. Rev. B* **1**, 2472 (1970).
[134] P.H. Dederichs, *Phys. Rev. B* **4**, 1041 (1971).
[135] H. Trinkhaus, *Phys. Status Solidi B* **51**, 307 (1972).

defects to be present in equal amounts so that the actual Huang scattering will be a superposition of the scattering from individual variants. Nevertheless, much can be inferred about the nature of the defect dipole tensor merely from the symmetry of the Huang scattering.

Interstitial defects are particularly good candidates for this type of investigation because of the large associated anisotropic strain fields, the resulting X-ray or neutron diffuse intensity of which can be analyzed in some detail. Recently, it has been possible to determine the structure of self-interstitials in cubic crystals [$\langle 100 \rangle$ dumbbells (split interstitials) in fcc, $\langle 110 \rangle$ dumbbells in bcc[123,136]], and the preferred interstitial site (octahedral, tetrahedral) for small interstitials in transition metals.[137,138] Substitutional solutions have also yielded interesting results, in particular the discovery[124] of a trigonal displacement field around Bi atoms in lead.

15. Discussion of Elastic Interactions

The strictly harmonic theory presented in the previous sections is very convenient and is the only tractable one for concentrated solid solutions. It was assumed that the Kanzaki forces **f** were independent of the local displacements **u**, an assumption which is not likely to be valid if displacements are large. For such cases, Kanzaki[81] proposed another force model in which **f** varies linearly with **u**, which is equivalent to saying that the force constants $\varphi_{\alpha\beta}$ are altered (by $\delta\varphi_{\alpha\beta}$) by the presence of the defect force field. Under this more general assumption, the Fourier method becomes less attractive than a calculation in direct space by means of the so-called defect Green's function **G**, related to the perfect crystal Green's function

$$\mathbf{G}_0(0, p) = \mathscr{F}^{-1} \Phi^{-1}(h) \tag{15.1}$$

through the Dyson equation[100,139]

$$\mathbf{G} = \mathbf{G}_0 + \mathbf{G}_0 \delta\varphi \mathbf{G}. \tag{15.2}$$

---

[136] W. Schilling, P. Ehrhart, and K. Sonnenberg, in "Fundamental Aspects of Radiation Damage in Metals" (M.T. Robinson and F.W. Young, eds.), ERDA CONF-751006-P1, p. 470, and references cited therein, particularly papers by P.H. Dederichs (p. 187), H.G. Haubold (p. 268), and P. Ehrhart (p. 302) in same conference proceedings. Gatlinburg, Tennessee, 1975. (Sponsored by: U.S. ERDA, National Science Foundation, Oak Ridge National Lab.)
[137] V.A. Somenkov, *Ber. Bunsenges. Phys. Chem.* **76**, 733 (1972), and references cited therein.
[138] G. Bauer, E. Seitz, H. Horner, and W. Schmatz, *Solid State Commun.* **17**, 161 (1975).
[139] R. Bullough and V.K. Tewary, in "Interatomic Potentials and Simulation of Lattice Defects" (P.C. Gehlen, J.R. Beeler, Jr. and R.I. Jaffe, eds.), p. 155. Plenum, New York, 1972.

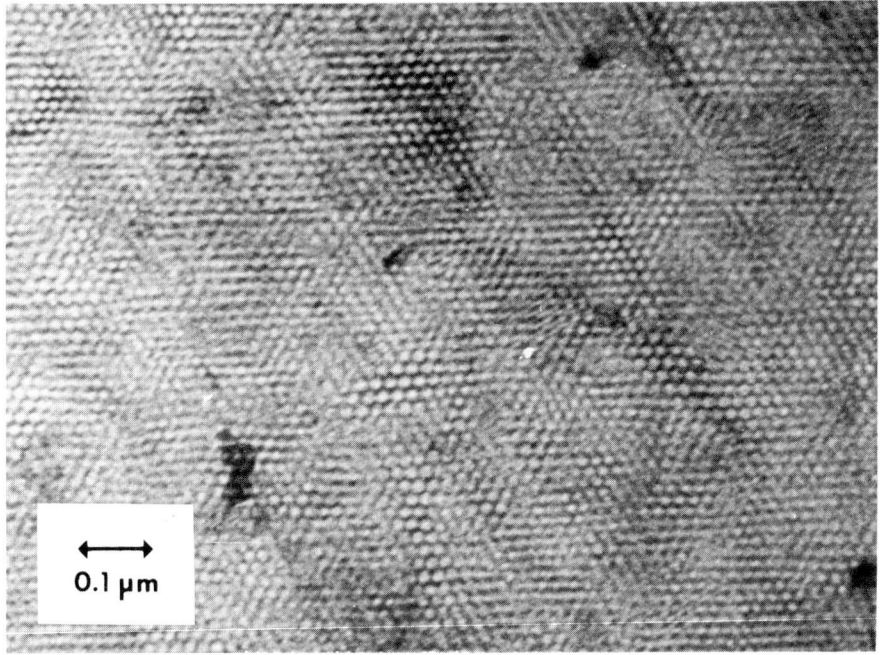

FIG. 20. Superlattice of voids in electron irradiated fluorite, (111) orientation. [L. T. Chadderton, E. Johnson, and T. Wohlenberg, *Phys. Scr.* **13**, 127 (1976)].

The defect Green's function method is suitable when only a few defects are considered (dilute solutions). This was extended by Tewary and Bullough[140] to the problem of the stability of periodic arrays of voids created in fcc and bcc metals by high-energy particle irradiations. An excellent example of void periodicity in irradiated fluorite is shown in Fig. 20.[141] It appears that the use of the defect Green's function gives a value of the void radius to void lattice parameter ratio which is closer to the experimental one than that given by the use of the perfect Green's function, or equivalently by the simple lattice statics formulation described above.[142,143]

The simple theoretical model of Section 13 does appear to work quite well in the case of elastic interactions of coherent precipitates or clusters, since that model predicts an equilibrium cluster separation slightly greater

[140] V.K. Tewary and R. Bullough, *J. Phys. F* **2**, L69 (1972).
[141] L.T. Chadderton, E. Johnson, and T. Wohlenberg, *Phys. Scr.* **13**, 127 (1976).
[142] A.M. Stoneham, *in* "The Physics of Irradiation-Produced Voids" (R.S. Nelson, ed.), AERE Harwell Rep. R7934, p. 319. 1975.
[143] A.M. Stoneham, *in* "Fundamental Aspects of Radiation Damage in Metals" (M.T. Robinson and F.W. Young, eds.), ERDA-Conf-751006P1, p.1221. Gatlinburg, Tennessee, 1975. (Sponsored by: U.S. ERDA, National Science Foundation, Oak Ridge National Lab.)

than a cluster characteristic dimension, in apparent agreement with experimental observations, as exemplified by Fig. 17. There are important factors which none of these elastic theories take into account, however; one of these is the fact that the diameter/spacing ratio cannot be determined independently but is in fact partially fixed by the precipitate/matrix volume fraction. Furthermore, voids and other clusters arise by nucleation and diffusional growth, so that cluster spacings and distributions must be at least partially governed by the requirement of minimizing diffusion field overlaps. To arrive at a satisfactory theory of defect cluster morphologies, one should therefore solve nonlinear partial differential diffusion equations in elastically anisotropic media, a formidable undertaking.

The formulas given in the previous sections have been extended by Khachaturyan,[144] Hoffman,[145] Tewary,[146] and others to the case of defects on interstitial sublattices. In these calculations it is assumed that the internal energy is primarily of elastic nature, so that approximate $V_{ij}(h)$ functions can be calculated by models based, in part, on continuum elasticity. It is then possible to predict interstitial ordered structures, and even to describe interstitial phase transitions such as those occurring in certain hydrogen bcc transition metal systems[147–149] and the more hypothetical one associated with the Martensitic transformation in the C–Fe system.[150,151] Some of these applications will be briefly described in later sections. Lattice statics methods have also been used to explain the periodic spacing of defect planes in certain nonstoichiometric oxides.[152]

Linear elasticity has been used thus far. Clearly, one should consider the important case of precipitate (cluster) elastic moduli differing from those of the matrix. Completely general theories are not available, but Eshelby[107] has shown, for example, that the elastic attraction of centers of dilatation in a matrix of different moduli decreases as $1/r^2$, rather than $1/r^3$ as predicted by the harmonic formula (13.15). Anharmonic potentials are also required to produce the partial splitting of dislocation cores in bcc predicted by Vitek,[153] for example. Such extensions of the theory are beyond the scope of this article.

[144] A.G. Khachaturyan, *Sov. Phys.—Solid State (Engl. Transl.)* **9**, 2249 (1968).
[145] D.W. Hoffman, *Acta Metall.* **18**, 819 (1970).
[146] V.K. Tewary, *J. Phys. F* **3**, 1275 (1973).
[147] See, for example, B. Stalinsky, *Ber. Bunsenges. Phys. Chem.* **76**, 724 (1972).
[148] G. Alefeld, *Ber. Bunsenges. Phys. Chem.* **76**, 746 (1972), and references cited therein.
[149] H. Wagner and H. Horner, *Adv. Phys.* **23**, 239 (1974).
[150] G.V. Kurdjumov and A.G. Khachaturyan, *Met. Trans.* **3**, 1069 (1972).
[151] G.V. Kurdjumov and A.G. Khachaturyan, *Acta Metall.* **23**, 1077 (1975).
[152] A.M. Stoneham and P.J. Durham, *J. Phys. Chem. Solids* **34**, 2127 (1973).
[153] V. Vitek, L. Lejček, and D.K. Bowen, in "Interatomic Potentials and Simulation of Lattice Defects" (P.C. Gehlen, J.R. Beeler, Jr., and R.I. Jaffe, eds.), p. 493. Plenum, New York, 1972.

V. Free Energy Models

Obtaining a convenient formula for the free energy is the central problem of solid solution theory. Formally, the Helmholtz free energy F is obtained by taking the logarithm of the appropriate partition function; however, even in the simple case of the nearest-neighbor Ising model of an equiatomic binary solution on a simple cubic lattice, no rigorous closed-form solution is available. Approximate methods must therefore be used in all practical cases. Various expansion methods are known which give very reliable results near a phase transition, for instance for the values of the so-called critical exponents. For the type of problems treated in this article, however, one is less concerned about correct behavior at the critical point than about such matters as predicting ordered structures, phase diagrams, and nonequilibrium processes. It is thus preferable to use approximate free energy functions which are analytic in the configuration variables and which can therefore be continued inside the metastable or unstable regions. Such free energy models generally yield "classical" (therefore incorrect) critical exponents. The Helmholtz free energy alone will be considered; the Gibb's free energy is not required here since it is assumed, for simplicity, that atomic rearrangements on lattice sites leave the lattice parameters of the average lattice invariant.

16. Configurational Free Energy

In Section 2, an ensemble of systems was set up, and configuration variables were defined by summing, over all possible states of the $\mathcal{M}$ systems, the products of values of occupation variables for the different clusters considered, weighted by the probability of finding the systems in the various states, Eq. (2.2). The probability of finding a system in given energy state is given formally by Eq. (2.2). The energy E(state) actually contains both vibrational and configurational contributions, but it will be assumed here for simplicity that the vibrational part factors out of the partition function in the form of a product of single-atom vibrational partition functions, it being assumed that the vibrational energy levels are independent of the occupation of neighboring sites. With these simplifying assumptions, the vibrational part contributes only to the configuration-independent portion of the free energy, not to the free energy of mixing, and consequently needs not be considered in what follows.

The configurational partition function can be written as

$$\mathscr{Z} = \sum_{\text{states}} e^{-E(\text{state})/k_B T} = \sum_{\{\sigma\}} g\{\sigma\} e^{-E\{\sigma\}/k_B T}, \qquad (16.1)$$

where $g\{\sigma\}$ is the statistical weight of the configuration defined by the set $\{\sigma\}$ of occupation variables $\sigma_i(p)$ first defined in Section 2. The energy E is for example the one given by Eq. (6.36), to which appropriate chemical potentials should be added if a grand canonical ensemble is used. Equation (16.1) can be rewritten as

$$\mathscr{L} = \sum_{\{\sigma\}} e^{-F\{\sigma\}/k_B T} \tag{16.2}$$

with the *nonequilibrium free energy function* F defined by

$$F\{\sigma\} = E\{\sigma\} - TS\{\sigma\} \tag{16.3}$$

where the configurational entropy is given by

$$S\{\sigma\} = k_B \ln g\{\sigma\}. \tag{16.4}$$

The *equilibrium* Helmholtz free energy is

$$F_{eq} = -k_B T \ln \mathscr{L}. \tag{16.5}$$

An important additional simplification is now introduced: the sum in the partition function formula (16.2) is replaced by its maximum term. In effect, fluctuations about the most probable state are thereby neglected. The equilibrium free energy is then obtained approximately by minimizing the nonequilibrium free energy function:

$$F_{eq} \cong \min_{\{\sigma\}} F\{\sigma\}. \tag{16.6}$$

The use of $F\{\sigma\}$ away from equilibrium is merely justified by the insight gained in treating nonequilibrium problems. As mentioned in Section 2, complete sets of $\{\sigma\}$ operators are unavailable, so that the configuration variables retained in F will actually be the cluster probabilities introduced in Eq. (2.1), and used to characterized the state of order of a solution. The minimization in Eq. (16.6) will thus be carried out with respect to these cluster probabilities; such is the basis of the cluster variation method (CVM) developed by Kikuchi.[14] In general, the larger the cluster retained, the more accurate will be the description of the state of order, and the more reliable will be the free energy function. (See, however, a counterexample in Section 20.)

The lowest cluster in the hierarchy is the single lattice point. When it is used alone, along with the zeroth (mean-field) approximation in the internal energy, the Bragg–Williams model is obtained, as shown in Section 17. Other mean-field models, including the Landau theory are developed in Sections 18 and 19, while the general cluster variation method is described in Section 20.

17. Generalized Bragg–Williams Model

The Bragg–Williams (BW) model is derived from a state-of-order description based on single-site averaging, i.e., on the "point" cluster. As was done in Section 2,b, an ensemble of lattices is set up, as shown in Fig. 1, such that the dimensions of the supercells, containing N lattice sites, display the essentials of the defect configurations expected. Consider the "grid" (of $\mathcal{M}$ points) through point (p) of the sample crystal. We seek the number of ways $g_\mathcal{M}$ of distributing

$$\mathcal{M}_i(p) = c_i(p)\mathcal{M} \tag{17.1}$$

defects (or atoms) of type i, for $i = 0, 1, \ldots, n$, over the $\mathcal{M}$ sublattice sites in such a way that these defects have correct fractional distribution $c_i(p)$, the latter average being determined by the procedure leading to Eq. (2.11). The result is the familiar one

$$g_\mathcal{M}(p) = \frac{\mathcal{M}!}{\{\bullet\}} \tag{17.2}$$

in which the convenient CVM notation[14]

$$\{\bullet\} \equiv \prod_{i=0}^{n} [c_i(p)\mathcal{M}]! \tag{17.3}$$

for the "point cluster" combinatorial factor has been used. Similar expressions are obtained for all other of the N points of the sample lattice, so that the total weight factor per system of the ensemble g is given by the $\mathcal{M}$th root of the product over all N points:

$$g = \left[\prod_p g_\mathcal{M}(p)\right]^{1/\mathcal{M}}. \tag{17.4}$$

For consistency, the internal energy must also depend only on single-site averages $c_i(p)$, i.e., on the concentration derivations

$$\gamma_i^\circ(p) \equiv \langle\gamma_i(p)\rangle_0 = c_i(p) - \bar{c}_i, \tag{17.5}$$

the symbol ($^\circ$) denoting microcanonical averaging, as in Eq. (2.11) or (11.20). From Eq. (6.36), the expectation value of the configurational energy is then given by

$$\langle E\rangle_0 = \frac{1}{2}\sum_{i,j=0}^{n}\sum_{p,p'} \nu_{ij}(p' - p)\langle\gamma_i(p)\gamma_j(p')\rangle_0, \tag{17.6}$$

and the configurational entropy is given by putting (17.2) and (17.4) into Eq. (16.4). The free energy for a sample lattice, defined by (16.3), thus takes the form

$$F = \sum_{i>j} \left\{ N w_{ij}^{\circ} \bar{c}_i \bar{c}_j + \sum_{p \neq p'} w_{ij}(p' - p) \gamma_i^{\circ}(p) \gamma_j^{\circ}(p') \right\} \quad (17.7)$$

$$+ k_B T \sum_{i=0}^{n} \sum_{p} c_i(p) \ln c_i(p),$$

in which Stirling's approximation has been used for the logarithm of the factorials. In keeping with the point-cluster approximation, the zeroth approximation has been introduced in deriving Eq. (17.7), which is to say that the average of the product in Eq. (17.6) has been replaced by the product of the averages γ_i° after removing the self-term ($p = p'$) given by Eq. (2.7). The double sum over defect types has been rewritten in the strictly off-diagonal manner of reduction scheme (6.39), but with a factor of $\frac{1}{2}$ introduced for convenience:

$$w_{ij}(p) = v_{ij}(p) - \frac{v_{ii}(p) + v_{jj}(p)}{2}. \quad (17.8)$$

As in Eq. (6.22), the coefficient of the self-term is

$$w_{ij}^{\circ} \equiv -w_{ij}(0) = +\sum_{p}' w_{ij}(p) \quad (17.9)$$

with a change of sign also introduced for convenience.

Equation (17.7) is the (nonequilibrium) free energy for a generalized Bragg–Williams model having arbitrary number of components $(n + 1)$, arbitrary number of sublattices (N) and arbitrary range of pair interactions.[3] Each lattice point (p) of the sample (or average) lattice is in a sense associated with an "average" atom of "composition" $c_i(p)$ ($i = 0, 1, \ldots, n$), thus regarded as a "microsystem"[4] of local entropy

$$s(p) = -k_B \sum_{i=0}^{n} c_i(p) \ln c_i(p), \quad (17.10)$$

and local internal energy

$$e(p) = \sum_{i=0}^{n} \varphi_i(p) \gamma_i^{\circ}(p) \quad (17.11)$$

where φ_i is the (nonlocal) potential

$$\varphi_i(p) = \sum_{j \neq i} \sum_{r} w_{ij}(r) \gamma_j^{\circ}(p + r). \quad (17.12)$$

Equation (17.12) shows that the microsystem at (p) is subjected to a potential averaged over all neighboring microsystems (sublattices).

Hence, the generalized BW model is a typical mean (or self-consistent, or Weiss) field model.

Well-known special cases may be derived easily from the general model (17.7), for example.

a. Regular Solution Model

This model is useful for the qualitative description of clustering solutions deviating little from the disordered one. It is derived from Eq. (17.7) by putting

$$c_i(p) = \bar{c}_i \quad \text{(all } p, \text{all } i). \tag{17.13}$$

For convenience, let us also write

$$c_A = \bar{c}_0, \quad c_B = \bar{c}_1, \quad c_C = \bar{c}_2, \ldots \tag{17.14}$$

and, for binary solution AB

$$c_B = c, \quad c_A = 1 - c, \quad w_{AB}^\circ = w. \tag{17.15}$$

The binary regular solution model is then

$$F^{(0)}(c, T) = Nwc(1 - c) + Nk_B T[c \ln c + (1 - c) \ln (1 - c)] \tag{17.16}$$

where the notation $F^{(0)}$ signifies that the expression on the right is actually the zeroth term in a Taylor's expansion of Eq. (17.7) (see Section 17,b). The ternary (ABC) regular solution model is

$$F^{(0)} = N(w_{AB}^0 c_A c_B + w_{BC}^0 c_B c_C + w_{CA}^0 c_A c_C)$$
$$+ Nk_B T[c_A \ln c_A + c_B \ln c_B + c_C \ln c_C], \tag{17.17}$$

and so on. The shape of the free energy curve $F^0(c)$ (17.16) at $k_B T/w = 0.8$ ($w > 0$) is shown in the upper portion of Fig. 21. The lower portion shows the derived miscibility gap (see Section 23). In favorable cases, experimentally determined miscibility gaps can be fitted quite well by assuming that the pair parameters $w_{AB}\cdots$ depend on temperature and average concentration,[154,155] as we know they must from the considerations of Section 6,a. In this way, various classes of subregular solution models are obtained. Note that in the regular solution model the sample system need contain only a single lattice point (microsystem) of composition $\bar{c}_i$.

---

[154] A.H. Hardy, *Acta Metall.* **1**, 202 (1953).
[155] L. Kaufman and H. Bernstein, "Computer Calculations of Phase Diagrams." Academic Press, New York, 1970.

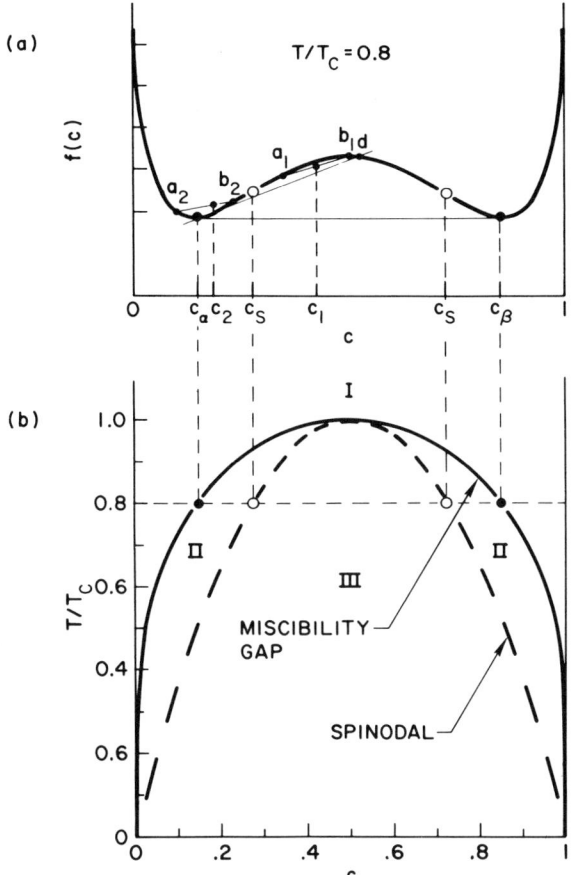

FIG. 21. Regular solution model. (a) Free energy curve at $T/T_c = 0.8$. (b) Miscibility gap and spinodal curve. Regions I, II, III are stable, metastable, and unstable, respectively.

b. Bragg-Williams Order-Disorder Models

In contradistinction to the regular solution models, at least two sublattices (microsystems) are required to describe order-disorder transitions. Concentrations $c_i(p)$ on the various sublattices are then usually given as linear functions of a set of long-range order parameters ξ, the number of which is one less than the number N of sublattices retained.

As an example, consider $L2_0$ ordering which occurs in bcc CuZn, for example. Two simple cubic sublattices, generally denoted α and β, are required with concentrations

$$c_0(p=0) \equiv c_A(\alpha) = c_A + c_A \xi$$
$$c_0(p=1) \equiv c_A(\beta) = c_A - c_A \xi \qquad (17.18)$$

and likewise for the second component. It follows from Eqs. (17.18) and (17.5) that

$$\gamma_A°(\alpha) = \gamma_B°(\beta) = c_A \xi$$
$$\gamma_B°(\alpha) = \gamma_A°(\beta) = -c_A \xi. \qquad (17.19)$$

The familiar free energy function for $L2_0$ ordering[156] then follows immediately from Eq. (17.7). At the critical temperature T_c we have

$$\left(\frac{d^2 F}{d\xi^2}\right)_{T_c} = 0. \qquad (17.20)$$

This is an example of a *transition of the second kind* (or second-order), defined as one for which the order parameter approaches zero continuously as the critical point (T_c) is approached from below.[157]

An example of a transition of the *first kind* is offered by $L1_2$ ordering which occurs in Cu_3Au. The long-range order parameter is conveniently defined by

$$c_0(p=0) \equiv c_A(\alpha) = c_A + 3c_A \xi$$
$$c_0(p=1,2,3) \equiv c_A(\beta) = \tfrac{1}{3}(c_A - 3c_A \xi), \qquad (17.21)$$

α being one simple cubic sublattice, and β a grouping of the other three which together make up the fcc lattice of the disordered solution. The familiar formula and free energy curves for $L1_2$ BW ordering are obtained by putting (17.21) into (17.5) and (17.7). The plot of ξ vs T shows a discontinuous jump in order parameter at the transition temperature (T_t), indicating the first-order character of the transformation. It is important to note in this context that the true character, first or second order, of a given transformation cannot be inferred from that of the corresponding BW model, as will be discussed below.

BW models have been used in slightly more complex cases, for instance in the Fe–Al system where both first- and second-neighbor interactions are required,[158,159] in ternary ordering systems such as $AgAuZn_2$ and

---

[156] T. Muto and Y. Takagi, *Solid State Phys.* **1**, 194 (1955).
[157] L.D. Landau and E.M. Lifshitz, "Statistical Physics," p. 430. Addison-Wesley, Reading, Massachusetts, 1958.
[158] C.R. Houska, *J. Phys. Chem. Solids* **24**, 95 (1963).
[159] S.V. Semenovskaya, *Phys. Status Solidi B* **64**, 291 (1974).

related systems,[160] and for the ω displacive transformation,[161] for example. All of these applications are special cases of the general formula (17.7).

c. Method of Concentration Waves

For binary systems one may write the thermodynamic potential as

$$G = \sum_p [e(p) - Ts(p) + \mu c(p)] \qquad (17.22)$$

where μ is a chemical potential introduced by the requirement of conserving atomic species, and where the local energy and entropy are given by Eqs. (17.10–17.12) specialized to binaries. Equilibrium sublattice concentrations are given by imposing the condition

$$\partial G/\partial c(p) = 0 \qquad (17.23)$$

yielding the solution[5,162]

$$c(p) = \frac{1}{\exp\left\{\left[\sum_{p'} v(p'-p)\gamma^\circ(p') + \mu\right]\Big/k_B T\right\} + 1} \qquad (17.24)$$

where the pair interaction parameter v, defined by Eq. (6.38), has been used instead of w, defined by Eq. (17.8), to conform to general practice of binary order–disorder theories. Khachaturyan[162] notes that Eq. (17.24) has the form of a Fermi–Dirac distribution function, which results from an analog of the Pauli exclusion principle, namely, that a lattice site can only be occupied either by an A or by a B atom. This analogy cannot be extended to multicomponent systems, however.

The "solution" (17.24) for $c(p)$ is implicit, because the right-hand side contains other concentrations $c(p')$ through the deviations $\gamma^\circ(p')$. Hence one must solve a set of nonlinear integral (summation) equations, just as in the more conventional treatment of BW models. Instead of defining sublattices *a priori*, as is usually done, Khachaturyan suggests the concentration wave approach,[5,54] setting

$$c(p) = \bar{c} + \sum_s \xi_s \sum_{h_s} \zeta_s(h_s) e^{i\mathbf{k}(h_s)\cdot\mathbf{x}(p)} \qquad (17.25)$$

where ξ_s denotes the amplitude of concentrations waves of wave vectors

---

[160] Y. Murakami, N. Nakanishi, and S. Kachi, *Acta Metall.* **19**, 93 (1971); Y. Murakami, S. Kachi, N. Nakanishi, and H. Takehara, *ibid.* p. 97.
[161] D. de Fontaine and R. Kikuchi, *Acta Metall.* **22**, 1139 (1974).
[162] A.G. Khachaturyan, *Phys. Met. Metallog. (Engl. Transl.)* **13**, 493 (1962).

$\mathbf{k}(h_s)$ belonging to the same star s, and where ζ_s are coefficients which determine the contribution of the particular wave $\mathbf{k}(h_s)$ to the amplitude ξ_s, the latter serving as an order parameter. The LRO parameters ξ_s are not completely defined by Eq. (17.25): additional conditions must be imposed on the ζ_s coefficients, either the normalization condition

$$\sum_{h_s} |\zeta_s(h_s)|^2 = 1 \qquad (17.26)$$

used by Landau and Lifshitz in their study of phase transformations of the second kind,[157] or the requirement that the parameters ξ_s be equal to unity for maximum degree of order at stoichiometry. Without actually solving the transcendental equations (17.24), Khachaturyan[5] then uses the CW method, based on Eq. (17.25), to predict possible ordered structures in fcc and bcc built up from CW with wave vectors $\mathbf{k}(h_s)$ located at the special points listed in Tables II and III (see Section 8). In principle, the state of order of a solution at any temperature and average concentration can be determined by solving (17.24) for equilibrium set of order parameters ξ_s.

Badalyan and Khachaturyan[163,164] have attempted to improve the accuracy of the BW model by using a linked cluster expansion such as proposed by Brout[165] and Horwitz and Callen.[166] The choice of terms retained by Khachaturyan and Badalyan appears to be somewhat arbitrary, however.

18. Taylor and Fourier Expansions

Much qualitative and some quantitative information can be gained by expanding a mean-field free energy in powers of the configuration variables $\gamma_i^\circ(p)$ or of their Fourier transforms $\Gamma_i^\circ(h)$.

a. Taylor's Expansion

The Taylor's expansion often need not be carried out beyond fourth-order terms:

$$F \cong F^{(0)} + F^{(2)} + F^{(3)} + F^{(4)}. \qquad (18.1)$$

If the expansion is performed on the generalized BW free energy given

---

[163] D.A. Badalyan and A.G. Khachaturyan, *Sov. Phys.—Solid State (Engl. Transl.)* **12**, 346 (1970).
[164] D.A. Badalyan and A.G. Khachaturyan, *Sov. Phys.—Solid State (Engl. Transl.)* **14**, 2270 (1973).
[165] R. Brout, "Phase Transitions." Benjamin, New York, 1965.
[166] G. Horwitz and H. Callen, *Phys. Rev.* **124**, 1757 (1961).

by Eq. (17.7), one obtains the following terms:

$$F^{(0)} = N \sum_{i>j} w_{ij}^0 \bar{c}_i \bar{c}_j + Nk_B T \sum_{i=0}^{n} \bar{c}_i \ln \bar{c}_i \tag{18.2}$$

$$F^{(2)} = \frac{1}{2} \sum_{i,j=0}^{n} \sum_{p,p'} f_{ij}''(p'-p) \gamma_i^\circ(p) \gamma_j^\circ(p') \tag{18.3}$$

$$F^{(3)} = \frac{1}{3!} \sum_{i=0}^{n} f_i''' \sum_p [\gamma_i^\circ(p)]^3 \tag{18.4}$$

$$F^{(4)} = \frac{1}{4!} \sum_{i=0}^{n} f_i^{IV} \sum_p [\gamma_i^\circ(p)]^4 \tag{18.5}$$

with coefficients

$$f_{ij}''(r) = \begin{cases} k_B T/\bar{c}_i & \text{for } i=j, r=0 \\ w_{ij}(r) & \text{for } i \neq j, r \neq 0 \\ 0 & \text{for } i=j, r \neq 0 \text{ and } i \neq j, r=0 \end{cases} \tag{18.6}$$

$$f_i''' = -k_B T/\bar{c}_i^2 \tag{18.7}$$

$$f_i^{IV} = k_B T/\bar{c}_i^3. \tag{18.8}$$

b. Fourier Expansion

The zeroth term $F^{(0)}$, Eq. (18.2), will be recognized as the free energy of an $(n+1)$ component regular solution model, binary and ternary examples of which were given in Eqs. (17.16) and (17.17), respectively. The quadratic or "harmonic" term $F^{(2)}$ is the relevant one for stability analysis. For that purpose, it is convenient to simplify it by a Fourier transformation:

$$F^{(2)} = \frac{N}{2} \sum_h \sum_{i,j=1}^{n} F_{ij}(h) \Gamma_i^{\circ*}(h) \Gamma_j^\circ(h). \tag{18.9}$$

It is essential that only independent concentration waves appear in $F^{(2)}$, hence the "host" CW have been eliminated from Eq. (18.9) by the reduction scheme (6.38):

$$F_{ij} = \mathscr{F}(f_{ij}'' - f_{i0}'' - f_{0j}'' + f_{00}'') \tag{18.10}$$

yielding,[3] by Eq. (18.6), the diagonal elements

$$F_{ii}(h) = -2W_{i0}(h) + k_B T \left(\frac{1}{\bar{c}_0} + \frac{1}{\bar{c}_i} \right), \tag{18.11}$$

and the off-diagonal elements

$$F_{ij}(h) = W_{ij}(h) - W_{i0}(h) - W_{0j}(h) + \frac{k_B T}{\bar{c}_0}, \qquad (18.12)$$

where $W_{ij}(h)$ is the Fourier transform of the pair potential $w_{ij}(r)$. Third- and fourth-order terms have Fourier representations which are fairly complicated, as seen in the binary example given in Eq. (18.15), below.

The harmonic portion of the free energy can be written in matrix form as

$$F^{(2)} = \frac{N}{2} \sum_h \Gamma^*(h) \mathbf{F}(h) \Gamma(h), \qquad (18.13)$$

Γ^* being the transpose of the n-component CW amplitude vector Γ (1 × n matrix) and $\mathbf{F}$ being an $n \times n$ Hermitian matrix. The stability of a solid solution with respect to a small-amplitude CW of given wave vector $\mathbf{k}(h)$ is guaranteed as long as the matrix $\mathbf{F}$ is positive definite. Instability sets in when the determinant of $\mathbf{F}$ vanishes:

$$\det(\mathbf{F}) \equiv \Delta(h; T, \bar{c}_1, \ldots, \bar{c}_n) = 0. \qquad (18.14)$$

For given wave vector, this equation generally defines n surfaces in (T, $\bar{c}$) phase diagram space; these are the stability limits, sometimes called spinodal surfaces, along each of which one of the (real) eigenvalues of $\mathbf{F}$ vanishes.[167] Above a stability limit, the corresponding eigenvalue is positive, indicating stability with respect to that mode of decomposition of the solution; below, the eigenvalue is negative, indicating instability. Although all $\Delta = 0$ surfaces are real, some may lie in negative temperature regions, in which case these are therefore not physically meaningful. Decomposition directions in composition space at a stability surface are given by the corresponding eigenvectors. Examples will be given in later sections. Here, we illustrate some of these ideas for the case of binary solid solutions.

c. Binary Systems

Instead of starting with a BW model, one may simply write down a phenomenological mean-field free energy expansion in Fourier space as follows[49,168]:

---

[167] D. de Fontaine, *J. Phys. Chem. Solids* **34**, 1285 (1973).
[168] A.G. Khachaturyan, "The Theory of Phase Transformation and Structure of Solid Solutions," p. 46. Nauka, Moscow, 1974 (in Russian).

$$F = F^{(0)} + \frac{N}{2} \sum_h F''(h) |\Gamma^\circ(h)|^2$$

$$+ \frac{N}{3!} \sum_{h,h',h''} F'''(h, h', h'') \Gamma^\circ(h) \Gamma^\circ(h') \Gamma^\circ(h'') \delta(h + h' + h''; g)$$

$$+ \frac{N}{4!} \sum_{h,h',h'',h'''} F^{IV}(h, h', h'', h''') \Gamma^\circ(h) \Gamma^\circ(h') \Gamma^\circ(h'') \Gamma^\circ(h''')$$

$$\times \delta(h + h' + h'' + h'''; g) + \cdots \quad (18.15)$$

with as yet undetermined coefficients. In the BW approximation, the quadratic coefficient F'' is simply given by F_{11} in Eq. (18.11). Hence, the stability condition (18.14) reduces here to

$$k_B T_0 = 2W(h)\bar{c}(1 - \bar{c}) \quad (18.16)$$

or, in the notation of Section 6,

$$k_B T_0 + V(h)\bar{c}(1 - \bar{c}) = 0. \quad (18.17)$$

In these equations, T_0 is the temperature at which instability sets in for the CW wave considered. According to the BW model, the stability limit, or spinodal, is always represented in the phase diagram $(T, \bar{c})$ by a parabola, symmetric about $\bar{c} = \frac{1}{2}$. For $h = (000)$, the classical "clustering spinodal"[169,170] is recovered from Eqs. (18.16) or (18.17). For other wave vectors, in particular those of the BZ boundary special points (Tables II and III), various "ordering spinodals" are obtained.[49]

In the BW case, the third-order coefficient has the simple form

$$F''' = k_B T \left(\frac{1}{\bar{c}_0^2} - \frac{1}{\bar{c}_1^2} \right) \quad (18.18)$$

obtained from Eq. (18.7). It follows that the third-order term in (18.15) vanishes identically for the symmetric composition $\bar{c} = \frac{1}{2}$ and hence a second-order transition is always expected in BW models at the 50/50 composition. The fourth-order coefficients F^{IV} is always positive according to Eq. (18.8).

19. Landau–Lifshitz Theory

Consider an ordering reaction in a binary solid solution for which the Fourier spectrum of the ordered phase is given by the set $\mathbf{k}(h_\kappa)$, the

[169] J.W. Cahn, *Acta Metall.* **9**, 795 (1961).
[170] J.W. Cahn, *Trans. AIME* **242**, 166 (1968).

index κ designating the various component waves, generally belonging to more than one star. In the free energy (18.15), perform the substitution

$$\Gamma^\circ(h) \to \xi\zeta(h_\kappa) \equiv \xi\zeta_\kappa \tag{19.1}$$

where, as in Eq. (17.25), ξ is the order parameter, and where the temperature-independent coefficients ζ_κ are fixed by the symmetry of the ordered phase. For this set of wave vectors, the expansion (18.15) takes the form

$$F = F^{(0)} + A(P, T)\xi^2 + B(P, T)\xi^3 + C(P, T)\xi^4 \tag{19.2}$$

with pressure and temperature-dependent coefficients

$$A = \frac{N}{2} \sum_\kappa F''(h_\kappa) |\zeta_\kappa|^2 \tag{19.3}$$

$$B = \frac{N}{3!} \sum_{\kappa,\kappa',\kappa''} F'''(h_\kappa, h_{\kappa'}, h_{\kappa''}) \zeta_\kappa \zeta_{\kappa'} \zeta_{\kappa''}$$
$$\times \delta(h_\kappa + h_{\kappa'} + h_{\kappa''}; g) \tag{19.4}$$

and a similar expression (involving four waves) for C. Equation (19.2) was postulated by Landau[157,171] as an approximate phenomenological expression for the free energy. It suffers from the usual shortcomings of a mean-field theory, which Landau was well aware of. Its use is warranted by its simplicity and by the remarkable symmetry rules which may be derived from it. A simplified account of the Landau and Lifshitz symmetry rules will now be given. A more general treatment is given in these authors' classical text *Statistical Physics*,[157] in the original references,[171,172] and in various other textbooks.[173–175]

In keeping with the general framework of this article, only those transformations which leave the average lattice invariant are considered. Just above the transition point, the order parameter ξ is zero by definition, and different from zero just below. Hence the effect of the transition is to remove certain symmetry elements from those of the symmetry group G_0 of the disordered (high-symmetry) phase. It follows that the group G of symmetry elements of the ordered phase must be a subgroup of G_0, G and G_0 being one of the 230 space groups (tabulated in the International

---

[171] L.D. Landau, *Phys. Z. Sowjetunion* **11**, 26 (1937).

[172] E.M. Lifshitz, *J. Phys. (Moscow)* **7**, 61 and 251 (1942).

[173] M.A. Krivoglaz and A.A. Smirnov, "The Theory of Order-Disorder in Alloys." Macdonald, London, 1965.

[174] A.P. Cracknell, "Magnetism in Crystalline Materials," p. 86. Pergamon, Oxford, 1975.

[175] G. Ya. Lyubarskii, "The Application of Group Theory in Physics." Pergamon, Oxford, 1960.

Tables for Crystallography[55]). Such is the first Landau and Lifshitz (LL) rule.

The second and third rules provide necessary (but not sufficient) conditions for transitions of the second kind, i.e., those for which the order parameter goes to zero continuously as the transition temperature T_c is approached from below. At the critical temperature, one must have[157]

$$B(P, T_c) = 0 \tag{19.5}$$

and

$$A(P, T) \begin{cases} > 0 & \text{for } T > T_c \\ = 0 & \text{for } T = T_c \\ < 0 & \text{for } T < T_c. \end{cases} \tag{19.6}$$

The latter inequalities are satisfied by setting

$$A(P, T) = \alpha(P)(T - T_c) \tag{19.7}$$

where $\alpha > 0$ depends on pressure only. Note that the second-order coefficient

$$F''(h) = \frac{k_B T}{\bar{c}(1 - \bar{c})} + V(h) \tag{19.8}$$

is of the form (19.7) when $T_0 = T_c$ is substituted into (19.8) from (18.17).

Let h_κ denote a CW, the amplitude of which vanishes above T_c but has finite value below T_c. Hence $\mathbf{k}(h_\kappa)$ is an ordering wave for this transition, along with all wave vectors of its star, the star being the set of wave vectors which transform into one another by the symmetry operations of G_0. Clearly, the value of $A(P, T)$ must be minimum at T_c for the wave vectors of this star, otherwise a phase transition would have occurred for other waves at higher temperatures. Certain wave vectors may accidentally minimize $A(P, T)$ for a certain temperature and pressure, and for a particular average concentration, but as these parameters are altered, the waves minimizing the second-order coefficient are expected to change. Therefore, stable ordered structures (in particular: stable with respect to long-period composition modulations)[5] must have CW spectra consisting entirely of waves whose wave vectors are located at *special points,* i.e., at those points in k-space where any function having the symmetry of the reciprocal lattice has extrema by virtue of symmetry requirements alone. As was explained in Section 8, these points are those at which symmetry elements intersect at a point; they are listed in Tables II and III for fcc and bcc lattices, respectively. Thus, second-order transitions may occur only for ordered structures described by "special" wave vectors. Such is the second LL symmetry rule, the so-called *Lifshitz criterion* recently discussed in detail by Michelson.[176]

The third rule concerns the necessary vanishing of the third-order coefficient $B(P, T)$. Again, we are not interested in its accidental vanishing, but in a general symmetry rule guaranteeing this property. The condition is easily discovered: it is seen, by Eq. (19.4), that B vanishes identically whenever

$$h_\kappa + h_{\kappa'} + h_{\kappa''} \neq g, \qquad (19.9)$$

i.e., if it is *impossible* to combine wave vectors of the star of ordering waves in such a way that they add up (vectorially) to a reciprocal lattice vector. This is the third LL symmetry rule.

To summarize, the three LL symmetry rules for second-order transitions are:

(I) The space group of symmetry elements of the ordered structure characterized by the local concentration $c(p)$ must be a subgroup of the space group of the disordered solution.

(II) The ordering wave vectors must be located at the special points of the Brillouin zone of the disordered phase, where symmetry elements intersect at a point.

(III) It must not be possible to find combinations of three members of the star of ordering wave vectors satisfying rule II which will sum (vectorially) to a reciprocal lattice vector.

These criteria are depicted symbolically as nested sets (solid lines) in the Venn diagram of Fig. 22.[49] Sets represented by dashed lines will be discussed later. The set J represents all possible ordered structures, and sets I, K, L represent ordered structures satisfying criteria I, II, III, respectively. Some examples will now be given for ordering transitions belonging to set I (where no lattice discontinuities are introduced). Discussions of the ordered structures are given in Section 22. A survey of experimentally observed structures was given by Guttman.[177]

$L2_0$ *ordering:* Occurring in near equiatomic bcc lattices, resulting in a CsCl structure upon complete ordering. The ordering waves are $\langle 100 \rangle$, which belong to the set of special points of Table III. Hence the LL II criterion is satisfied. Criterion III is also satisfied since three $\langle 100 \rangle$ waves cannot add up to a reciprocal lattice vectors; for example

$$[100] + [010] + [001] = [111] \qquad (19.10)$$

which is not a reciprocal lattice point in bcc. Hence $L2_0$ is candidate for a second-order transition, as is found to be the case experimentally in βCuZn, or AgCd.

[176] A. Michelson, *Phys. Rev. Phys. Rev. B* **16**, 577, 585 (1977).
[177] L. Guttman, *Solid State Phys.* **3**, 145 (1956).

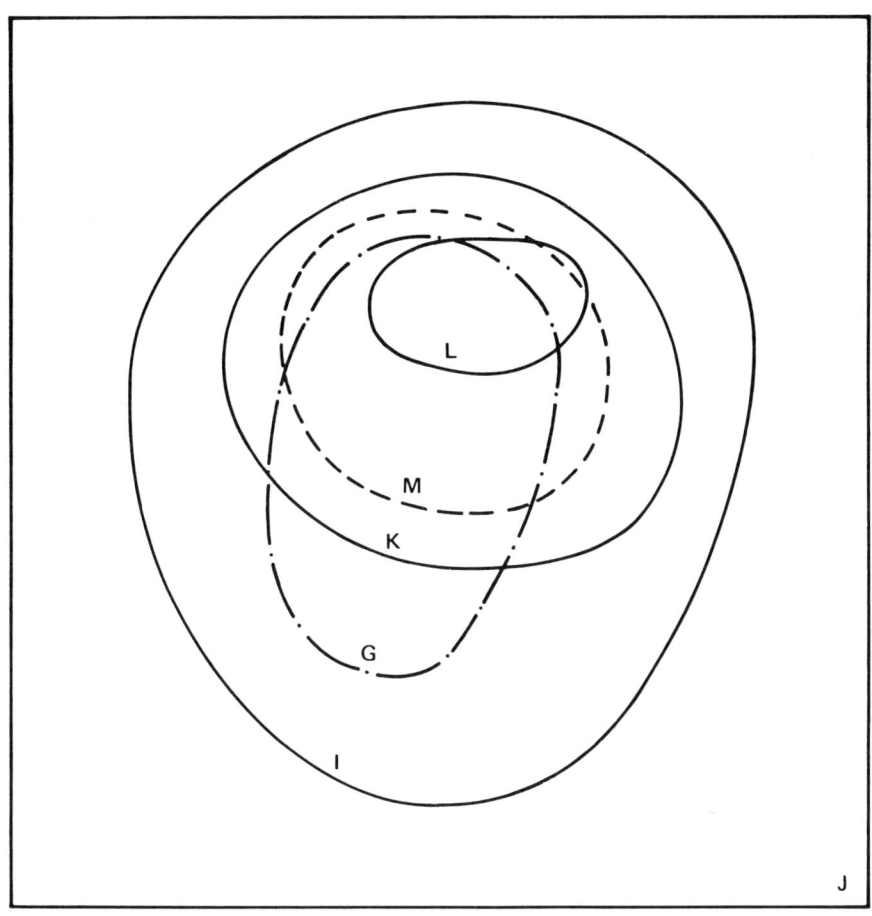

J ☐ : ALL ORDERED PHASES
I ——— : LLI; REPLACIVE ORDERING
K ——— : LLII; KHACHATURYAN "SPECIAL-POINT" CUBIC STRUCTURES
L ——— : LLIII; LIFSHITZ "SECOND-KIND" CUBIC STRUCTURES
M --- : CLAPP – MOSS CUBIC STRUCTURES
G —·— : RICHARDS – ALLEN – CAHN GROUND STATE STRUCTURES

FIG. 22. Venn diagram classification of ordered structures [D. de Fontaine, *Acta Metall.* **23**, 553 (1975)].

$L1_2$ ordering: Occurring in fcc lattices. The ordering waves are again $\langle 100 \rangle$, and Criterion II is again satisfied, but not LL III, since [111], from Eq. (19.10), *is* an fcc reciprocal lattice point. Hence $L1_2$ ordering *must* be a first-order transition, as found experimentally in Cu_3Au, Ni_3Mn, etc. The same conclusion holds for *$L1_0$ ordering* at the equiatomic composition. Thus, according to the LL rules, CuAu ordering must be first-order, as observed experimentally, whereas according to the BW model, it is second-order. This discrepancy will be briefly discussed in Section 24 in the CVM context. An interpretation, based on elastic energy considerations, is given by Kajitani and Cook.[178]

$D0_{22}$ ordering: Occurs in fcc, for example in Ni_3V[179] and in metastable Ni_3Mo.[180] The ordering waves are $\langle 1\frac{1}{2}0 \rangle$ and $\langle 210 \rangle$, the latter belonging to the star of $\langle 100 \rangle$. Hence two stars are involved and the transition *cannot* be of second kind. Indeed, though LL II is satisfied LL III is not[49]:

$$[1\tfrac{1}{2}0] + [1\tfrac{1}{2}0] + [210] = [420] \qquad (19.11)$$

which is an fcc reciprocal lattice vector. Surprisingly, Lifshitz[172] considered this ordering reaction to be a possible candidate for a second-order transition, apparently failing to recognize that the [210] wave, in addition to [1½0], was required to describe the structure correctly. The oversight probably arose because Lifshitz defined the order parameter with respect to wave vectors of a single star only, as in Eq. (17.25), whereas no such restriction is imposed on ξ as defined by Eq. (19.1).

$D1_a$ ordering: Occurs in Ni_4Mo (fcc disordered). The ordering waves are $q[420]$ where

$$q = \tfrac{1}{5}, \tfrac{2}{5}, \tfrac{3}{5}, \tfrac{4}{5}. \qquad (19.12)$$

Thus, LL II is not satisfied. Furthermore, $D1_a$ ordering must be driven, not by the minimum in the harmonic portion of the free energy, but by the third-order term B being negative. Anharmonic interactions of three ordering waves is indeed possible since, for example,

$$\tfrac{1}{5} + \tfrac{1}{5} + \tfrac{3}{5} = 1. \qquad (19.13)$$

$L1_1$ ordering: Occurs in equiatomic (fcc) CuPt. The ordering waves are $\langle \tfrac{1}{2}\tfrac{1}{2}\tfrac{1}{2} \rangle$, hence LL II is satisfied, and so is III since

$$[\tfrac{1}{2}\tfrac{1}{2}\tfrac{1}{2}] + [\bar{\tfrac{1}{2}}\tfrac{1}{2}\tfrac{1}{2}] + [\tfrac{1}{2}\bar{\tfrac{1}{2}}\tfrac{1}{2}] = [001], \text{ etc.} \qquad (19.14)$$

which is not a reciprocal lattice vector. Hence $L1_1$ ordering can be second-order. However, CuPt ordering is strongly first-order, which Ka-

---

[178] T. Kajitani and H.E. Cook, *Acta Metall.* **26**, 1371 (1978).
[179] L.E. Tanner, *Phys. Status Solidi* **30**, 685 (1968).

jitani and Cook[178] explain on the basis of a strong negative fourth-order contribution to the free energy expansion (18.1), due to an anharmonic "shape change" on ordering. The Landau expansion (18.1) for CuPt must then be carried out to sixth order, resulting in a first-order transition at the equiatomic composition.

The LL symmetry rules are extremely helpful in determining possible ordering mechanisms: one needs only the knowledge of the positions in k-space of the ordering waves.[49] For example, if the wave vectors of the ordering transformation place it in set L of Fig. 22, the reaction may proceed by a continuous ordering process occurring homogeneously throughout the crystal by continuous growth in amplitude of the ordering waves, with no discernible nucleation event. In set K (but outside L), ordering must proceed at equilibrium by third-order anharmonic interactions of ordering waves obeying the Lifshitz criterion (II), hence by the accidental local interaction of three high-amplitude waves with correct phase relationship. This can only occur in isolated regions, and thus requires a nucleation event. Away from equilibrium however, continuous ordering may occur in set K, as will be explained in Section 24. In set I (but outside K), the transformation mechanism must be nucleation and growth of the equilibrium ordered phase, with no continuous ordering of the equilibrium structure possible even far away from equilibrium. Heterogeneous nucleation at imperfections (grain boundaries, dislocations, etc.) may of course compete with homogeneous nucleation. Outside of J, heterogeneous nucleation must be the dominant mechanism since the lattice structure is destroyed on ordering, and this can best be initiated on preexisting lattice discontinuities. Such is the case for equilibrium orthorhombic Ni_3Mo, a distorted hexagonal phase, since no hexagonal point group can be a subgroup of cubic point group m3m (that of the disordered fcc lattice), and hence LL I is not obeyed.[49] Indeed, nucleation of Ni_3Mo is via a very complex heterogeneous mechanism.[180,181]

Landau-type phenomenological expansions, such as that of Eq. (19.2), are often used in connection with displacive transformations, particularly for the case of so-called *soft-mode transitions,* i.e., those for which a phonon frequency goes to zero at the transition temperature.[182–185] If the

---

[180] G. van Tendeloo, R. De Ridder, and S. Amelinckx, *Phys. Status Solidi A* **27**, 457 (1975).
[181] G. van Tendeloo, J. van Landuyt, P. Delavignette, and S. Amelinckx, *Phys. Status Solidi A* **25**, 697 (1974); G. van Tendeloo, P. Delavignette, J. van Landuyt, and S. Amelinckx, *ibid.* **26**, 299 (1974).
[182] J.D. Axe, J. Harada, and G. Shirane, *Phys. Rev. B* **1**, 1227 (1970).
[183] G. Shirane, *in* "Structural Transitions and Soft Modes" (E.J. Samuelson, E. Anderson, and J. Feder, eds.), p. 217, and other papers in symposium proceedings. Universitäts Forlaget, Oslo, 1971.
[184] S.M. Shapiro, J.D. Axe, G. Shirane, and T. Riste, *J. Phys. Rev. B* **6**, 4332 (1972).
[185] E. Pytte, *Comments Solid State Phys.* **5**, 41 and 57 (1973).

appropriate symmetry rules are obeyed, the coefficient $A(P, T)$ of the harmonic term may vanish at $T = T_c$, which means that the energy (or frequency) of a displacive wave of wave vector $\mathbf{k}(h)$ goes to zero, indicating lattice instability with respect to that mode, just as in the replacive case described by Eq. (19.8) or (18.17). A particularly interesting application of the Landau theory was made by McMillan to the transitions in TaS_2 and similar compounds.[43] In this formulation, rotations as well as more usual lattice displacements, both commensurate and incommensurate with the average lattice are predicted in successive temperature ranges.

20. Cluster Variation Method

The general principle of constructing a free energy function F based on cluster probabilities was outlined in Section 16: $F(\tau_{ijk\cdots l})$ is defined as an analytic function in the variables $\tau_{ijk\cdots l}$ denoting the frequency of occurrence of a cluster consisting of the indicated atomic species $i, j, \ldots$, occupying the sites of a *basic figure* containing say m neighboring lattice sites. These *cluster variables* $\tau_{ij\cdots l}$ are the configuration variables for the system, and they define the *state of order* (see Section 2); it is with respect to these that the free energy function must be minimized at equilibrium, as indicated symbolically in Eq. (16.6).

This cluster variation method[14] (CVM) appears not to have been used as yet to its full potential. Many of the early practical shortcomings of the CVM have now been overcome: schemes for the systematic improvement of accuracy have been devised,[15,186,187] the limitation to first-neighbor interactions in the internal energy has been lifted,[15,188] and a new computational algorithm which guarantees convergence, the natural iteration technique,[189] has been developed. The author believes, therefore, that the CVM will find increasing use in solid solution theory, hence the rather extensive coverage given below. For more detailed treatments, the interested reader is referred to the cited papers of Kikuchi, who originated and developed the CVM. See also the review by Burley.[190]

The major difficulties encountered in CVM calculations are: (a) the derivation of suitable weight factors $g(\tau_{ij\cdots l})$, which occur in Eqs. (16.1) and (16.4), and (b) the computation of equilibrium values of cluster frequencies $\tau_{ij\cdots l}$. These topics are examined in subsections (a) and (b),

[186] M. Kurata, R. Kikuchi, and T. Watari, *J. Chem. Phys.* **21**, 434 (1953).
[187] S.K. Aggarwal and T. Tanaka, *Phys. Rev. B* **16**, 3963 (1977).
[188] R. Kikuchi and C.M. van Baal, *Scr. Metall.* **8**, 425 (1974).
[189] R. Kikuchi, *J. Chem. Phys.* **60**, 1071 (1974).
[190] D.M. Burley, *in* "Phase Transitions and Critical Phenomena (C. Domb and M.S. Green, eds.), Vol. 2, p. 329. Academic Press, New York, 1972.

below; a third subsection (c) is devoted to a comparison of transition temperatures T_t predicted by various approximations.

a. CVM Configurational Entropy

Various schemes have been proposed for deriving approximate combinatorial formulas for the weight factor g, for successively larger basic figures.[14,15,191,192] The one outlined below was recently proposed by Kikuchi[193]; resulting formulas are identical to those derived earlier by other methods.

One begins with the simplest possible case: that of a one-dimensional Ising chain with nearest-neighbor (n.n.) interactions. An ensemble of $\mathcal{M}$ chains of N_1 lattice sites is set up, as was done with sample systems in the BW approximation described in Section 17. To conform with the usual CVM notation, let

$$x_i \equiv c_i(p) \qquad (20.1)$$

be the fractional distribution of i at all homologous lattice sites (p) of the various chains of the ensemble. In this simple example, sublattices need not be defined, so that all points (p) are equivalent, and the argument (p) need not be retained. Consider at first only a binary system AB. Hence, the only "point cluster" variables are x_A and x_B. Nearest-neighbor pair frequencies will likewise be denoted by y_{ij}, specifically: y_{AA}, $y_{AB} = y_{BA}$, y_{BB}.

Assume that A and B atoms (or spins $+1$, -1) have been distributed among the $\mathcal{M}$ chains up to point (p') such that both the x_i and y_{ij} have specified fractional distribution. Consideration of pair frequencies is essential since the internal energy is assumed to be given by a sum of pair interactions, as explained at some length in Section III. It is now required to distribute A and B among the sites of column (p) (Fig. 23), in such a way that the new cluster frequencies at (p) have correct fractional distribution. To a randomly selected set of $y_{AA}\mathcal{M}$ A points (a subset of all $x_A\mathcal{M}$ A points) and $y_{BA}\mathcal{M}$ B points of column (p'), add $x_A\mathcal{M}$ A symbols in corresponding positions on column (p). This forms $y_{AA}\mathcal{M}$ AA pairs and $y_{BA}\mathcal{M}$ AB pairs, and this can be performed in

$$g_A(p) = \frac{(x_A\mathcal{M})!}{(y_{AA}\mathcal{M})!(y_{BA}\mathcal{M})!} \qquad (20.2)$$

---

[191] J. Hijmans and J. de Boer, *Physica (Utrecht)* **21**, 471, 485, and 499 (1955); **22**, 408 (1956); J. Hijmans, *ibid.* p. 429.
[192] J.A. Barker, *Proc. R. Soc. London, Ser. A* **216**, 45 (1953).
[193] R. Kikuchi, private communication.

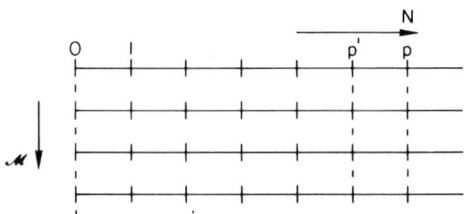

FIG. 23. Ensemble of $\mathcal{M}$ linear chains used to derive Eq. (20.4).

ways.[14,190] Likewise, B symbols can be placed on (p) in

$$g_B(p) = \frac{(x_B \mathcal{M})!}{(y_{AB}\mathcal{M})!(y_{BB}\mathcal{M})!} \quad (20.3)$$

ways. The combined number of ways of distributing A and B on (p) is given by the product of these two factors: $g_A g_B$. In general, for an $(n+1)$-component chain

$$g_{\mathcal{M}}(p) = \prod_{i=0}^{n} g_i(p) = \frac{\{\bullet\}}{\{\bullet\!\!-\!\!\bullet\}} \quad (20.4)$$

with shorthand CVM notation

$$\{\bullet\} = \prod_{i=0}^{n} (x_i \mathcal{M})! \quad (20.5)$$

$$\{\bullet\!\!-\!\!\bullet\} = \prod_{i,j=0}^{n} (y_{ij}\mathcal{M})!, \quad (20.6)$$

a convenient symbolism easily generalized to basic figures larger than the "point" and "pair." The lower-approximation "point" CVM formula was given above in the BW approximation: Eq. (17.2). Since all points of the chain are assumed to be equivalent, the weight factor for all N_1 points is given by the N_1th power of $g_{\mathcal{M}}(p)$ [Eq. (20.4)]. In cases involving ordered structures (antiferromagnet), sublattices must be introduced by setting $p = \alpha, \beta, \ldots$ as explained in Section 17,b. For a single chain, the $\mathcal{M}$th root must be taken so that, finally, the required weight factor to be used in the configurational entropy (16.4) is

$$g(1 - \text{dim. chain}) = \left[\frac{\{\bullet\}}{\{\bullet\!\!-\!\!\bullet\}}\right]^{N_1/\mathcal{M}}. \quad (20.7)$$

This formula is exact, and the free energy derived from it is the exact one for the linear chain Ising model with n.n. interactions.[14] Formula (20.7) will now be generalized to lattices in two and three dimensions, treated as extensions of the linear chain.

Consider an ensemble of $\mathcal{M}$ two-dimensional lattices containing $N_1 \times N_2$ square unit cells (Fig. 24). Symbols A, B, C, ... (atomic species, defects) have been placed on lattice sites up to column (p') with correct fractional distribution for both x_i and y_{ij}. Symbols must now be placed on column (p) with correct fractional distribution. The total number of ways of doing this is obtained by considering the lattice as a one-dimensional chain, each of the N_1 sites of which is occupied by a "string" of N_2 points, the frequencies of occurrence of various types of strings being

$$\tilde{x}_I; \quad I = 1, 2, \ldots, \mathcal{N} \tag{20.8}$$

where

$$\mathcal{N} = (n + 1)^{N_2} \tag{20.9}$$

a very large number, generally, but assumed to be greatly surpassed by the number $\mathcal{M}$ of systems in the ensemble. Likewise, let $\tilde{y}_{IJ}$ denote the fraction of n.n. double strings or "ladders." By Eq. (20.4), the weight factor at (p) is:

$$g_\mathcal{M}(p) = \frac{\prod_{I=1}^{\mathcal{N}} (\tilde{x}_I \mathcal{M})!}{\prod_{I,J=1}^{} (\tilde{y}_{IJ} \mathcal{M})!} \equiv \frac{\left\{ \begin{matrix} \bullet \\ \vdots \\ \bullet \end{matrix} \right\}}{\left\{ \begin{matrix} \bullet - \bullet \\ \vdots \\ \bullet - \bullet \end{matrix} \right\}}. \tag{20.10}$$

Formula (20.10) is exact; if evaluated explicitly, it would lead to Onsa-

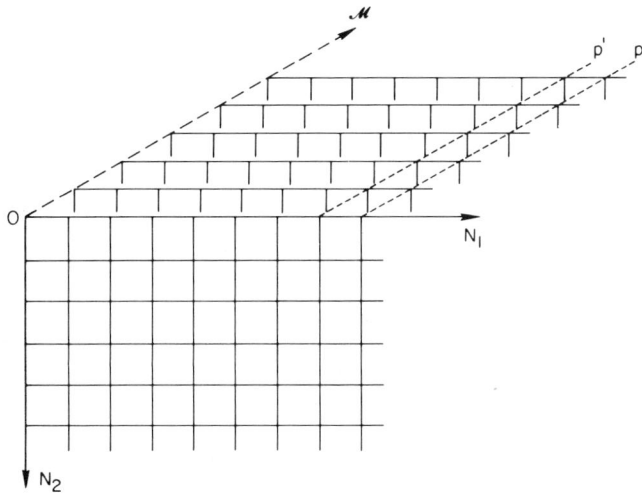

FIG. 24. Ensemble of $\mathcal{M}$ $N_1 \times N_2$ two-dimensional square lattices used to derive Eq. (20.15).

ger's exact solution of the two-dimensional Ising model.[194] Here, however, we are interested in simple approximations suitable for extension to higher-dimensionality systems.

To approximate (20.10) in terms of lower cluster variables, one notes that $g(p)$ may be written as the ratio $g_\mathcal{M}(\text{ladder})/g_\mathcal{M}(\text{string})$ where

$$g_\mathcal{M}(\text{string}) = \frac{\mathcal{M}!}{\{\bullet\!\!-\!\!\bullet\!\!-\!\!\bullet\!\!-\!\!\bullet\!\!-\!\}} \tag{20.11}$$

and

$$g_\mathcal{M}(\text{ladder}) = \frac{\mathcal{M}!}{\{\begin{smallmatrix}\bullet\!\!-\!\!\bullet\!\!-\!\!\bullet\!\!-\!\!\bullet\\ \bullet\!\!-\!\!\bullet\!\!-\!\!\bullet\!\!-\!\!\bullet\end{smallmatrix}\}}, \tag{20.12}$$

both of these formulas being generalizations of Eq. (17.2): Eq. (20.11) gives the total number of ways of distributing $\tilde{x}_1 \mathcal{M}$ chains of type 1, $\tilde{x}_2 \mathcal{M}$ chains of type 2, ..., $\tilde{x}_\mathcal{N} \mathcal{M}$ chains of type $\mathcal{N}$ among $\mathcal{M}$ points, while Eq. (20.12) has similar meaning for ladders. The crucial approximation is now made of discarding the fractional string distributions $\tilde{x}_l$ themselves, requiring merely that each string, of length N_2, have correct fractional distribution of points and pairs. Hence, by Eq. (20.4), we take

$$g_\mathcal{M}(\text{string}) \cong \left[\frac{\{\bullet\}}{\{\bullet\!\!-\!\!\bullet\}}\right]^{N_2}. \tag{20.13}$$

Likewise, for ladders with correct fractional distribution of pairs and n.n. squares we take

$$g_\mathcal{M}(\text{ladder}) \cong \left[\frac{\{\bullet\!\!-\!\!\bullet\}}{\{\begin{smallmatrix}\bullet\bullet\\ \bullet\bullet\end{smallmatrix}\}}\right]^{N_2}. \tag{20.14}$$

Hence $g(p)$ is approximately given by the ratio of expressions (20.14) and (20.13). The required formula for the two-dimensional square lattice with n.n. square as basic figure is thus

$$g(2-\text{dim. sq. lattice}) \cong \left[\frac{\{\bullet\!\!-\!\!\bullet\}^2}{\{\begin{smallmatrix}\bullet\bullet\\ \bullet\bullet\end{smallmatrix}\}\{\bullet\}}\right]^{N_1 N_2/\mathcal{M}} \tag{20.15}$$

obtained by following the procedure leading to Eq. (20.7). Formula (20.15) is precisely that of the Kramers-Wannier approximation of the square lattice.[14,195] The formula is not exact because correct two-dimensional correlations are no longer respected in going from (20.11) to (20.13), and from (20.12) to (20.14).

The method given here has straightforward extension to three dimen-

[194] L. Onsager, *Phys. Rev.* **65**, 117 (1944).
[195] H.A. Kramers and G.H. Wannier, *Phys. Rev.* **60**, 252 and 263 (1941).

sions, which is essential for treating real systems. Thus, the simple cubic lattice is treated as a one-dimensional chain, each of the N_1 points of which is associated with a square lattice of $N_2 \times N_3$ points. By repeating the above arguments, one is lead successively to the following approximations:

$$g_\mu(\rho) = \frac{\{\boxed{}\}}{\{\boxed{}\}} \cong \left[\frac{\{\boxed{\square\square\square}\}}{\{\boxed{\square\square\square}\}}\right]^{N_2} \cong \left[\frac{\{\square\}/\{\boxdot\}}{\{\leftrightarrow\}/\{\square\}}\right]^{N_2 N_3} \left[\frac{\{\leftrightarrow\}/\{\square\}}{\{\cdot\}/\{\leftrightarrow\}}\right]^{N_2 N_3} \quad (20.16)$$

Hence

$$g\text{ (simple cubic)} \cong \left[\frac{\{\square\}^3 \{\cdot\}}{\{\boxdot\} \{\leftrightarrow\}^3}\right]^{N_1 N_2 N_3 / \mathcal{M}} \quad (20.17)$$

is the required formula for the simple cubic lattice with n.n. cube as basic figure.

Kikuchi[193] has devised an ingenious procedure to reduce Eq. (20.17) to one valid for the more useful case of the fcc lattice: since the sc lattice is made up of two interpenetrating fcc lattices, one simply assigns "vacancies" (i.e., "defect" $i = 0$) to one of the sublattices, the remaining n components of the initial $(n + 1)$-component system now being distributed over the other fcc sublattice. The basic figure for species $i = 1, 2, \ldots, n$ is now the n.n. tetrahedron, as shown in Fig. 25 (points 1-2-3-4). Likewise, the square clusters of (20.17) go into pairs (face diagonals), the pairs into points, and the points into a product of two terms: $\mathcal{M}!$ for "vacancies" and $\{\bullet\}$ for occupied sites, of which there are $N = N_1 N_2 N_3 / 2$. The weight factor is then found to be

$$g(\text{fcc; tetrahedron}) \cong \left[\frac{\{\bullet\text{---}\bullet\}^6 \mathcal{M}!}{\{\bullet\diamondsuit\bullet\}^2 \{\bullet\}^5}\right]^{N / \mathcal{M}}. \quad (20.18)$$

It is also possible to derive a very convenient simplified version of the

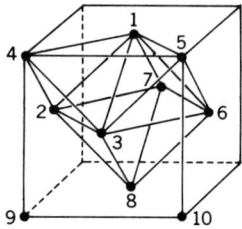

FIG. 25. Tetrahedron (1-2-3-4), octahedron (1-2-3-6-7-8) and double-tetrahedron (1-2-3-4-5-6) clusters used in formula (21.1) [J. M. Sanchez and D. de Fontaine, *Phys. Rev. B*, **17**, 2926 (1978)].

above formulas which uses the n.n. pair as basic figure[189.]

$$g(\text{pair}) \cong \left[\frac{\{\bullet\}^{2\omega-1}}{\{\bullet\!-\!\bullet\}^{\omega}(\mathcal{M}!)^{\omega-1}} \right]^{N/\mathcal{M}}. \tag{20.19}$$

This formula is valid for either sc, fcc, or bcc lattices, or for square or triangular 2-dimensional networks of N lattice points; one needs only specify the number $2\omega = z^{(1)}$ of nearest neighbors.

All of the above formulas are limited to nearest-neighbor cluster variables. This means that, for example, only a limited number of order-disorder reactions can be treated successfully: those for which the internal energy can be approximated by a sum of n.n. pair interactions. If second and third neighbors could be included, almost all fcc and bcc "coherent" ordering reactions could be handled by the CVM in principle. Fortunately, a formula for the g factor in fcc has recently been given[188] which uses as basic figure the double tetrahedron depicted in Fig. 25 (points 1-2-3-4-5-6), which is seen to contain first-, second-, and third-neighbor pairs. A better cluster combination is the double-tetrahedron/octahedron (points 1-2-3-6-7-8 in Fig. 25),[15] for which the g factor is given in Eq. (21.1), below.

The method outlined above for constructing combinatorial weight factors can be extended to larger basic figures, for example, to the 9-point cube in bcc,[196] and the 13- and 14-point cubes in fcc.[15] The number of cluster variables in an $(n + 1)$-component solution, $(n + 1)^m$, increases very rapidly with the number of m of sites in the basic cluster, however.

b. CVM Free Energy Computations

The free energy is obtained as explained in Section 16, with an appropriate weight factor selected for the entropy formula (16.4). The internal energy is written down as a sum of cluster energies with the restriction that the basic figures used in the internal energy must be contained in the set of basic figures used in the entropy formula. For systems of fixed composition, the minimization must be performed on the "grand potential"

$$G = F - N \sum_{i=0}^{n} \mu_i x_i \tag{20.20}$$

where μ_i are chemical potentials.

To proceed, let us consider a definite example: that of an fcc solution

---

[196] J.M. Sanchez, Ph.D. Dissertation, School of Engineering and Applied Science, University of California, Los Angeles (1977).

in the n.n. tetrahedron approximation. Let z_{ijkl} be the tetrahedron clusters variables, with i, j, k, l denoting the atomic species occupying the four vertices of the tetrahedron. Cluster variables are normalized so that

$$\sum_{i,j,k,l} z_{ijkl} = 1 \qquad (20.21)$$

the sums, here and in what follows, running over all atomic species (i, j, k, $l = 0, 1, \ldots, n$). Lower cluster probabilities can be obtained from the tetrahedron variables by the identities

$$x_i = \sum_{j,k,l} z_{ijkl} \qquad (20.22)$$

$$y_{ij} = \sum_{k,l} z_{ijkl} . \qquad (20.23)$$

It is convenient to write the internal energy in terms of the highest cluster variables:

$$E = N \sum_{i,j,k,l} w_{ijkl} z_{ijkl} \qquad (20.24)$$

where the w's denote the energy of the corresponding tetrahedron clusters. These cluster energies can be calculated on the basis of a sum of pairs, as was done in Section III, or distinct clusters may be given distinct energies even if they are made up of the same pairs. In this way, anharmonic (many-body) interactions can easily be introduced in the internal energy.

By means of Eqs. (20.24) and (20.18), the grand potential (20.20) may be written explicitly as

$$G = N \sum_{i,j,k,l} w_{ijkl} z_{ijkl} + N k_B T [5 \sum_i x_i \ln x_i - 6 \sum_{i,j} y_{ij} \ln y_{ij} \qquad (20.25)$$
$$+ 2 \sum_{i,j,k,l} z_{ijkl} \ln z_{ijkl}] - N \sum_i \mu_i x_i + N\lambda [1 - \sum_{i,j,k,l} z_{ijkl}]$$

where λ is a Lagrangian multiplier introduced to satisfy the normalization condition (20.21). One then substitutes x_i and y_{ij} from Eqs. (20.22) and (20.23) into (20.25), and minimizes with respect to z_{ijkl} to yield the equilibrium condition

$$z_{ijkl} = Y_{ijkl}^{1/2} X_{ijkl}^{-5/8} \exp[-(\tfrac{1}{2} w_{ijkl} - \tfrac{1}{8}\mu_{ijkl} - \tfrac{1}{2}\lambda)/k_B T] \qquad (20.26)$$

where

$$\mu_{ijkl} = \mu_i + \mu_j + \mu_k + \mu_l . \qquad (20.27)$$

Equation (20.26) is implicit since the X_{ijkl} and Y_{ijkl} are products of point and pair cluster variables. The nature of these products differ depending upon whether one is investigating the disordered phase (D) or an ordered phase (L1$_2$, for example):

$$X_{ijkl}(D) = x_i x_j x_k x_l$$
$$Y_{ijkl}(D) = y_{ij} y_{ik} y_{il} y_{jk} y_{jl} y_{kl}, \qquad (20.28)$$

or

$$X_{ijkl}(L1_2) = x_i{}^\alpha x_j{}^\alpha x_k{}^\alpha x_l{}^\beta$$
$$Y_{ijkl}(L1_2) = y_{ij}^{\alpha\alpha} y_{ik}^{\alpha\alpha} y_{il}^{\alpha\beta} y_{jk}^{\alpha\alpha} y_{jl}^{\alpha\beta} y_{kl}^{\alpha\beta}, \qquad (20.29)$$

where the superscripts denote the two sublattices α and β of the ordered L1$_2$ structure. Equations (20.29) are derived from slightly modified versions of Eqs. (20.22) and (20.23), appropriate for the ordered structure.

The value of λ is obtained by combining condition (20.21) and Eq. (20.26):

$$e^{-\lambda/2k_B T} = \sum_{i,j,k,l} Y_{ijkl}^{1/2} X_{ijkl}^{-5/8} \exp[-(\tfrac{1}{2} w_{ijkl} - \tfrac{1}{8} \mu_{ijkl})/k_B T] \qquad (20.30)$$

which has the form of a sum over states, limited to tetrahedron clusters. At equilibrium, the parameter λ gives the value of the free energy per lattice point, appropriate for the particular level of the CVM approximation used. The sum (20.30) differs, however, from the conventional partition function through the presence of the factor XY, without which the cooperative nature of the problem would be lost.[189]

The structure of Eqs. (20.26)–(20.30) suggests a numerical solution algorithm which is far preferable to the usual Newton–Raphson (NR) method of solving simultaneous nonlinear equations. This algorithm, called by Kikuchi the natural iteration (NI) method,[189] consists of performing the following operations: (1) select fixed values of the chemical potentials μ_i, (2) start with initial guess values of the point variables x_i and of the pair variables y_{ij} (which may be approximated initially by the products $x_i x_j$), (3) obtain a value of λ by Eq. (20.30), (4) obtain values of the tetrahedron variables z_{ijkl} explicitly by Eq. (20.26), (5) insert the latter values in Eqs. (20.22) and (20.23), or in their modified versions in case of ordered phases, (6) insert the new values of x_i, y_{ij} into (20.28) or (20.29) and reiterate the whole process from step (2) onward. When suitable convergence has been attained, new sets of values of the chemical potentials may be selected, and the whole cycle (2–6) is repeated.

Generally, the number of iterations required for the NI method exceeds that required for NR iteration, but this slight disadvantage is more than offset by the fact (a) that the NI method is explicit, so that there are no matrix inversions required, and (b) that the natural iteration method is guaranteed to converge to equilibrium states even for initial guess values of the lower cluster variables which are far from the correct equilibrium ones. This is a particularly remarkable feature of NI which makes CVM calculations with large basic figures quite attractive. With NR iteration, such computations would not be practical due to the extreme sensitivity of NR convergence on the choice of initial values.

To conclude this subsection, new developments in the CVM now allow extensive and accurate computations to be made on systems of considerable complexity. For example, equilibrium states have recently been calculated for binary[197–200] and ternary[201,202] fcc solutions in the tetrahedron approximation, and for the displacive ω transformation in the 8-point bcc primitive unit cell approximation.[203] The number of simultaneous equations to be solved in these three cases was, respectively, $2^4 = 16$, $3^4 = 81$, $3^8 = 6561$. Application to the calculation of phase diagrams will be given in Section 23.

c. Comparison of Cluster Variation and Bragg–Williams Methods

A good measure of the accuracy of free energy calculations is provided by the numerical values of the predicted transition temperatures. A comparison of transition temperatures predicted by various approximate methods is given in Table V, adapted from Kikuchi,[14] Burley,[190] and Sanchez and de Fontaine.[15] Values listed are for the transition parameter

$$\tau = 4k_B T_t / |z^{(1)} V_1| \tag{20.31}$$

where T_t is the transition temperature, V_1 is the n.n. pair interaction parameter, as defined in Section III, and $z^{(1)}$ is the first-shell coordination number. For the Ising ferromagnet, isomorphous to the AB binary solution, and $L2_0$ ordering in bcc, the transition is second-order, while those listed for fcc are first-order for $L1_2$ in both BW and CV models, first-order for $L1_0$ (which is correct) in the tetrahedron CV method, and second-order in the BW approximation. Note however that the pair CVM approximation gives no transition at all for $L1_2$ and $L1_0$. The first-order

[197] R. Kikuchi and H. Sato, *Acta Metall.* **22**, 1099 (1974).
[198] C.M. van Baal, *Physica (Utrecht)* **64**, 571 (1973).
[199] D. de Fontaine and R. Kikuchi, in "Applications of Phase Diagrams in Metallurgy and Ceramics," N.B.S. Special Publication 496 (G.C. Carter, ed.), p. 999 (1978).
[200] R. Kikuchi and D. de Fontaine, in "Applications of Phase Diagrams in Metallurgy and Ceramics," N.B.S. Special Publication 496 (G.C. Carter, ed.), p. 967 (1978).
[201] R. Kikuchi, *Acta Metall.* **25**, 195 (1977).
[202] R. Kikuchi, D. de Fontaine, M. Murakami, and T. Nakamura, *Acta Metall.* **25**, 207 (1977).
[203] J.M. Sanchez and D. de Fontaine, *Acta Metall.* **26**, 1083 (1978).

TABLE V. CALCULATED NORMALIZED TRANSITION TEMPERATURES

A. General results[a]

Transition point in units of $k_B T/|z^{(1)}V_1|$ with $V_j = 0$ ($j \geq 2$)	Phase separation (Ising ferromagnet)					Ordering		
	2-D Square lattice	2-D Triangular lattice	3-D Simple cubic	BCC Also $L2_0$ ordering	Phase separation	FCC		
						$L1_0$		$L1_2$
Coordination $z^{(1)}$	4	6	6	8	12	12		12
Point approximation (Reg. sol., B–W)	1.0	1.0	1.0	1.0	1.0	0.3333		0.2734[b]
CV pair (Bethe approximation)	0.7212	0.8222	0.8222	0.8690	0.9142	0		0
CV square or triangle	0.6057	0.6525	0.7683	0.8454	—	—		—
CV cube or tetrahedron	—	—	0.7628	0.8113	0.8354	0.1577[b]		0.1604[b]
Exact (or best known)	0.567	0.6062	0.7522	0.7944	0.8163	NA[c]		NA[c]

[a] R. Kikuchi, Phys. Rev. **81**, 988 (1951); D.M. Burley in "Phase Transitions and Critical Phenomena" (C. Domb and M.S. Green, eds.), Vol. 2, p. 339. Academic Press, New York, 1972.
[b] First-order transitions; all others second-order.
[c] NA: not available.

B. More recent results for Ising ferromagnet[a]

Approximation	$k_B T_c/12V_1$
Tetrahedron	0.83544
Double tetrahedron	0.84045
Octahedron-tetrahedron	0.83394
Double tetrahedron-octahedron	0.82981
High T expansion	0.81627

[a] J.M. Sanchez and D. de Fontaine, Phys. Rev. B **17**, 2926 (1978).

transitions are determined by finding the intersections of curves representing equilibrium values of G vs μ_i in both disordered and ordered phases. The second-order transitions are determined by setting to zero the determinant of the matrix of second derivatives of the free energy with respect to independent cluster variables.

In two-dimensional lattices for which exact solutions are known,[194,204] the highest CVM approximation predicts transition temperatures which differ from the exact values by about 7%, whereas the BW approximation is off by more than 60%. In three-dimensional phase separation cases (spin $\frac{1}{2}$ ferromagnet), the highest CVM approximation (double-tetrahedron/octahedron) predicts T_c values which differ by only 1.5% from those predicted by the best available series expansions, whereas the BW approximation is off by about 25%. In the β-brass case, with $T_c = 468°C$, this would mean an error of 15°C for the CVM, and 185°C in the BW approximation. In the fcc ordered cases, the BW approximation fails even more dramatically: the transition temperatures are off by more than 70%, the order of the transformation is predicted incorrectly for $L1_0$, and the calculated phase diagram for CuAu-like systems is not even qualitatively correct, as will be shown in Section 23.

Further comparison is given in Fig. 26[204a] for the case of the two-dimensional square lattice: it is seen that T_c from the CVM square approximation comes much closer to the exact Onsager solution than that from the Bethe pair approximation, and also that the heat capacity at constant volume c_v has a more reasonable behavior in the vicinity of the critical temperature. As a final example, Fig. 27 compares experimental data for the LRO parameter ξ as a function of temperature with the curve calculated for both $L1_0$ (CuAu, Fig. 27a) and $L1_2$ (Cu$_3$Au, Fig. 27b) ordering by the tetrahedron CVM approximation.[197] The first-order character of the $L1_0$ transition is predicted correctly, and agreement is generally quite good.

21. Discussion of Free Energy Models

Free energy functions which depend on the configurational variables used to characterize the state of order of a solid solution were derived in Section 16–20. These functions are analytic and may be used away from equilibrium, the equilibrium free energy then being given by that function of the configurational variables which corresponds to the most probable

---

[204] C.J. Thompson, "Mathematical Statistical Mechanics," Appendix D, p. 229. Macmillan, New York, 1972.

[204a] H.E. Stanley, "Introduction to Phase Transitions and Critical Phenomena," p. 15. Oxford Univ. Press, London and New York, 1971.

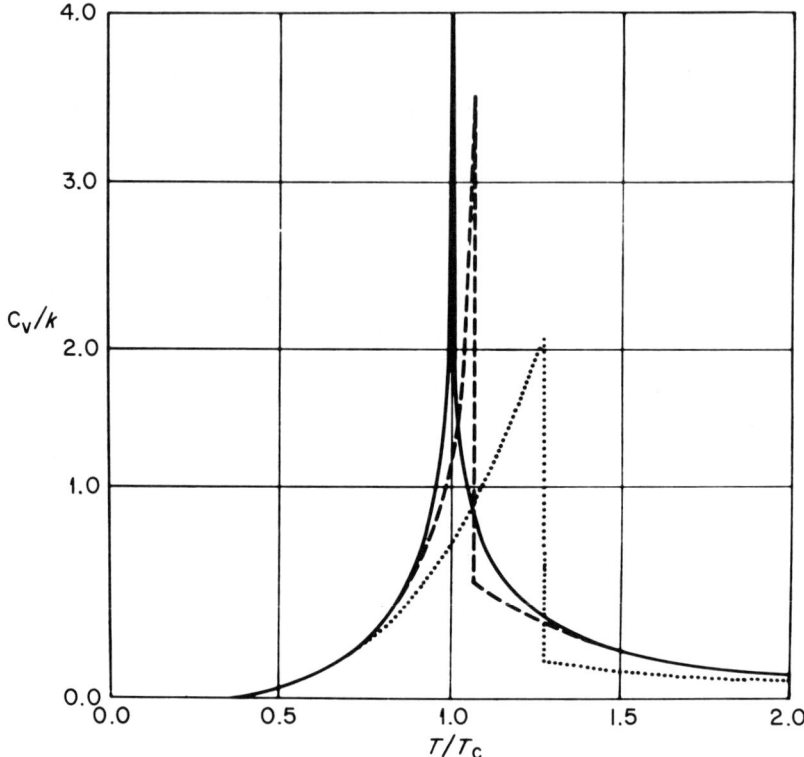

FIG. 26. Specific heat for the two-dimensional Ising model (square lattice) as obtained from the exact solution of Onsager (solid curve), from the Bethe approximation equivalently, the CVM pair approximation (dotted curve), and the Kramers–Wannier and Kikuchi CVM square approximation (broken curve) [H. E. Stanley, "Introduction to Phase Transitions and Critical Phenomena," p. 15. Oxford Univ. Press, London and New York, 1971].

state. With these stated assumptions, critical exponents are not expected to be predicted accurately. Additional approximations must also be introduced in order to make the mathematical models tractable, and these approximations can introduce further nonphysical effects or artifacts which must be examined in each case.

Two basic descriptive elements are introduced: the wave concept and the cluster concept. The first (wave) is associated with the Fourier representation, is essentially mean-field, is particularly suited to harmonic effects, and is applicable to arbitrary range of interatomic interactions; in fact the mean-field approximation is exact in the limit of infinite-range interactions.[165] The second (cluster) is complimentary to the first: it is essentially a direct space formulation, the mean-field is used, only after the first cluster correlations have been treated (almost) correctly, many-body effects are easily incorporated, only near-neighbor interactions can be treated. Which alternative of this *wave–cluster duality* is selected depends upon the nature of the application. Generally speaking, qualitative aspects of a problem are best handled by the wave formulation, quantitative aspects by the cluster formulation. However, elastic interactions cannot readily be incorporated within the cluster framework.

The CVM appears to hold much promise whenever fairly reliable numerical results are required for realistic problems. New developments have made the method particularly attractive in cases heretofore regarded as untractable. For example, the statistical weight formula for the double-octahedron/tetrahedron approximation in the fcc lattice looks rather for-

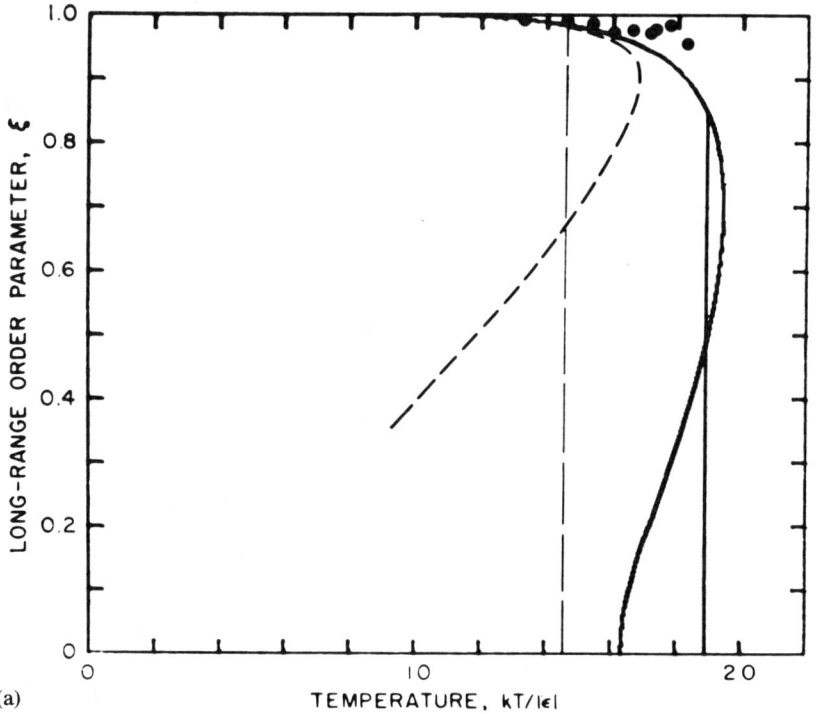

FIG. 27. Long-range order parameter for fcc binary alloy plotted against reduced temperature, CVM tetrahedron approximation. (a) A–B type alloy (dashed curve is from quasichemical calculation). (b) A_3B-type alloy. Transition temperatures are indicated by vertical lines, experimental points by dots [R. Kikuchi and H. Sato, *Acta Metall.* **22**, 1099 (1974)].

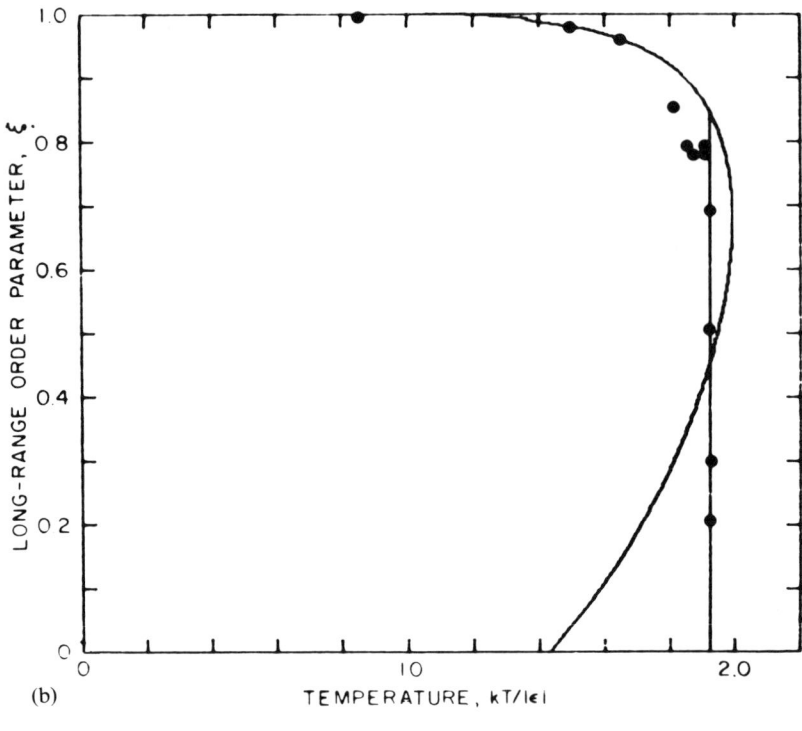

(b)

FIG. 27. (Continued)

midable:

$$g(\text{double-tetra./octa.}) = \frac{\{\text{tetra}\}^{10} \{\text{octa}\}^{12} \{\text{point}\}}{\{\text{triangle}\}^{16} \{\text{oct}\}^{6} \{\text{oct}\}^{6}} \tag{21.1}$$

yet its application to the calculation of the Ising ferromagnet critical temperature required very little computational effort,[15] far less than is required for more precise theoretical methods. More importantly, the CVM, in the above approximation, allows the incorporation in the energy expression of second- and third-neighbor pair interactions, and even many-body (up to six-point) interactions, in a perfectly straightforward manner. This is a significant result, since the stability of experimentally determined ordered structures can only be demonstrated theoretically by appealing to higher than first-nearest-neighbor pair interactions (see Section 22). Furthermore, multiplet correlations are obtained directly as a by-product of the CVM, as described in Section 25,d.

Mean-field theories predict second-order transformations which can occur at equilibrium in a completely continuous manner, i.e., properties such as local concentration may vary continuously with time and homogeneously throughout the sample as the transformation progresses. This is an oversimplification and does not take into account the essentially discrete (1, 0) occupation of lattice sites in replacive transformations, nor the partially discrete character of displacive transformations. Monte Carlo[205] or molecular dynamics[206] simulation of such critical phenomena invariably suggest a microdomain picture of the transforming system just below T_c. This does not mean that a nucleation event is required in higher-order transitions, since true critical phenomena are characterized by a transition temperature which has the property

$$T_t = T_c = T_0 \tag{21.2}$$

where T_0 is the instability (or spinodal) temperature, and the interfacial free energy vanishes at $T = T_0$. Hence the critical nucleus at $T_0 (= T_c)$ is infinitely large, infinitely diffuse, and infinitely shallow in the classical description.[207] The actual critical nucleus at T_c is far from the classical one and appears to consist of interpenetrating, continuous, infinitely ramified (or branching) domains in which one may recognize microdomains by taking plane sections.[208] It is believed[209] that the existence of these microdomains gives rise to the so-called central phonon peak observed in soft-mode transitions, which mean-field Landau-type theories cannot account for. Although the Landau theory is not suitable for quantitative work pertaining to critical phenomena, the associated symmetry rules are of fundamental importance for the study of the crystallographic aspects of phase transformations,[210] and appear to retain their validity despite being based on an incorrect Taylor's expansion.[176]

The main theoretical concepts have now been introduced. The remainder of this article will be primarily devoted to applications of the models presented.

VI. Stable and Unstable States

Three general topics will be treated under this heading: coherent ordered ground states (Section 22), coherent phase diagrams (Section 23),

---

[205] N. Ogita, A. Ueda, T. Matsubara, H. Matsuda, and F. Yonezawa, *J. Phys. Soc. Jpn.* **26S**, 145 (1969).
[206] T. Schneider and E. Stoll, *Phys. Rev. B* **13**, 1216 (1976).
[207] J.W. Cahn and J.E. Hilliard, *J. Chem. Phys.* **31**, 688 (1959).
[208] A. Sur, J.L. Lebowitz, J. Marro, and M.H. Kalos, *Phys. Rev. B* **15**, 3014 (1977).
[209] H.E. Cook, M. Suezawa, T. Kajitani, and L. Rivaud, *J. Phys. (Paris)* **38**, C7-430 (1977).
[210] See, for example, N. Boccara, "Symétries Brisées." Hermann, Paris, 1976.

and stability limits, i.e., those surfaces which separate phase diagram regions where a given structure is stable (or metastable) and where it is unstable with respect to arbitrary small composition perturbations (Section 24). As was stated in the Introduction, the word "coherent" refers to those processes of atomic rearrangement which conserve the average lattice. The determination of ground states makes use of the internal energy formulas of Section III, the derivation of phase diagrams makes use of the free energy models of Section V, particularly in the cluster formulation, and the determination of stability limits (or spinodals) makes use of the concentration wave concept also introduced in Section V.

22. Ordered Ground States

Ground states are generally defined as equilibrium states at zero absolute temperature. Hence the ground-state atomic configurations are simply those which minimize the internal energy E. Here, we are interested in those ordered structures which can be obtained by rearrangement of A, B, C··· atoms (or defects) on lattice sites, i.e., the coherent ordered ground states belonging to set I (and included subsets) of the Venn diagram of Fig. 22. No fluctuations are allowed about this minimum energy, and it follows that the cluster averages defined by Eq. (2.1) may be replaced by microcanonical averages, as in Eq. (2.11). In particular, the pair frequencies may now be written

$$\rho_{ij}(r) \equiv \langle \sigma_i(p)\sigma_j(p+r) \rangle = \frac{1}{\mathcal{M}} \sum_{\text{syst.}} \sigma_i(p)\sigma_j(p+r), \qquad (22.1)$$

independent of (p) because of translation symmetry. If the number of sites N is large in each of the $\mathcal{M}$ systems of the ensemble, and if boundary effects are eliminated (for example by periodic boundary conditions), then Eq. (22.1) may be simplified to

$$\rho_{ij}(r) = \frac{1}{N} \sum_p \sigma_i(p)\sigma_j(p+r). \qquad (22.2)$$

Likewise, the pair correlation function becomes, from Eq. (2.4),

$$q_{ij}(r) = \frac{1}{N} \sum_p \gamma_i(p)\gamma_j(p+r) = \rho_{ij}(r) - \bar{c}_i\bar{c}_j. \qquad (22.3)$$

The ground-state problem is then that of determining those atomic (defect) configurations which minimize the configurational energy given either in the k-space representation (6.41) or in the real space representation (6.42), for given average concentrations $\bar{c}_i$, and for given pair interaction parameters $v_{ij}(r)$, where (r) runs over the first few neighbor coordination shells. Here again, both "wave" and "cluster" methods are available, the former being used in conjunction with Eq. (6.41), the latter with (6.42). In what follows, we assume that only pair interactions contribute significantly to the configurational energy E.

a. "Wave" Method

The concentration wave method was used primarily by Khachaturyan[5,54] and by Clapp and Moss.[7] Consider first binary systems (A-B) for which the energy is given by Eq. (6.43) or by

$$E = \frac{N}{2} \sum_h V(h)Q(h) \qquad (22.4)$$

where the positive function Q is the Fourier transform of the pair correlation function (22.3). It is clear that E will surely be minimized for those ordered structures which are generated by concentration waves $\Gamma(h)$ with wave vectors $\mathbf{k}(h)$ located exclusively at the absolute minima of $V(h)$. As discussed previously in Section 8, meaningful structures are associated with special-point waves (Tables II and III). The location of $V(h)$ minima in interaction parameter space was given in Figs. 8, 9, 10, 11a–d, in which second- and third-neighbor pair interactions in fcc and bcc lattices were considered. From such considerations, Clapp and Moss[7] were able to determine a number of ordered structures at stoichiometry AB and A_3B.

Khachaturyan did not specifically seek ground-state structures, but described all ordered structures which could be generated by combinations of fcc and bcc special-point concentration waves, the wave vectors of which were given in Tables II and III. Most of the resulting structures are shown in Figs. 28 and 29.[5] In these figures, a set of three structures such as a_k, b_k, c_k, for given index k, were generated by a particular set of special-point waves, a_k representing substitutional, b_k interstitial, and c_k reciprocal space structures. Crystal structures are described in Tables VI and VII. Body-centered cubic derivatives a_2 and a_3 (Fig. 28) are generated by special-point waves $\langle\frac{11}{22}0\rangle$, and $\langle\frac{11}{22}0\rangle$, $\langle 100\rangle$ combined, respectively. These structures have not been observed experimentally in substitutional solutions and are not listed in Table VI. Their interstitial counterparts b_2 and b_3 have been observed, however.[211] Diffraction patterns (Figs. 28, 29; c_k) are included since they can be particularly useful for structure determination from electron diffraction patterns, for example.

[211] M.P. Usikov and A.G. Khachaturyan, *Sov. Phys.—Crystallogr.* (*Engl. Transl.*) **13**, 910 (1969).

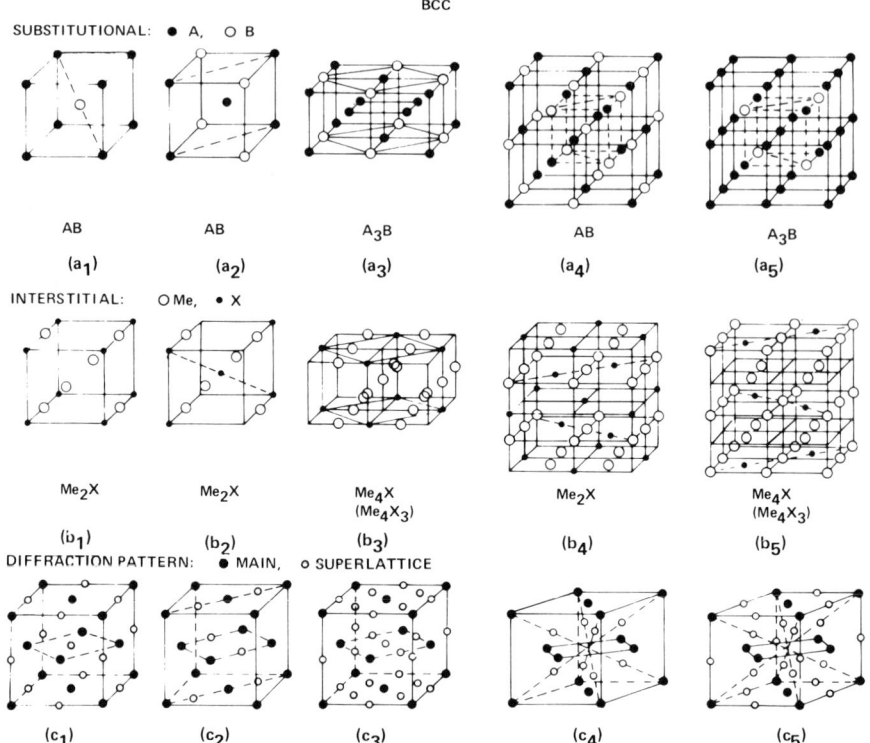

FIG. 28. Ordered structures generated by combinations of bcc special-point concentration waves. a_1-a_6: substitutional structures (described in Table VI), b_1-b_6: interstitial structures; c_1-c_6: corresponding diffraction patterns [A. G. Khachaturyan, *Phys. Status Solidi B* **60**, 9 (1973)].

Substitutional structures are discussed further in the next subsection. Of particular interest are the interstitial ordering phases obtained by Khachaturyan[5] as follows: the ordering is assumed to take place between empty and filled (X) sites of one of the sublattices of interstitial sites, the host lattice (Me) remaining unchanged. Hence, any substitutional structure generates a corresponding interstitial one, and in some cases two distinct interstitial phases when the roles of empty and filled sites are interchanged. Octahedral interstitial sites form an fcc sublattice in the fcc host, so that a single set of interstitial structures are obtained from the substitutional ones. In the bcc host lattice however, there are three bcc sublattices of octahedral sites, and six of tetrahedral sites. In Fig. 28, only a single octahedral bcc lattice was allowed to be occupied by interstitial solute.

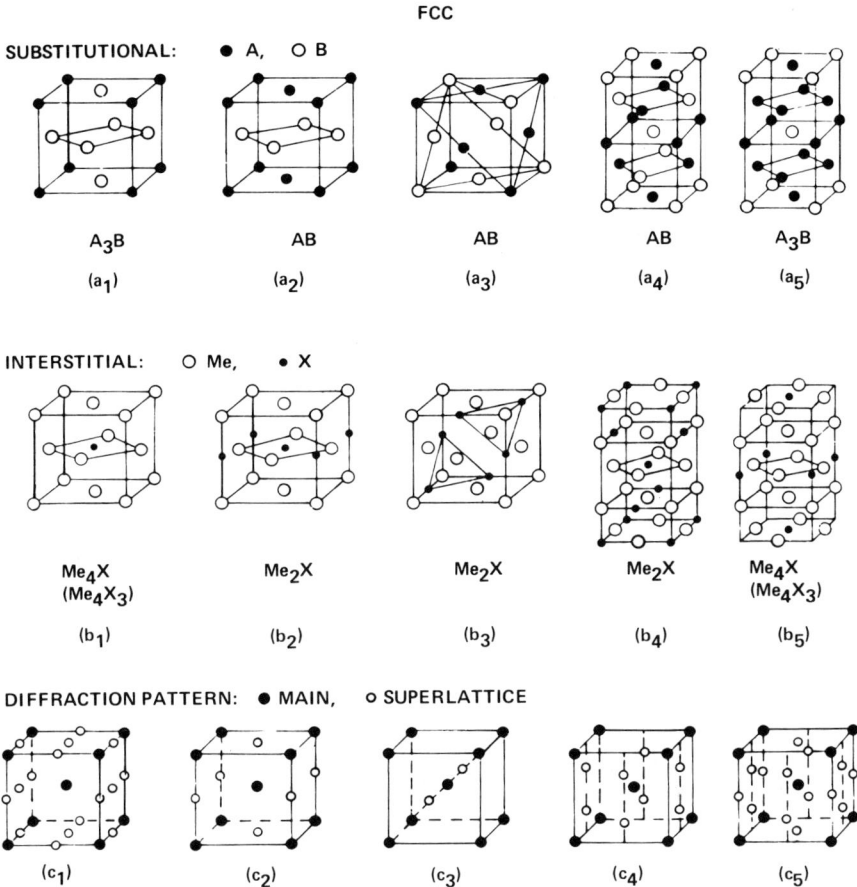

FIG. 29. Ordered structures generated by combinations of fcc special-point concentration waves. a_1–a_5: substitutional structures (described in Table VII); b_1–b_5: interstitial structures; c_1–c_5: corresponding diffraction patterns [A. G. Khachaturyan, *Phys. Status Solidi B* **60**, 9 (1973)]. See also Fig. 34.

Khachaturyan and Shatalov[212] later lifted this restriction and treated the multiple-sublattice ordering problem in considerable detail. Their method can be briefly described as follows: let indices

$$i, j = \begin{cases} 1, 2, 3 & \text{(octahedral)} \\ 1\text{–}6 & \text{(tetrahedral)} \end{cases} \qquad (22.5)$$

denote sublattices. An interstitial site can be represented by two indices:

[212] A.G. Khachaturyan and G.A. Shatalov, *Acta Metall.* **23**, 1089 (1975).

i, indicating the sublattice, p, indicating the position on the particular sublattice. The energy can then be written in the multicomponent notation (6.41) or

$$E = \frac{N}{2} \sum_h \Gamma^*(h) \mathbf{V}(h) \Gamma(h) \tag{22.6}$$

in the matrix notation of Eq. (18.13). This quadratic form in the concentration amplitudes Γ_i can be diagonalized to yield

$$E = \frac{N}{2} \sum_h \sum_{l=1}^{3 \text{ or } 6} \lambda_l(h) |\bar{\Gamma}_l(h)|^2 \tag{22.7}$$

where λ_l are the eigenvalues of the symmetric matrix $\mathbf{V}$, and where $\bar{\Gamma}_l$ are normal concentration modes which mix the sublattices in the appropriate manner. In Eq. (22.7), l can be regarded as a polarization index, related to the eigenvectors of $\mathbf{V}$. Low-energy ordered states were found by seeking special wave vectors $\mathbf{k}(h)$ located at the minima of the lowest-lying eigenvalue $\lambda_l(h)$. Ordered interstitial structures were then constructed from the corresponding $\Gamma_l(h)$ concentration waves, with as little contribution from other waves, $\Gamma_{l'}(h')$ say, as possible.

It was assumed by Khachaturyan and Shatalov[212] that the major contribution to the internal energy was of elastic nature. Hence the components of the k-space energy matrix were obtained from the elastic energy formula (11.3), the Born–Huang force constants and elastic constants for tantalum being used as an illustrative example. The distortion caused by an interstitial was specified by a distortion tensor $\eta_{\alpha\beta}$ of tetragonal symmetry, with tetragonality ratio t defined by Eq. (11.11). Various interstitial ordered phases are predicted for various ranges of t in good agreement with available experimental data pertaining to Ta hydrides and oxides.

Despite its successes, the wave method cannot guarantee ground states. For one thing, concentrations amplitudes Γ do not yield (1, 0) occupation of lattice sites unless auxiliary conditions are applied.[7] Furthermore, ordered structures may be found which have lowest possible energy, although their ordering k-vectors neither lie at special points nor at individual minima of $V(h)$. Such structures cannot readily be discovered by k-space methods. Hence, real-space "cluster" methods (to be described below) must be regarded as the only reliable ones for the problem at hand. Actually, the wave method is more useful in the context of dynamic and kinetic problems (see Sections 25 and 26).

b. "Cluster" Method

Ground states have been fully determined for the case of binary systems only. The problem thus consists in finding those configurations of

A and B atoms (or 1 and 0 symbols) on the sites of a given lattice which minimize the internal energy E expressed in terms of pair correlations $q_{11}(r)$, with given pair interaction parameters $v_{11}(r)$, and average concentration $\bar{c} \equiv \bar{c}_1$. Equation (6.42) must therefore be used in the reduction scheme (6.38). Because of the high symmetry of the lattices investigated (bcc, fcc), it is preferable to sum over coordination shells s, rather than over individual lattice vectors $\mathbf{x}(r_s)$ of the shells. It is also more convenient to use pair frequencies ρ rather than correlations q. The energy formula may then be written as

$$E = \frac{N}{2}\left[\sum_{s=1}^{m} V_s \rho_s - V_0 \bar{c}\right] \qquad (22.8)$$

where, by Eq. (22.2),

$$\rho_s = z_s \rho_{11}(r_s) \qquad (22.9)$$

is the average number of (1, 1) sth-neighbor pairs (i.e., B–B solute atom pairs) in the crystal, z_s being the corresponding coordination number. In Eq. (22.8) the pairwise energy parameters are given by

$$V_s = v_{11}(r_s) = v^{AA}(r_s) + v^{BB}(r_s) - 2v^{AB}(r_s) \qquad (22.10)$$

[where the (0, 1) notation of Eq. (6.38) has been replaced by the more familiar (A, B) notation] and V_0 is given by

$$V_0 = \sum_{s=1}^{m} z_s V_s \qquad (22.11)$$

where the sum is extended to the maximum number m of coordination-shell interactions retained.

Richards and Cahn[71] used a heuristic method to find the bcc and fcc ordered ground states under the assumption of first- and second-neighbor interactions. Phase diagrams were obtained which show the domains of existence of the ordered structures as a function of the average concentration $\bar{c}$ and of the pair interaction parameter ratio V_2/V_1. These diagrams are given in Figs. 30 and 31 (see corrections to the fcc ground states given by Allen and Cahn[71a]). Explanation of structures is to be found in Tables VI and VII. In Figs. 30 and 31, the stoichiometric composition for a given structure is indicated by a thin line if it occurs in the middle of a single-phase field, and by a heavy line if at the limit of a two-phase field. The extension of the field of a more symmetrical structure on the grounds of entropy is indicated by a dashed line.[71,71a]

To prove that the ordered structures indicated in Figs. 30 and 31 were indeed the lowest possible energy states, Allen and Cahn[72] rewrote Eq. (22.8) in terms of cluster frequencies. The basic figure for the cluster

TABLE VI. BINARY (A–B) ORDERED GROUND STATES FOR THE bcc LATTICE

Formula	Structurbericht	Int. tables	Symmetry class	Examples	Special points	Figures
AB	$L2_1$	Pm3m	Simple cubic	CsCl, CuZn	$\langle 100 \rangle$	$28a_1$
AB	B32	Fd3m	f.c. cubic	NaTl	$\langle \frac{1}{2}\frac{1}{2}\frac{1}{2} \rangle$	$28a_4$
A_3B	Do_3	Fm3m	f.c. cubic	Fe_3Al	$\langle \frac{1}{2}\frac{1}{2}\frac{1}{2} \rangle, \langle 100 \rangle$	$28a_5$

TABLE VII. BINARY (A–B) ORDERED GROUND STATES FOR THE fcc LATTICE

Formula	Structurbericht	Int. tables	Symmetry class	Examples	Special pts.	Kanamori symbol	Figures
AB	$L1_0$	P4/mmm	s. tetragonal	CuAu I	$\langle 100 \rangle$	$(2, 3, 4, 6; \frac{1}{2})$	$29a_2$, 34a
AB		$I4_1/amd$	b.c. tetragonal		$\langle 1\frac{1}{2}0 \rangle$	$(2, 2, 8, 2; \frac{1}{2})$	$29a_4$, 34b
AB	$L1_1$	R$\bar{3}$m	Rhombohedral	CuPt	$\langle \frac{1}{2}\frac{1}{2}\frac{1}{2} \rangle$	$(3, 0, 6, 6; \frac{1}{2})$	$29a_3$, 34c
A_2B		Immm	b.c. orthorhombic	Pt_2Mo	—	$(1, 1, 6, 1; \frac{1}{3})$	34d
A_2B		B2/m (C2/m)	s.c. monoclinic		—	$(1, 1, 6, 1; \frac{1}{3})$	34e
A_5B_2	$mC14^a$	C2/m	base-c. monoclinic	Au_5Mn_2	—	$(\frac{1}{2}, \frac{3}{2}, 5, 1; \frac{2}{7})$	34f
A_3B	$L1_2$	Pm3m	s. cubic	Cu_3Au	$\langle 100 \rangle$	$(0, 3, 0, 6; \frac{1}{4})$	$29a_1$, 34g
A_3B	$D0_{22}$	I4/mmm	b.c. tetragonal	Al_3Ti	$\langle 1\frac{1}{2}0 \rangle, \langle 010 \rangle$	$(0, 2, 4, 2; \frac{1}{4})$	$29a_5$, 34h
A_4B	$D1_a$	I4/m	b.c. tetragonal	Ni_4Mo	—	$(0, 1, 4, 0; \frac{1}{5})$	34i
A_5B		B2/m	s.c. monoclinic		—	$(0, 0, 4, 1; \frac{1}{6})$	34j

[a] Pearson symbol: base-centered monoclinic with 14 atoms per unit cell [W. B. Pearson, "The Crystal Chemistry of Metals and Alloys," p. 347. Wiley (Interscience), New York, 1972].

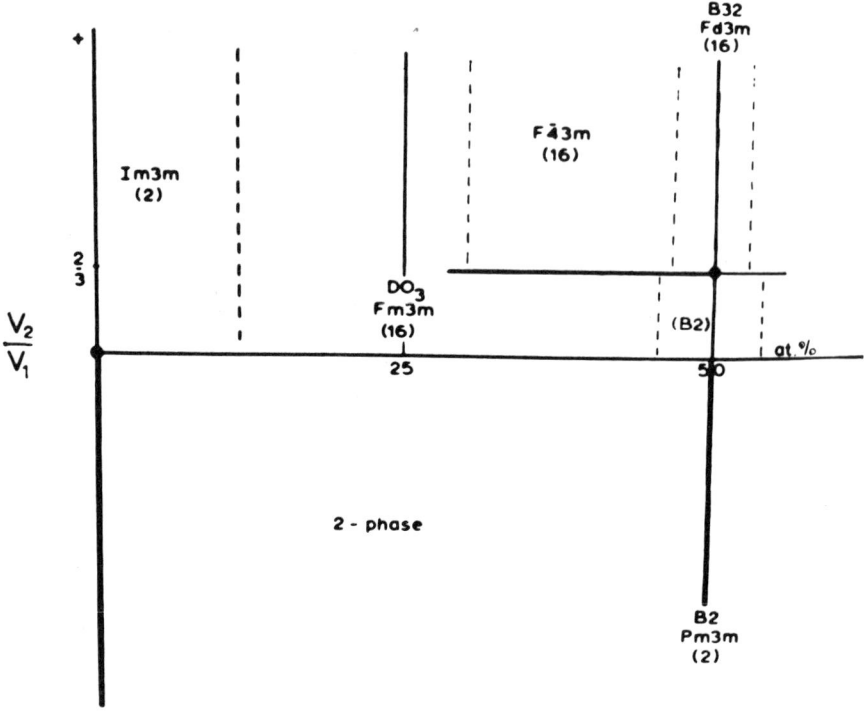

FIG. 30. Bcc ground-state diagram showing the state of lowest energy, in the first- and second-neighbor pair interaction model, as a function of average concentration $\bar{c}$ and ratio V_2/V_1 [M. J. Richards and J. W. Cahn, *Acta Metall.* **19**, 1263 (1971)].

(these terms were defined in Section 20) must contain pairs of lattice sites which include the highest pair interaction considered. For the bcc lattice, Allen and Cahn used the nearest and next-nearest neighbor tetrahedron, and for the fcc lattice they used the regular octahedron whose edges join the centers of the cubic unit cell. Minimization of the energy was obtained by a method borrowed from linear programming. In this way, the structures discovered by Richards and Cahn could be proved rigorously to be absolute ground states, except in the fcc case for a narrow range of energy parameters.

Unbeknown to Cahn and co-workers, Kanamori had already published an important paper entitled "Magnetization Process in an Ising Spin System."[213] In this paper, Kanamori derived a set of inequalities which must necessarily hold for geometrical reasons between linear combina-

[213] J. Kanamori, *Prog. Theor. Phys.* **35**, 66 (1966).

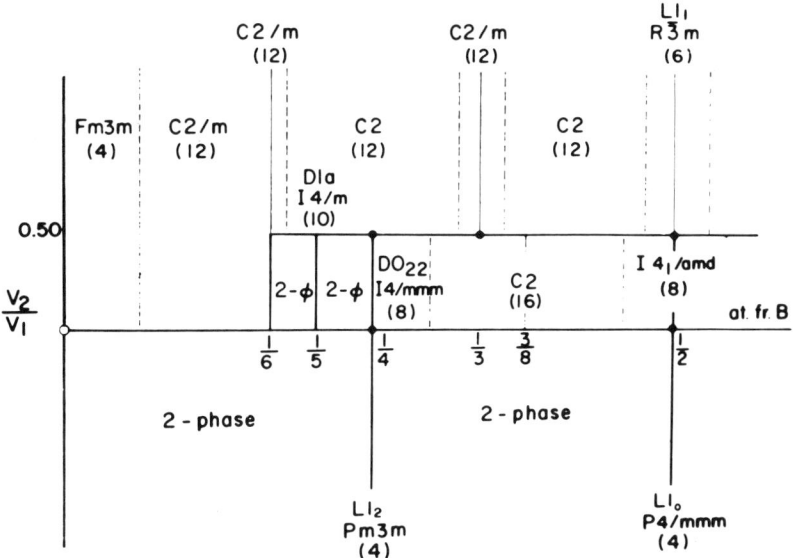

Fig. 31. Fcc ground-state diagram showing the state of lowest energy, in the first- and second-neighbor pair interaction model, as a function of average concentration $\bar{c}$ and ratio V_1/V_2 [M. J. Richards and J. W. Cahn, *Acta Metall.* **19**, 1263 (1971); see also correction by S. M. Allen and J. W. Cahn, *Scripta Metall.* **7**, 1261 (1973)].

tions of pair frequencies ρ_s and the average concentration $\bar{c}$. With the help of these inequalities, Kanamori was able to determine rigorously all bcc and fcc ground states of magnetization under the assumption of first- and second-nearest-neighbor spin interactions. Since the Ising model is isomorphous to the binary solution, the problem of Richards, Cahn and Allen was effectively solved.

The difficulty of Kanamori's method lies in the derivation of the inequalities. A systematic technique for obtaining these has been given by Kaburagi and Kanamori[214] which makes use of a basic figure of lattice sites, each site of the figure being used as a "peephole" through which the occupation of lattice sites may be counted as the basic figure is translated through the lattice. For first- and second-neighbor interactions, the basic figures were the same as those used by Allen and Cahn, except that the nearest-neighbor tetrahedron had to be included for the fcc lattice, in addition to the octahedron, in order to cover all ranges of pair interactions.

The Kaburagi-Kanamori method, which is really a cluster method like

[214] M. Kaburagi and J. Kanamori, *Prog. Theor. Phys.* **54**, 30 (1975).

that of Allen and Cahn, is extremely powerful and versatile. It is particularly transparent when handled in the manner recently proposed by Kudō and Katsura,[73] a method which works particularly well in cases of pair interactions limited to first and second coordination shells. These authors consider the $(m + 1)$-dimensional hyperspace of coordinates $\rho_1, \ldots, \rho_s, \ldots, \rho_m, \bar{c}$. Points in this space which satisfy the Kanamori inequalities are found to lie within a convex polyhedron, called the configuration polyhedron, such as the one shown in Fig. 32 for the (ideal) hcp structure with first- and second-neighbor interactions.[73] Since, by Eq. (22.6), the energy is linear in the pair and "point" ($\bar{c}$) variables, the locus of constant energy is a hyperplane in the $(\rho_1, \ldots, \rho_m, \bar{c})$ space, the direction of its normal being set by the ratios $V_1/V_0, \ldots, V_m/V_0$. For fixed energy parameters, the energy E can be decreased (or increased) by translating the hyperplane parallel to itself. Finally, one vertex (or a subspace spanned by more than one vertex) of the configuration polyhedron must be reached, for example point IX in Fig. 33. One then tries to distribute A and B atoms on lattice points so as to satisfy

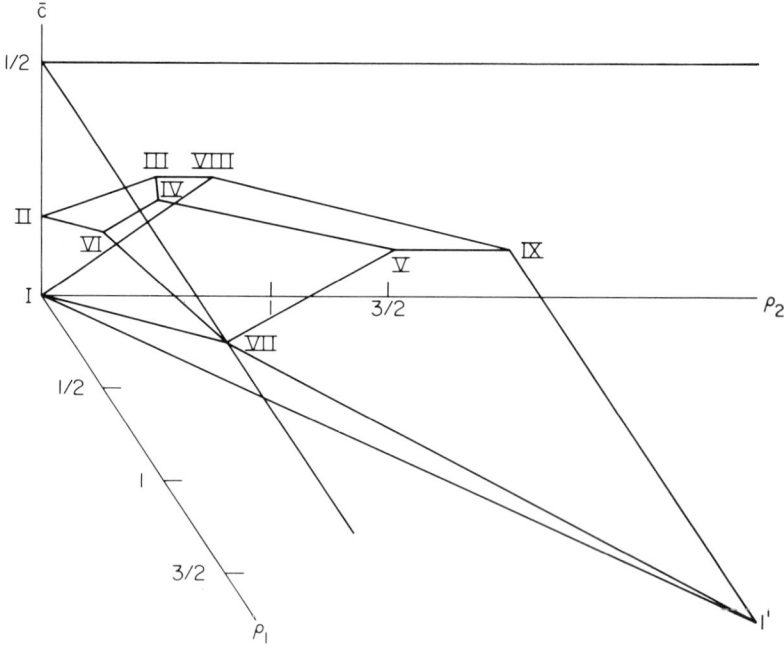

FIG. 32. Configuration polyhedron for the ideal hcp structure in first- and second-neighbor pair interaction model. Coordinates are ρ_1, ρ_2, $\bar{c}$, defined in Eqs. (22.8) and (22.9) [T. Kudō and S. Katsura, *Prog. Theor. Phys.* **56**, 435 (1976)].

pair and point frequencies, which are the known coordinates of the vertex in question. If the attempt is successful, the correct ground state has been found for the specified ratio of interaction parameters. In the case of point IX, the resulting structure is that of MgCd, with stoichiometric composition $\frac{1}{2}$. If the attempt at constructing a structure is not successful, this means that new inequalities must be sought, and a "tighter" polyhedron must be constructed. Compositional disorder is not allowed in this model: at fixed average concentration, the trace in $\bar{c}$ = const. of a constant-energy plane will generally (for $m = 2$) intersect an edge of the configuration polyhedron, indicating a ground state made up of a mixture of the two ordered phases which correspond to the two vertices which the edge joins. In exceptional cases, the ground state may consist of a mixture of three or more ordered phases if the energy plane is parallel to a face of the polyhedron. In this context, Kudō and Katsura[73] have demonstrated a general inverse relationship between the configuration polyhedron in $(\rho_1, \ldots, \rho_m, \bar{c})$-space and the corresponding *phase diagram* which indicates domains of stability of the various ordered structures in $(V_1, \ldots, V_m)$-space.

With such techniques, much progress has been made in the determination of ground-state structures, in particular for the following cases: simple cubic with up to second[213] and third[214,215] neighbor interactions, bcc with up to second,[71,72,213] third,[216] and fourth[63] neighbor interactions, fcc with up to second[71,72,213] and fourth[62,63] neighbor interactions (with $|V_1| \gg |V_2|, |V_3|, |V_4|$), (ideal) hcp with up to second-neighbor interactions[73] and many lower-dimensionality lattices (see Kudō and Katsura[73]). Corresponding phase diagrams have been constructed.

The method of the Kanamori inequalities confirms the findings of Cahn and co-workers.[71,72] The bcc ground states (for up to second-neighbor interactions) are those given in Fig. 30 and Table VI. It is interesting to note that all bcc ground states are special-point structures in the sense of Khachaturyan.[5] This is not the case for fcc structures. Khachaturyan's a_2 and a_3 structures (Fig. 28) are ground states only in the body-centered tetragonal lattice with up to third-neighbor interactions.[216] This is probably why these structures have not been found experimentally in substitutional bcc solutions. In interstitial solutions, however, the predominant interactions are with the host lattice, and the elastic distortions cause the effective pair interactions to propagate to high coordination shells. Hence the b_2 and b_3 structures (Fig. 28), which are indeed observed experimentally, may thus be ground states of interstitial ordering.

Kanamori[62,63] has recently tackled the extremely complex problem of

[215] S. Katsura and A. Narita, *Prog. Theor. Phys.* **50**, 1426 and 1750 (1973).
[216] A. Narita and S. Katsura, *Prog. Theor. Phys.* **52**, 1448 (1974).

fcc ground states with up to fourth-neighbor interactions. The basic figure used for the derivation of the required inequalities was the cuboctahedron shown in Fig. 33. Some of the many resulting ordered ground-state structures discovered thus far are shown in Kanamori and Kakehashi[62,63] and the most common ones are given here in Fig. 34 and Table VII. The Kanamori symbols have arguments

$$S(p_1, p_2, p_3, p_4; \bar{c}) \qquad (22.12)$$

where

$$p_s = \rho_s/\bar{c} = z_s P_{11}(r_s), \qquad (s = 1, 2, 3, 4), \qquad (22.13)$$

the latter equality resulting from Eq. (2.6). Again, the ground-state structures of Cahn and co-workers are confirmed, including the side-centered monoclinic A_2B and A_5B structures (denoted by the symbol C2/m in Fig. 36). Of course, many additional ordered structures were found by Kanamori because of the longer-range interactions considered, in particular the Au_5Mn_2 structure indicated in Table VII. By including fourth-neighbor interactions, it is also possible to lift certain degeneracies which resulted from considering only first- and second-neighbor interactions.

In actual practice, ground states for a given binary system may be quite different from those predicted by theory. For example, stoichiometric Ni_3Mo at equilibrium is not the expected $D0_{22}$ structure, which is observed for Ni_3V, but a distorted hcp derivative which is completely incoherent with the parent fcc solid solution. Hence the Ni_3Mo ordered ground state belongs to the complement of set I in the Venn diagram of Fig. 22, and its existence cannot be predicted by existing order–disorder models. Still, the study of coherent ground states is of interest even in

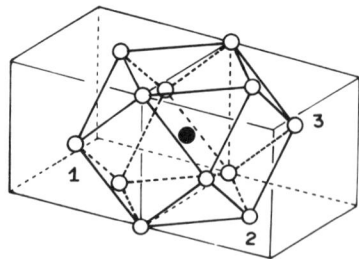

FIG. 33. Central lattice point and its twelve nearest neighbors (cuboctahedron) used in derivation of Kanamori inequalities in fcc lattice with up to fourth-neighbor pair interactions [M. Kaburagi and J. Kanamori, *Prog. Theor. Phys.* **54**, 30 (1975)]. Also basic cluster used by Clapp [P. C. Clapp, *Phys. Rev. B* **4**, 255 (1971)] for determining cluster probabilities in fcc solid solutions above the transition, see Section 25.

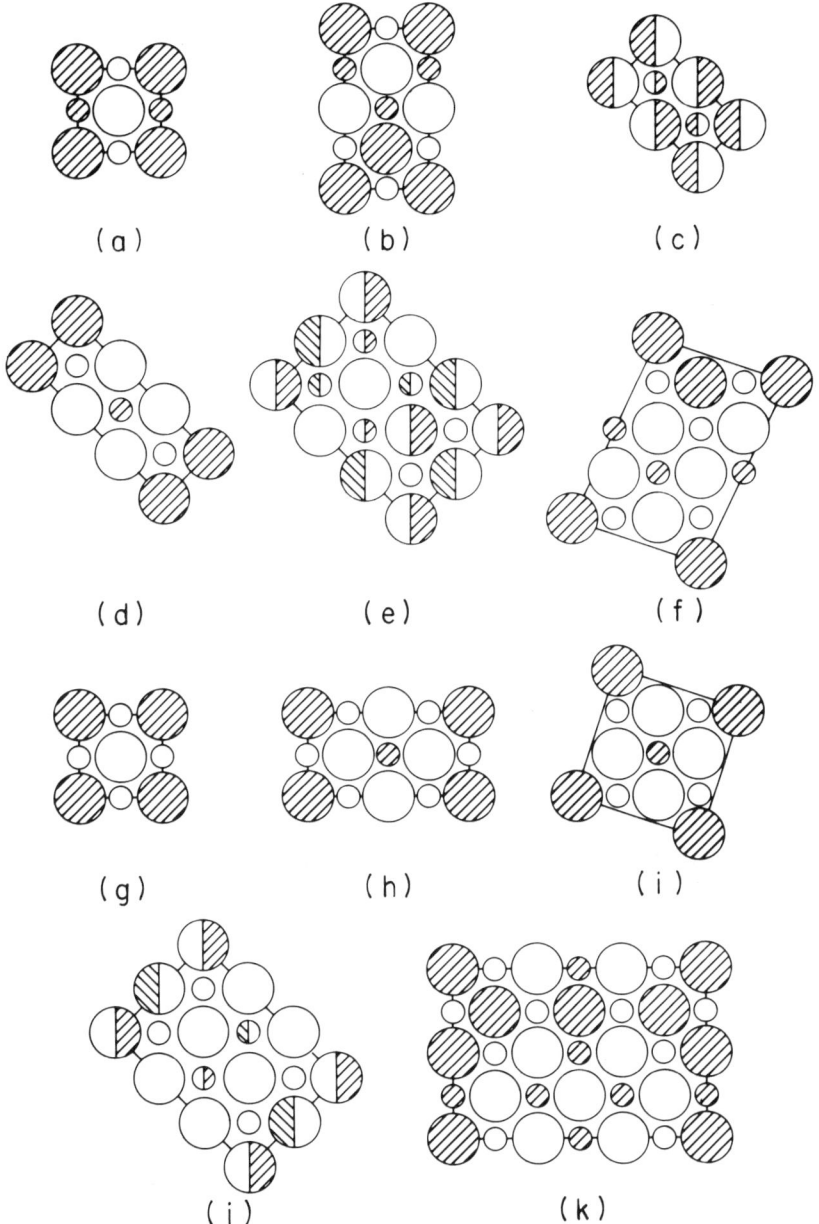

FIG. 34. Most common ground-state structures in the fcc lattice with up to fourth-neighbor pair interactions [J. Kanamori and Y. Kakehashi, *J. Phys. (Paris)* **38**, C7–274 (1977)]. Description of crystal structures given in Table VII. Structure (k) is an example of complex ordered structure A_7B_5 discovered in this way, with Kanamori symbol $S(\frac{1}{5}, 2, \frac{28}{5}, 2; \frac{5}{12})$, see Eq. (22.12). Unit cells of ordered structures are shown projected on (001) plane. Large circles indicate lattice sites on the $(0, 0, n)$ planes, with n integer, small circles on the $(0, 0, n + \frac{1}{2})$ planes. Open circles represent A atoms, closed represent B atoms. Half-shaded circles represent ABAB . . . sequence along (001) row. Left-shaded and right-shaded sequences are out of phase by one fcc lattice parameter along (001).

such cases because the $D0_{22}$ structure has been observed experimentally[180] at the Ni_3Mo composition as a metastable intermediate product phase obtained by rapid quenching of the parent solid solution. This will be discussed further in Section 26.

The ordered ground-state structures derived by Cahn and co-workers[71,72] are those represented symbolically by set G in Fig. 22.[49] This set must be a subset of I since only coherent structures are considered, and must contain elements in the complement of K since some of the Cahn ordered structures (Table VI and VII) are generated by concentration waves whose wave vectors do not obey the Lifshitz criterion. Ni_4Mo and Ni_2Pt are typical examples of this class. The set of ordered structures derived by Clapp and Moss by a "wave" method (set M) must lie entirely within set K, since all of these structures are special-point ones; but M must be a subset of K, since not all of Khachaturyan's special-point structures can be ground states, at least within the assumption of first- and second-neighbor interactions only. Cu_3Au, Ni_3V are typical examples in set M. Certain ground-state structures may of course appear through higher-order transitions, and must thus belong to set L; CuZn is a typical example.

23. Coherent Phase Diagrams

The term "phase diagram" is used in this section in the familiar sense: that of a graphical representation of regions of stability of various phases in $(c_1, \ldots, c_n, T)$-space. These diagrams thus differ from the phase diagrams referred to in Section 22 by the fact that the interaction parameters are assumed to be known and that the influence of temperature on phase equilibrium is now taken into account. In keeping with the general limitations imposed in this article, only *coherent* phase diagrams will be considered, i.e., those diagrams which relate to crystalline phases that belong to the same parent average lattice. Such diagrams can in principle be derived theoretically, in contrast with phase diagrams involving incoherent phases which must be obtained empirically or at least semiempirically through the use of separate free energy functions for each phase considered.[155]

In addition to a pairwise (or many-body) internal energy expression, the theoretical construction of coherent phase diagrams requires a reliable configurational entropy formula. Not surprisingly then, the cluster variation method has proved to be the most useful one to date for that purpose. Equilibrium is determined by minimizing the grand potential G given by Eq. (20.20), which is equivalent to the familiar common tangent construction, with free energy F given by one of the models examined in Section V.

Three classes of diagrams will now be considered: (a) binary phase separation, (b) binary ordering, and (c) ternary systems.

a. Binary Phase Separation

For phase separation-type systems, the regular solution model often gives an adequate qualitative picture of coherent phase equilibrium. The free energy function is given by Eq. (17.6) and is represented in Fig. 21a. Figure 21b shows the resulting phase diagram which consists of a symmetrical miscibility gap with critical point at $c_c = \frac{1}{2}$ and $k_B T_c/w = \frac{1}{2}$. The loci of the common tangency points, which in this symmetrical case are simply the minima of the free energy (17.6), give the miscibility gap (MG).

Experimental MG's are generally not symmetrical but may be fitted to so-called subregular solution models obtained from Eq. (17.6) by allowing the effective pair interaction coefficient w to be concentration and temperature dependent, as it must be if the pseudopotential considerations of Section 6 are taken into account. A possible subregular free energy curve is shown in Fig. 35. Although in this way the internal energy contribution is made somewhat more realistic, the configurational en-

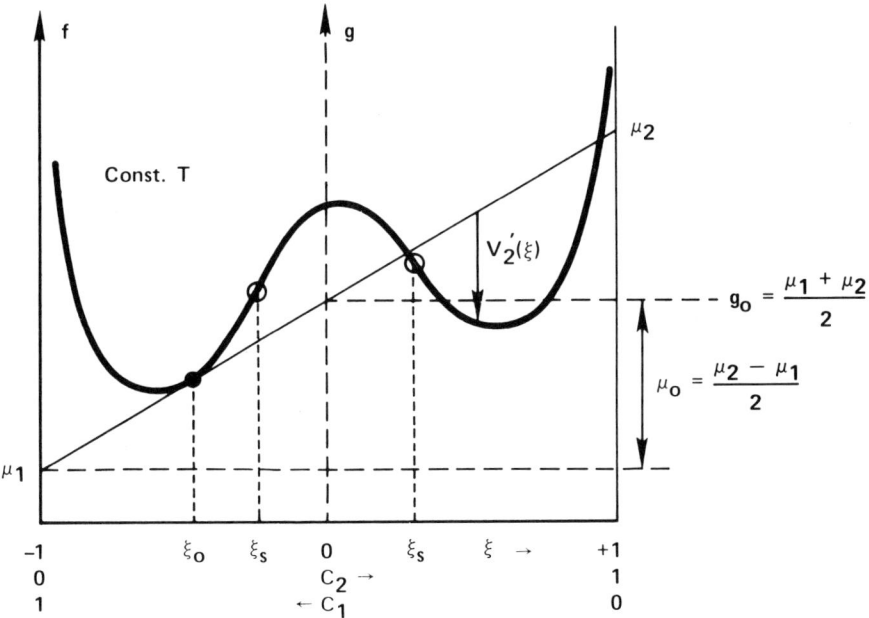

FIG. 35. Free energy $f(\xi)$ and grand potential $g(\xi)$ as a function of order parameter $\xi = c_2 - c_1$ [R. Kikuchi and D. de Fontaine, *Scr. Metall.* **10**, 995 (1976)].

tropy is still that of the "point" approximation of Eq. (17.2). It follows that the entropy is estimated too high and the free energy too low at a given temperature. Hence, in a fit to an experimental MG, the magnitude of the energy parameter w will be estimated too high by as much as 30 or 40%. It is thus worth bearing in mind that the values of the energy parameters deduced by fitting free energy models to phase diagram data depend sensitively on the entropy model chosen. The numerical values of regular or subregular pair interaction parameters therefore can have little theoretical significance.

Experimentally, MG's are constructed for bulk phases separated by incoherent interfaces. This represents the equilibrium situation since small coherent "precipitates" must coarsen to reduce interfacial energy. These must then eventually lose coherency since the replacement of coherency strain energy (proportional to volume) by incoherent interfacial energy (proportional to surface) is favored as soon as the volume/surface ratio of the precipitates (or defect clusters) exceeds a certain value which depends upon the system in question. For large bulk phases, equilibrium does not depend on the sizes and shapes of the coexisting incoherent phases. This property no longer holds for small coherent precipitates since, by Eq. (6.43) for example, the internal energy is seen to depend, through the concentration intensities $|\Gamma|^2$, on the actual location of all solute atoms. It follows that a different MG should be constructed for each distinct state of phase separation in the system, provided that these metastable states are sufficiently stable to be experimentally differentiated from one another. A classical example of this is that of a few at. % of Cu in solution in an Al matrix, as explained in Section 13.

Calculations can readily be performed whenever the solid solution can be approximated by an isotropic elastic continuum in which clusters of solute atoms cause only pure dilatational stress-free strains. This case was treated in Section 12,b where it was shown, by Eq. (12.19), that the continuum elastic energy $\hat{E}_e$ depended only on the integrated concentration deviation squared. The same dependence is found for the difference in internal energy ΔE between a solid solution containing concentration modulations and the uniform continuum. This can be seen in Eqs. (9.7) and (9.8) in which $\Omega(c)$ may be regarded as the bulk internal energy density of a fictitious solution of identical-size atoms. Actual size differences induce coherency strains during clustering so that the integral (12.19) must be combined to that of Eq. (9.7) to obtain the bulk contribution to the total energy difference

$$\Delta E_{\text{bulk}} = \frac{1}{2} \int_{V_N} (N_V \Omega^0 + 2\eta^2 Y_0)[c(\mathbf{x}) - \bar{c}]^2 \, d^3\mathbf{x}. \qquad (23.1)$$

Equivalently, Eq. (23.1) shows that the term $-2\eta^2 Y_0$ may simply be added to the coefficient of $\bar{c}(1 - \bar{c})$ in a regular or subregular solution model, the energy parameters of which have been determined by fitting the free energy to the incoherent MG. The effect of adding this isotropic continuum elasticity correction depresses the incoherent MG to lower temperatures as demonstrated originally by Cahn.[217]

A system where this type of calculation may be carried out is the much-investigated Al–Zn system. At about 10 or 20 at. % Zn, the fcc crystalline solution is practically isotropic and, in contrast to Cu atoms in Al, the Zn solute causes only isotropic distortions. Hence it is permissible to derive a metastable coherent fcc–fcc MG by the method outlined above. This was done by Lašek,[218] with the resulting coherent MG shown in Fig. 36.[219] The general validity of the concept of coherent phase (metastable) equilibrium appears to be well documented here by the excellent agreement with numerous observations, at least for the leaner Zn compositions. The low-Zn portion of the coherent MG is also referred to as the G.P. zone solvus, the zones in this case being small spherical Zn-rich fcc clusters.

b. Binary Ordering

Loss of coherency does not take place on ordering since one can hardly surround each A or B atom by an incoherent interface. Particularly in cases of lowering of rotational symmetry on ordering, however, such as a cubic phase going to an ordered tetragonal one, it is quite possible for the nature of the atomic bonding to change as ordering proceeds, thereby creating volume or shape changes, hence coherency stress, but these anharmonic effects cannot be treated fully by the models presented thus far, and will therefore not be discussed further. The paper by Kajitani and Cook,[178] mentioned earlier, is devoted to this topic.

The ordering problem is more complex than the phase separation problem for crystallographic reasons: the diversity of structures derived from fcc and bcc lattices alone is quite remarkable, as was explained in Section 22. It is of course essential to use a discrete, or lattice free energy formulation, whereas a continuum model was adequate for the phase separation case. From the comparison given in Table V (Section 20) it is clear that the cluster variation method is to be preferred over the Bragg–Williams method, particularly in $L1_0$ ordering where even the order of the phase transition is predicted incorrectly by the BW theory. In the fcc

---

[217] J.W. Cahn, *Acta Metall.* **10**, 907 (1962).
[218] J. Lašek, *Czech. J. Phys.* **15**, 848 (1965).
[219] D. de Fontaine, *in* "Treatise on Solid State Chemistry" (N.B. Hannay, ed.) Chapter 3, p. 129. Plenum, New York, 1975.

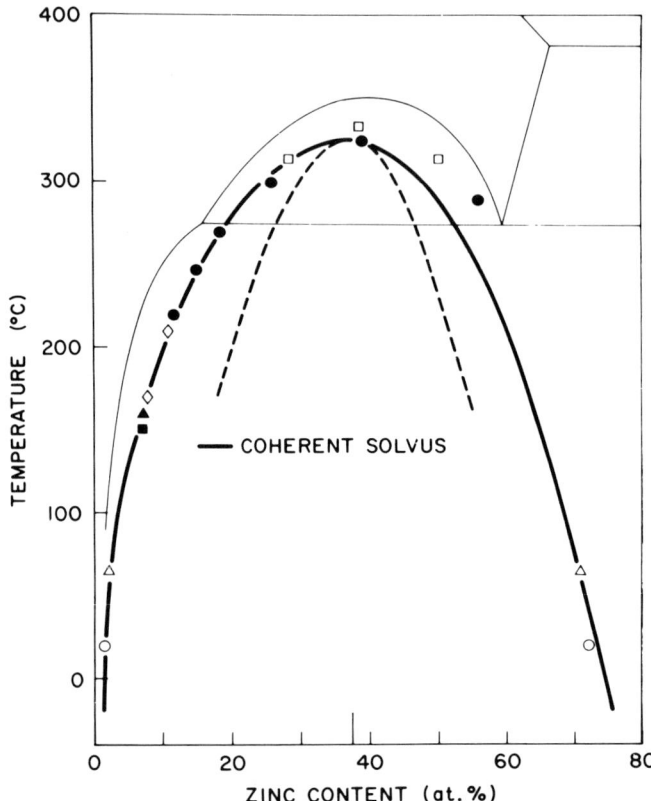

FIG. 36. Metastable fcc coherent miscibility gap calculated by Lašek [J. Lašek, *Czech. J. Phys.* **15**, 848 (1965)] (heavy line) for Al–Zn phase diagram (light line). Dashed line is coherent spinodal calculated by root-three rule [H. E. Cook and J. E. Hilliard, *Trans. Metall. Soc. AIME* **233**, 142 (1965)]. Experimental points from various sources, quoted in de Fontaine.[219]

case, at least the n.n. tetrahedron cluster must be used, but n.n. interactions alone cannot account for the variety of ordered structures described in Section 22. It is therefore expected that, for the fcc case, the CV method with the six-point double tetrahedron basic figure shown in Fig. 25 will provide adequate prototype coherent ordering phase diagrams for all systems of practical interest.

In the CV framework, phase equilibrium is determined mathematically by minimizing the grand potential (20.20), and the computations are carried out as outlined in Section 20,b. If the ordering transition is of higher order, the critical temperature is obtained by equating to zero the determinant of the matrix of second derivatives of the free energy with respect to independent cluster variables. If the ordering reaction is of first order, the transition temperature is obtained by calculating the temperature and concentrations at which the equilibrium grand potentials are the same in both ordered and disordered states, these states differing by the definitions of cluster variables, as was explained in connection with Eqs. (20.28) and (20.29), for example.

The BW model has produced bcc-based phase diagrams which are at least in semiquantitative agreement with experimental data.[220] For example, if magnetic interactions are included, a fairly realistic diagram is obtained[221] which compares favorably with the Fe-Al diagram determined experimentally. Both theoretical and experimental diagrams exhibit a tri-critical point, the implications of which have been discussed by Allen and Cahn.[222]

Progress in the theoretical determination of fcc-based ordering diagrams has been rather slow. The only system to be investigated systematically is Cu-Au, the experimentally determined phase diagram of which is shown in Fig. 37.[222a] By using the BW theory, Shockley in 1938[223] could obtain no better than the diagram shown in Fig. 38, clearly a very unsatisfactory result which shows that the BW model is not even qualitatively correct for the fcc case. The phase diagram calculated by Kikuchi and van Baal[224] by the n.n. tetrahedron CV method is shown in Fig. 39: good qualitative agreement is obtained, at least on the Cu-rich side. Since only a single (n.n.) pair interaction energy is used, all diagrams calculated in this fashion must be identical when plotted on a normalized temperature scale. Furthermore, all diagrams must be symmetrical about $c = \frac{1}{2}$ when only constant pair interaction parameters are included in the internal energy, no matter how many distant neighbor pairs are retained. To break the symmetry, either many-body interactions must be introduced in the internal energy expression, or concentration-dependent pair interaction parameters must be used. As mentioned in Section 20, many-body interactions can be treated very simply in the framework of the CV method. Following the suggestion made by van Baal,[198] tetrahedron interactions were recently incorporated into the model, resulting in the two-parameter fit shown in Fig. 40.[199] The dashed line is an instability locus, to be discussed in Section 24.

[220] G. Inden, *Z. Metallkd.* **66**, 577 (1975).
[221] S.V. Semenovskaya, *Phys. Status Solidi B* **64**, 291 (1974).
[222] S.M. Allen and J.W. Cahn, *Acta Metall.* **24**, 425 (1976).
[222a] M. Hansen, "Constitution of Binary Alloys." McGraw-Hill, New York, 1958.
[223] W. Shockley, *J. Chem. Phys.* **6**, 130 (1938).
[224] R. Kikuchi and C.M. van Baal, unpublished, based on results of van Baal[198] and Kikuchi[189].

Although the latter diagram represents considerable improvement over previous ones, it is still not completely satisfactory. Conspicuously absent is the distinction between CuAuI and CuAuII ordered phases which are present in the center of the experimental phase diagram (Fig. 37). A more complete theoretical determination of the CuAu phase diagram will require at least a double tetrahedron or tetrahedron–octahedron CVM calculation with many-body interactions. If a good fit can be obtained, an excellent understanding of atomic interactions, correlations and state of order (long- and short-range) will have been achieved.

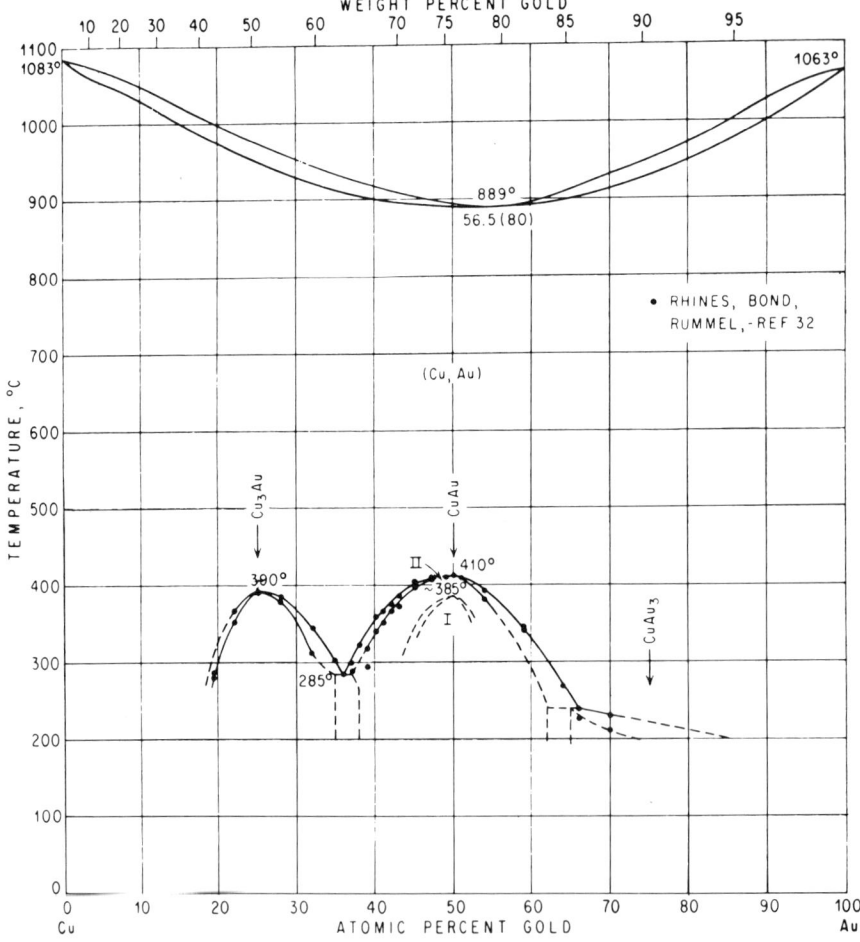

FIG. 37. Experimentally determined Cu–Au phase diagram [After M. Hansen, "Constitution of Binary Alloys." McGraw-Hill, New York, 1958].

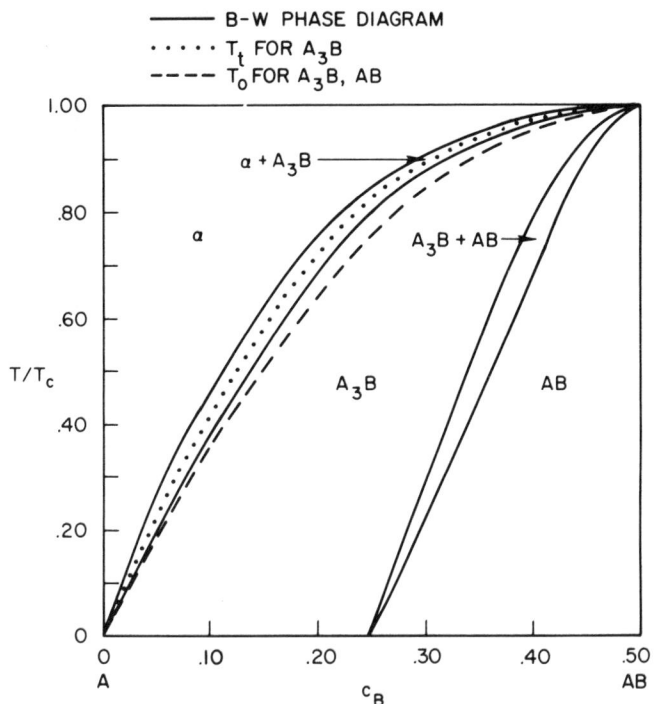

FIG. 38. Coherent ordering phase diagram in fcc lattice calculated from Bragg–Williams model, full curve [W. Shockley, *J. Chem. Phys.* **6**, 130 (1938)]. Locus of equality of chemical potentials in disordered (α) and ordered phases (A_3B), dotted curve. $\langle 100 \rangle$ instability, dashed curve.

c. Ternary Systems

Experimental and theoretical investigations have largely been confined to binary systems. Yet the study of ternary and higher systems seems to be well worth the added effort: all systems of technological interest are multicomponent and the theoretical complexities which appear when more than two components are considered are particularly challenging. The new features presented by multicomponents are well illustrated by ternary phase diagrams which will be presented here in a little more detail than is customary. An additional reason for examining ternaries is that binary systems should really be regarded as three-component systems with vacancies as the third component. Such an approach should be particularly valuable in the case of off-stoichiometric compounds when vacancies may undergo an ordering reaction, or in irradiated or rapidly quenched materials where excess vacancies may cluster, i.e., form

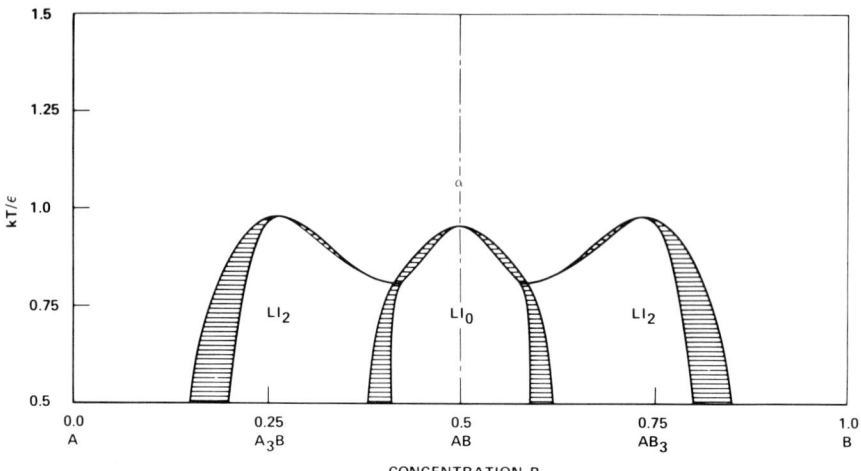

FIG. 39. Coherent ordering phase diagram in fcc lattice calculated by van Baal and Kikuchi with n.n. tetrahedron CVM [R. Kikuchi and C. M. van Baal, unpublished, based on results of van Baal[198] and Kikuchi[189]].

voids.[225] Understandably, the study of quaternary systems is even less advanced than that of ternaries; for the former, the representation of the phase diagram in four-dimensional space constitutes a major difficulty.[226]

For ternaries, composition is represented in an equilateral triangle and the temperature is represented on an axis normal to the composition (or Gibbs) triangle, the main features of ternary phase equilibrium being conveniently exhibited in triangular isothermal sections. Prototype (coherent) clustering ternaries have been calculated by both regular solution and CV methods, and the study of ternary ordering–clustering systems is in progress.

A general classification of ternary regular solution diagrams was presented in the two classic papers of Meijering.[227,228] The basic free energy formula used to construct the coherent miscibility gap phase diagrams was Eq. (17.17), while the classification itself was based on the instability criterion (18.14), with the origin (000) substituted for the wave vector $\mathbf{k}(h)$. By Eqs. (18.11) and (18.12), Eq. (18.14) becomes explicitly, in the present case[227]

$$(k_B T)^2 - 2H(k_B T) - L\bar{c}_1 \bar{c}_2 \bar{c}_3 = 0 \qquad (23.2)$$

[225] J.W. Corbett and L.C. Ianniello, *AEC Symp.* Ser. **26,** 1972.
[226] A. Prince, "Alloy Phase Equilibria." Am. Elsevier, New York, 1966.
[227] J.L. Meijering, *Philips Res. Rep.* **5,** 333 (1950).
[228] J.L. Meijering, *Philips Res. Rep.* **6,** 183 (1951).

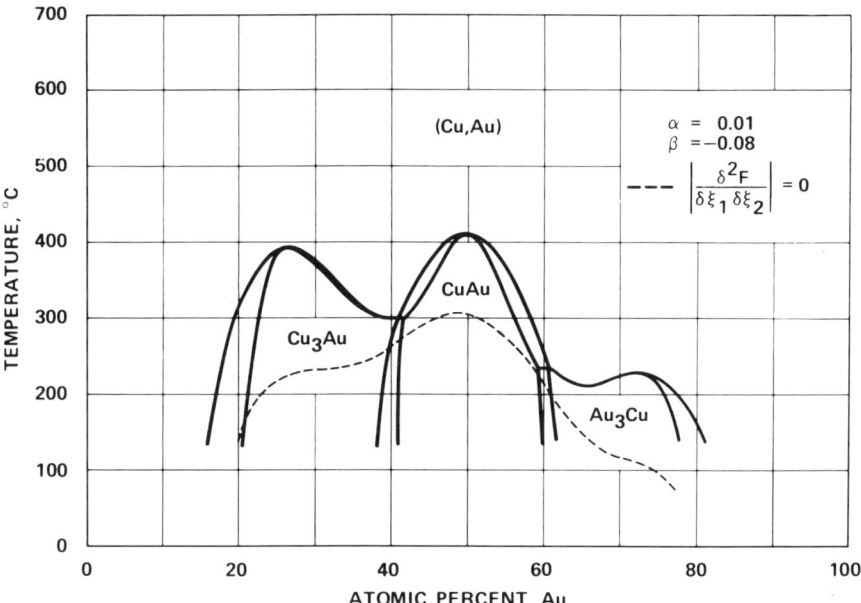

FIG. 40. Coherent ordering phase diagram in fcc lattice calculated with n.n. tetrahedron CVM including tetrahedral many-body interactions [D. de Fontaine and R. Kikuchi, in "Applications of Phase Diagrams in Metallurgy and Ceramics" N.B.S. Special Publication 496 (G. C. Carter, ed.) p. 999 (1978)]. Dashed line is ordering spinodal (see Section 24,b).

with

$$H = w_{12}\bar{c}_1\bar{c}_2 + w_{23}\bar{c}_2\bar{c}_3 + w_{31}\bar{c}_3\bar{c}_1 \tag{23.3}$$

and

$$L = w_{12}^2 + w_{23}^2 + w_{31}^2 - 2w_{12}w_{23} - 2w_{23}w_{31} - 2w_{31}w_{12} \tag{23.4}$$

where the w_{ij} are the values of $W_{ij}(h)$ at the origin and where numerical subscripts (1, 2, 3) have been used to conform to the notation generally used for ternaries. Equation (23.2), first derived by Prigogine,[229] represents the ternary regular solution instability limit or spinodal (see Section 24).

The existence of a binary critical point on the i–j side is guaranteed by the existence of the binary spinodal

$$k_B T = 2w_{ij}\bar{c}_i\bar{c}_j \tag{23.5}$$

which exists for finite positive temperatures provided that w_{ij} is positive.

[229] I. Prigogine, *Bull. Soc. Chim. Belg.* **52**, 115 (1943).

TABLE VIII. MEIJERING'S EIGHT CATEGORIES FOR TERNARY REGULAR SOLUTIONS

Pair potentials $w_{12} \geq w_{23} \geq w_{31}$			Class	Meijering categories[a]	
				A	B
−	−	−	I	$L^b < 0$	$L^b > 0$
+	−	−	II	$w_{12} > w_{23} - w_{31}$	$w_{12} < w_{23} - w_{31}$
+	+	−	III	$w_{12} > w_{23} - w_{31}$	$w_{12} < w_{23} - w_{31}$
+	+	+	IV	$w_{12} > w_{23} + w_{31}$	$w_{12} < w_{23} + w_{31}$

[a] J.L. Meijering, *Philips Res. Rep.* **5**, 333 (1950).
[b] The quantity L is given by Eq. (23.4).

Hence, four basic types of ternaries are expected, depending on whether 0, 1, 2, or 3 binary critical points (at $2k_B T_c = w_{ij}$, $\bar{c}_i = \bar{c}_j = \frac{1}{2}$) are present. Such is the basis for Meijering's classification[227] into the four basic categories I, II, III, IV, respectively. Each basic category is further broken down as indicated in Table VIII where it has been assumed, without loss of generality, that

$$w_{12} \geq w_{23} \geq w_{31}. \tag{23.6}$$

In category IA ($L < 0$), no phase separation occurs, in IB ($L > 0$) a ternary critical point appears, and an isothermal section below it shows a characteristic island MG which does not touch the binaries at any finite temperature. In category II, there is only one binary critical point which is lowered by the addition of a third component when $w_{23} - w_{31} < w_{12}$ (IIA) and raised in the opposite case (IIB). In category III, the higher of the two binary critical temperatures is lowered or raised under the same conditions (IIIA and IIIB, respectively). In category IV, three binary critical points exist; when $w_{12} > w_{23} + w_{31}$ the two lowest critical temperatures are raised by the addition of c_1 and c_2, respectively (IVA); they are lowered in the opposite case (IVB). The useful property concerning the raising or lowering of the binary critical temperature or, equivalently, concerning the expansion or contraction of the binary MG as the third component is added can be proved by using Eq. (23.2),[227,229] a property discovered empirically by Timmermans.[230]

The customary way of deriving ternary MG diagrams is by minimizing the free energy (17.17) under appropriate constraints.[155,218] Since the resulting simultaneous equations to be solved are transcendental, an algorithm based on Newton–Raphson iteration is generally used. Actually, it was found[201,202] to be both more accurate and numerically sim-

[230] J. Timmermans, *Z. Phys. Chem.* **58**, 129 (1907).

pler to use the CV method coupled with natural iteration (see Section 20). Typical isothermal sections calculated in the CVM pair approximation for "B" categories are shown in Fig. 41.[202] The miscibility gaps are obtained as loci of pairs of *binodal points*[228] joined by *tie-lines*. A pair of binodal points denotes two-phase equilibrium, represented geometrically by a tangent plane simultaneously touching the free energy surface at two points. Cases IIA and IIIA would show a MG contracting (rather than expanding) from the 1-2 binary.

An example of category IVA was calculated at the same temperature in the BW, CV pair, and CV tetrahedron approximations. The resulting three MG's are shown on the same diagram in Fig. 42a. In a sense, it is as if three different isothermal sections had been calculated by the same method, the lowest being the BW isotherm, the highest being the CV tetrahedron isotherm. This indicates that the BW approximation predicts the highest and the CV tetrahedron the lowest critical points, respectively, for the same energy parameters, in accordance with the rule whereby the lowest approximation gives the highest critical temperature and vice versa as discussed in Section 20,c and Table V. Upon lowering the temperature, a three-phase equilibrium appears in this system as shown in Fig. 42b. At still lower temperatures, the little MG, located on the short side of the three-phase triangle, must eventually touch the AC side, as required by the fact that all three pair interaction parameters are positive in category IV.

All ternary solution models that are limited to pair interactions in the internal energy have these two features in common: all binary MG's are symmetrical and ternary equilibria depend exclusively on the knowledge of the three binary pair parameters w_{ij}. Hence, once the three binary MG's are known, the ternary diagram is uniquely determined. This is clearly an unrealistic situation, as one expects, on theoretical grounds alone, that the values of pair parameters will depend on average or local composition. In the first alternative, the w_{ij} parameters are allowed to vary with average composition (and temperature), in the second alternative, the internal energy is allowed to depend on many-body (cluster) interactions. A ternary subregular solution fit to the experimental MG of the Cu-Ag-Au system and to tie-lines determined experimentally by Ziebold and Ogilvie[231] gave excellent results as shown in Figs. 43a and b.[232] This example illustrates the use of models in obtaining information on ternary systems, such as the spinodal surface and instability directions

---

[231] T.O. Ziebold and R.E. Ogilvie, *Trans. AIME* **239**, 942 (1967).

[232] M. Murakami, D. de Fontaine, J.M. Sanchez, and J. Fodor, *Thin Solid Films* **25**, 465 (1975).

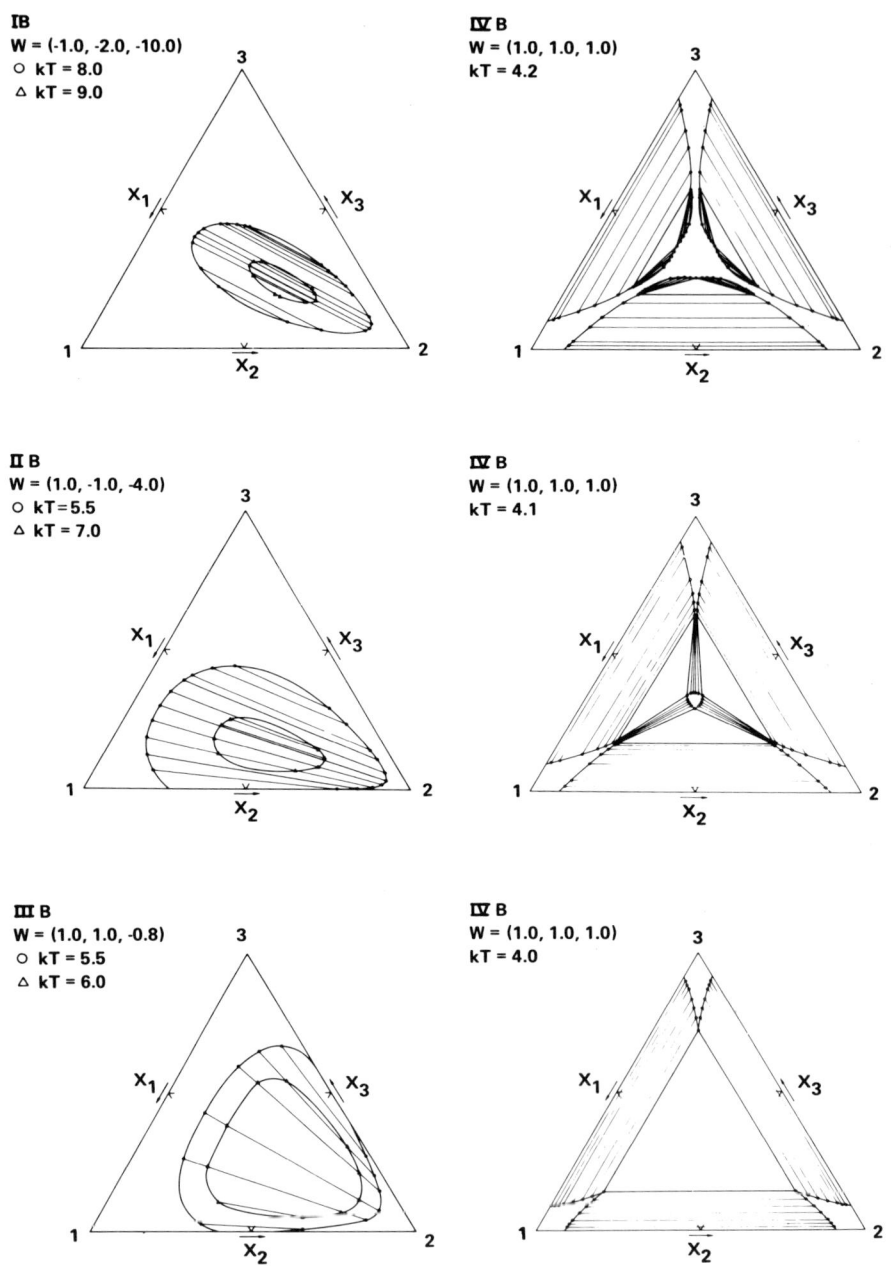

FIG. 41. Isothermal sections for unmixing-type ternary solutions calculated by CVM pair approximation [R. Kikuchi, D. de Fontaine, M. Murakami, and T. Nakamura, *Acta Metall.* **25**, 207 (1977)], illustrating Meijering categories I–IVB (see Table VIII).

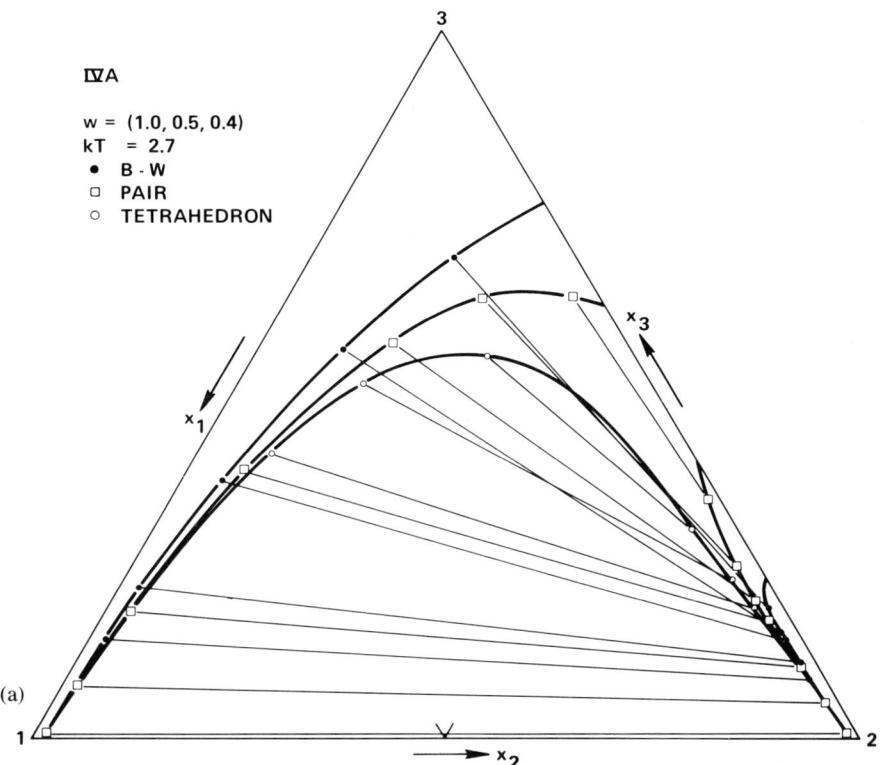

FIG. 42. Isothermal sections of ternary unmixing illustrating Meijering category IVA [R. Kikuchi, D. de Fontaine, M. Murakami, and T. Nakamura, *Acta Metall.* **25**, 207 (1977)]. (a) Comparison of calculations according to Bragg–Williams, pair CVM, and tetrahedron CVM approximations. (b) Appearance of small MG on one side of three-phase triangle.

(Fig. 49b, to be discussed in Section 24), even when knowledge of the full ternary is imperfect.

Thus far, almost no work, experimental or theoretical, has been done on ordering ternaries. As was mentioned in the binary context, for fcc systems the BW and regular solution models are not reliable. For these cases, the tetrahedron CVM must be used, so that the total number of cluster variables is $3^4 = 81$. Thus, it is not surprising that the study of ternary coherent ordering diagrams is in its infancy.[202,233] Nevertheless, the subject should pursued: in fact, all of Meijering's prototype phase diagrams for categories I, II, and III must be in error since, in each instance, at least one of the w_{ij} pair parameters must be negative, so that

[233] N.S. Golosov, L.E. Popov, and L. Ya. Pudan, *J. Phys. Chem. Solids* **34**, 1149 and 1157 (1973).

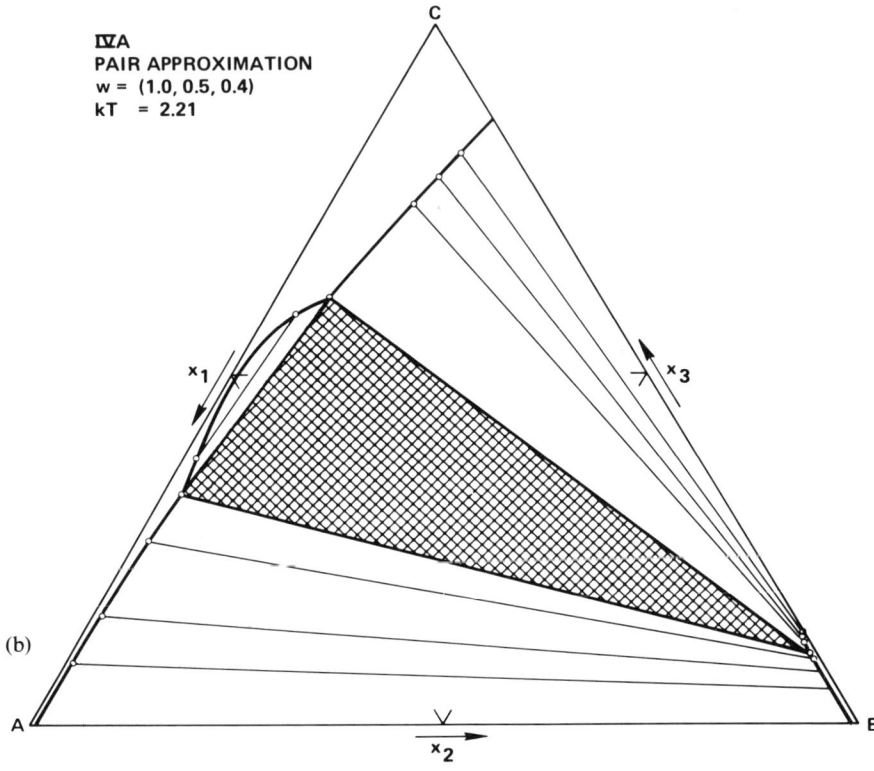

FIG. 42. (Continued)

ordered phases must appear at finite temperatures, and should be represented in the diagrams.

The theoretical calculations of coherent phase diagrams is far from being merely an academic pursuit: for coherent diagrams which have been determined experimentally (Cu–Au for example) a good fit by a CV procedure will yield accurate pair and many-body atomic interactions against which, say, pseudopotential calculations may be tested. Also, pair and other cluster probabilities may be derived theoretically and compared to experimental short-range order data and local atomic arrangements determined by diffuse intensity diffraction experiments. For systems where the incoherent diagram is known, an approximate computation of the related coherent diagram is essential for the study of phase transformations occurring away from equilibrium: at sufficient supersaturations or undercoolings, coherent phases are easier to nucleate than incoherent ones, so that the transforming system will usually evolve along a path which is dictated by the (metastable) coherent equilibria

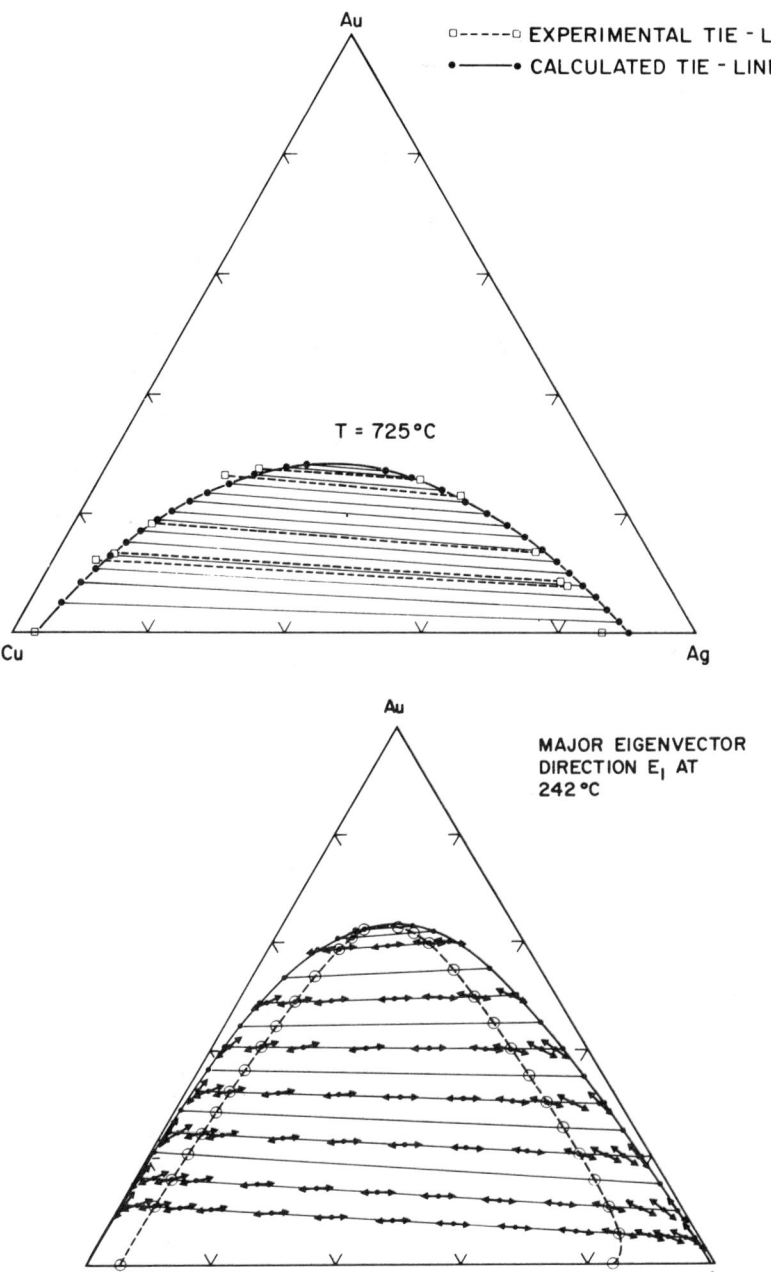

FIG. 43. Subregular solution model fit to the experimental ternary MG in Cu–Ag–Au system. (a) Experimentally determined tie-lines at 725°C [T. O. Ziebold and R. E. Ogilvie, *Trans. AIME* **239,** 942 (1967)]. (b) Calculated ternary MG (full line), spinodal (dashed line), and principal directions (double arrows) corresponding to largest negative eigenvalue [M. Murakami, D. de Fontaine, J. M. Sanchez, and J. Fodor, *Thin Solid Films* **25,** 465 (1975)].

rather than by the incoherent ones. Even if the solid evolves to practically complete equilibrium, the final microstructure observed will generally reflect the metastable stages through which it has passed, hence some knowledge of the coherent phase diagram is required for the rational control of microstructure. Finally, when phase diagrams are unknown or partially determined experimentally, calculations may provide first guesses or plausible extrapolations into uncharted regions.

24. STABILITY ANALYSIS

Knowledge of ordered ground states and phase diagrams is not sufficient for a complete understanding of microstructures resulting from solid state phase transformations: transformations of practical interest take place often considerably away from thermodynamic equilibrium. Only a complete kinetic study can reveal the nonequilibrium path followed by the system during such transformations; but this is outside the scope of the present article, and only a very brief account of kinetics will be given in Section 26.

A less ambitious undertaking, not involving the time coordinate, is that of stability analysis which concerns itself with the manner by which stability is initially lost at the start of the phase transformation. Gibbs[234] recognized that two basic types of instabilities could exist: (a) instability with respect to a local perturbation, large in degree but small in extent; and (b) instability with respect to a nonlocal perturbation, small in degree but large in extent. Elementary theories clearly distinguish between these two mechanisms, generally referred to as *nucleation and growth* and *spinodal decomposition*, respectively, although it is now recognized that these two mechanisms are neither in competition with one another, nor mutually exclusive. Rather, the modes of transformations available to a system form a continuous sequence, the mode selected depending on the type of system, its past history, and its distance from thermodynamic equilibrium.

A model which has realistic "continuous" features is that proposed by Cahn and Hilliard[65,207] to account for phase separation in binary systems. The basic equation is (9.10) which gives the free energy of an elastically isotropic solution containing arbitrary compositional inhomogeneities $c(\mathbf{x})$. Below the miscibility gap, or coherent solvus, the mean-field local free energy density used in Eq. (9.10) has a form like that shown in Fig. 35, for example, and it is possible to find a local concentration fluctuation $c(\mathbf{x})$ in (unstable) equilibrium with the metastable solution of uniform

[234] J.W. Gibbs, "Scientific Papers," p. 105 and 252. Dover, New York, 1961.

concentration c_0, say. Phase transformation proceeds by the growth, in size and amplitude, of this *critical* fluctuation or nucleus, naturally assumed to have spherical shape in isotropic systems. In contrast to classical theory, the nucleus has a smooth concentration profile which varies continuously from the value c_n at the nucleus center to the value c_0 far away from it. The equilibrium profile is obtained by minimizing the free energy functional (9.10) subject to the conservation constraint

$$\int (c - c_0) \, d\mathbf{x} = 0. \tag{24.1}$$

This is done by solving the Euler equation

$$\frac{\partial \mathscr{L}}{\partial c} - \nabla \cdot \frac{\partial \mathscr{L}}{\partial \nabla c} = 0 \tag{24.2}$$

in which the Lagrangian is

$$\mathscr{L}(c, \nabla c) \equiv f(c) + \kappa (\nabla c)^2 - 2\mu_0 (c - c_0), \tag{24.3}$$

μ_0 being the half difference of chemical potentials, as indicated in Fig. 35. Typical calculated profiles are shown in Fig. 44.[207] It is seen that close to the phase boundary (coherent MG), the Cahn–Hilliard nucleus tends to resemble the classical one exhibiting constant concentration inside and sharp interface (top curve). The nucleus concentration c_n is that which maximizes the tangent-to-curve distance indicated by the arrow in Fig. 35. As the average concentration approaches the spinodal concentrations (inflection points ξ_s in Fig. 35), the interface profile becomes increasingly diffuse and the concentration c_n at the center of the nucleus decreases. Finally, at the spinodal, the uniform solution itself is the critical nucleus: infinitely shallow and infinitely diffuse. The work of nucleation W is obtained by inserting the correct equilibrium profile into the integrand of

$$W = \int [f(c) - f(c_0) + \kappa (\nabla c)^2] \, d\mathbf{x}. \tag{24.4}$$

It is found[207] that W increases without limit as the equilibrium phase boundary is reached, again as in classical theory, but goes to zero at the spinodal, where classical theory incorrectly predicts a finite value given by $\frac{4}{3}\pi r^2 \sigma$, with r and σ the classical radius and surface tension, respectively.

The fact that stability is lost for a solution at or "below" (or "inside") the spinodal can be easily seen by referring to Fig. 21 which shows a regular solution free energy curve (a) and the resulting MG (b). The spinodal, or locus of the vanishing of the second derivative of the free energy with respect to concentration, is indicated by a dashed line. Inside

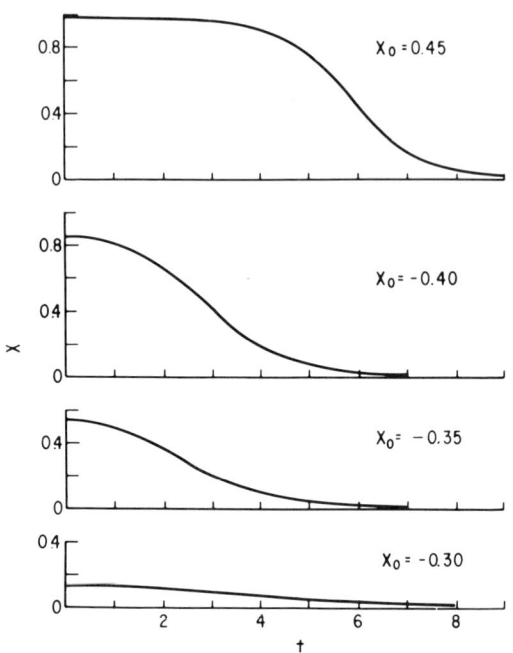

FIG. 44. Calculated reduced concentration profiles (X) of critical nuclei as a function of reduced distance t for various values of concentration from close to MG boundary (top) to close to spinodal (bottom) [J. W. Cahn and J. E. Hilliard, *J. Chem. Phys.* **31**, 688 (1959)].

the spinodal (region III), any incipient phase separation, $a_1 - b_1$ for example, is seen to lower the free energy: the uniform solution is thus inherently unstable, and no nucleation event with finite W is required. Outside the spinodal, but inside the MG (region II), infinitesimal concentration perturbations, $a_2 - b_2$ for example, increase the free energy so that large local departures from the uniform value c_2 are required, i.e., supercritical fluctuations. In region I, the uniform solution is stable.

The important but difficult subject of nucleation will not be pursued further here. The interested reader is referred to the recent article on nucleation by Binder and Stauffer[235] or, for nucleation in solids, to the one by Russell,[236] for example.

The spinodal is obviously an important concept, but it is not an equilibrium thermodynamic one, and depends on the existence of free energy

[235] K. Binder and D. Stauffer, *Adv. Phys.* **25**, 343 (1976).
[236] K.C. Russell, *in* "Phase Transformations" (H.I. Aaronson, ed.), p. 219. Am. Soc. Metals, Metals Park, Ohio, 1970.

curves with concave portions as in Figs. 21 or 35. Only convexity is allowed in the strict thermodynamic sense,[237] i.e., convex curve $f(c)$ from $c = 0$ to c_α in Fig. 21, then the common tangent to c_β, thence the curve to $c = 1$, so that concave portions and associated spinodals can only be justified by imposing artificial constraints, as argued recently by Reiss.[238] In practice, these constraints are kinetic ones: solid solutions may be quenched into the unstable region thereby retaining the quasi-uniform solution for indefinite periods of time. Theoretically, Kikuchi[239] has shown that if one performs an equilibrium cluster variation calculation in region I, then changes the value of the temperature T to one in region II or III, but without changing the cluster frequencies, one obtains an $f(c)$ curve as that of Fig. 21. Even an "exact" calculation with large clusters reproduces the central "hump" of the free energy curve for such a quenched system, although the curvature is not as pronounced and the spinodal points tend to move toward the phase boundary points c_α and c_β. Kikuchi[239] has also performed CVM calculations below T_c: large "humps" are found for small cluster approximations, but rapidly decay as larger clusters are included in the calculation; simultaneously, the common tangent is approached and the spinodal moves toward the MG.

It is seen, therefore, that the position, indeed the very concept of the spinodal is computationally model-dependent and experimentally path-dependent (a function of the severity of the quenching operation, and of state of order at the quenching temperature). Hence, statements made about spinodals in general must be taken with a great deal of caution. However, it would be a mistake to reject the notion altogether, as simple approximate free energy models are often found to give surprisingly good and useful results concerning thermodynamic systems displaced far from equilibrium.

With these words of caution, let us pursue in a more formal way the study of loss of stability by nonlocal small-amplitude perturbations for solid solutions inside the appropriate spinodals. What is required is an analytical function Φ which closely approximates the correct free energy in an equilibrium state in the vicinity of the transition. Of course, the function Φ is not expected to give correct critical exponents, but that is not its role: it is to be used away from equilibrium for the purpose of analyzing the modes of decomposition available to an unstable system. This analytical function is then expanded to second order in the config-

---

[237] H.E. Stanley, "Introduction to Phase Transitions and Critical Phenomena," p. 28. Oxford Univ. Press, London and New York, 1971.
[238] H. Reiss, *Scr. Metall.* **10**, 5 (1976).
[239] R. Kikuchi, *J. Chem. Phys.* **47**, 1664 (1967).

uration variables ξ_s which measure the departure of the solution from the disordered state:

$$\Delta F = \frac{1}{2} \sum_{s,s'} \varphi_{ss'} \xi_s \xi_{s'} \qquad (24.5)$$

where $\varphi_{ss'}$ are the coefficients of the expansion

$$\varphi_{ss'} = \left[\frac{\partial^2 \Phi}{\partial \xi_s \partial \xi_{s'}}\right]_{\xi=0} \qquad (24.6)$$

and ΔF represents the difference in free energy between the slightly perturbed system and the uniform (disordered) state. Equation (24.5) has the same form as (18.13) which is the second-order term of a Taylor's expansion of a generalized Bragg–Williams model. In Eq. (24.5), as in (18.13), the variables of the quadratic form must be long-range order parameters, or sublattice-averaged concentration deviations, the sublattice averaging in turn implying the zeroth or mean-field approximation. The variables ξ_s may be more general than the ones introduced in Sections 17 and 18, as the suffix s may refer to different components or to different linear combinations of cluster variables taken on different sublattices. It is only required that the variables ξ_s be linearly independent and that they vanish identically in the disordered state.

If the configuration variables are sublattice-averaged concentration derivations $\gamma_i^0(p)$, the quadratic form (24.5) may be diagonalized by a Fourier transform followed by an orthogonal transformation of the matrix $\Phi(h)$,

$$\Delta F = \frac{N}{2} \sum_h \Gamma^*(h) \Phi(h) \Gamma(h) = \frac{N}{2} \sum_h \sum_{l=1}^n \Lambda_l(h) |\bar{\Gamma}_l(h)|^2, \qquad (24.7)$$

where Λ_l are the eigenvalues of Φ, and $\bar{\Gamma}_l$ are what we may call *normal concentration modes*. If the approximate free energy function Φ used in (24.6) is taken from the generalized Bragg–Williams model introduced in Section 17, then Eq. (24.7) is identical to $F^{(2)}$ given by (18.13), the matrix elements being given by Eqs. (18.11) and (18.12). Whatever the model used, the brief discussion of stability given in Section 18 can be taken over without modification. Thus, the solution is stable to all small-amplitude perturbations as long as Φ is positive definite for all possible wave vectors $\mathbf{k}(h)$. For given $\mathbf{k}$, different compositional modes become unstable as the corresponding eigenvalue goes from positive to negative. Hence, stability limits in (T, c_i) phase diagram space are given by the vanishing of the determinant Δ of the matrix of the quadratic form, as in

Eq. (18.14). Usually, only the uppermost (in temperature) stability limit is plotted: it is the one which corresponds to that value of $\mathbf{k}(h)$ (actually, of all vectors of the star of $\mathbf{k}$) which makes the lowest eigenvalue $\Lambda_1(h)$ of Φ vanish at the highest possible temperature. Since this matrix and all of its eigenvalues must have the symmetry of the reciprocal lattice, generally only wave vectors at the special points (see Section 8) need be considered.

In the Bragg–Williams approximation of Sections 17 and 18, the configurational entropy contribution to the second-order term $F^{(2)}$ does not depend on the wave vector and the internal energy is temperature-independent. Consequently, the wave vector belonging to the uppermost stability surface is simply the one that can be deduced from a knowledge of pair interaction parameter ratios, for example with the help of diagrams of special-point minima such as shown in Fig. 10. As mentioned earlier, these diagrams are really more useful in the context of stability, rather than ground-state analysis. Hence, in this model, specified ratios of pair interaction parameters determine uniquely the type of special-point instability that will occur for all temperatures and average concentrations. It is then possible to group all binary solid solutions based on a given average lattice, into a few basic families, each family being characterized by the special-point ordering wave belonging to the uppermost stability curve. Members of the same family may of course exhibit various equilibrium ordered phases which are not manifestly related to the characteristic ordering wave of the family. Examples will be given below. If pair parameters are strongly temperature- and/or concentration-dependent, or if the configurational entropy is wave vector-dependent, a given binary solution may exhibit different uppermost instabilities in different regions of temperature and concentrations, but there appears to be little evidence for the existence of such hybrid systems.

This simple classification into stability families is possible because, in binary systems, the generalized Bragg–Williams model yields the simple formula (18.17) for the stability locus or spinodal: by inserting different wave vectors $\mathbf{k}(h)$ into the k-space potential $V(h)$ one obtains a set of nested parabolic loci T_0 vs $\bar{c}$ which can only intersect at absolute zero. The situation is quite different for multicomponent systems where stability surfaces belonging to different k vectors may intersect at finite temperatures. Nevertheless, multicomponent solutions may still be classified by regions according to characteristic ordering waves. When ΔF, Eq. (24.5), is obtained by expanding a cluster variation free energy to second order in LRO parameters, the physical meaning of the stability limits is not apparent. At least, it is not obvious that these limits may be identified with the loci along which special-point ordering waves just become unstable.

Stability analysis of special cases will now be described.

a. Binary Phase Separation

For phase separation systems, the equilibrium state is the fully phase-separated one, the associated "ordering wave" having wavelength of the order of the crystal's linear dimensions. The special-point wave vector is thus at the Brillouin zone center. In the Bragg–Williams approximation, the binary clustering spinodal is obtained by inserting $(h) = (000)$ in Eq. (18.17), $V(0)$ denoting the sum of pair interaction parameters over corresponding coordination shells. Such a regular solution spinodal was shown in Fig. 21 (dashed curve). Because of the singularity of the k-space potential V at the origin (Section 7), a unique clustering spinodal exists only if the elastic energy contribution to the total potential is very small, or if the solid solution is nearly elastically isotropic.

Experimental miscibility gaps do not generally exhibit such symmetry. The customary procedure is then to fit a subregular model free energy (for which pair parameters depend on average concentration and temperature, such as the one shown in Fig. 35, for example) to the incoherent MG. An elastic energy correction is then made, as discussed in connection with Eq. (23.1). The coherent spinodal is given by

$$\left[\frac{\partial^2 f_c}{\partial c^2}\right] \equiv f_I'' + 2\eta^2 Y(0) = 0 \quad (24.8)$$

where f_c is the coherent free energy density and f_I'' is the second derivative with respect to concentration of the incoherent (fitted) free energy. The elastic energy correction is (24.8) is the continuum (long wavelength) form appropriate for misfitting solute creating isotropic stress-free distortions $\eta_{\alpha\beta} = \eta\delta_{\alpha\beta}$ in cubic crystals. If the crystal is nearly isotropic, $Y(0)$ in (24.8) is simply the modulus Y_0 of Eq. (12.18). If the crystal is elastically anisotropic, the value of $Y(0)$ depends on the direction of approach to the origin of the BZ. Hence Y at $k = 0$ must be given by the exact formula (12.14) for $Y(\kappa)$ or its approximate but simpler form (12.17), where κ is the unit vector along **k**.

Thus, even when a unique coherent MG cannot be defined, a set of coherent spinodals may be constructed on the phase diagram, each belonging to concentration waves along various directions such as $\langle 100 \rangle$, $\langle 110 \rangle$, $\langle 111 \rangle$, for example.[217] In solids for which the anisotropy parameter A, defined by Eq. (11.13), is positive (the usual case), the $\langle 100 \rangle$ spinodal will be the uppermost stability limit, $\langle 111 \rangle$ the lowest, all others intermediate. For negative anisotropy, the reverse is true. If the distortion tensor η has lower symmetry than cubic, these simple rules no longer hold; compare, for instance, the long wavelength limits of the curves of Fig. 13 (cubic symmetry) and Fig. 14 (tetragonal symmetry).

A simple method for constructing a spinodal directly from the MG near the critical point has been suggested by Cook and Hilliard.[240] One starts from the Landau expression (19.2), with second-order coefficient A given by Eq. (19.7), with third-order coefficient $B \equiv 0$, since the expansion in this case is about a second-order unmixing critical point (c_c, T_c), and with fourth-order coefficient $C > 0$. The equilibrium (common tangent) concentrations c_e are obtained by setting the first derivative of the free energy equal to zero:

$$\xi_e \equiv c_e - c_c = \pm \sqrt{\alpha(T_c - T)/2C}. \tag{24.9}$$

The spinodal concentrations c_s are obtained by setting the second derivative equal to zero:

$$\xi_s \equiv c_s - c_c = \pm \sqrt{\alpha(T_c - T)/6C} = \xi_e/\sqrt{3}. \tag{24.10}$$

This so-called "root-three rule" (24.10) has been used for constructing an approximate spinodal curve for the nearly isotropic Al-Zn system in Fig. 36.

More realistic estimates of the spinodal concentrations place these much closer to the equilibrium MG concentrations, as may be seen by examining Fig. 5 of Kikuchi[239] which refers to Kikuchi's CVM calculation of free energy curves, or Fig. 1 of Sur et al.[208] which shows spinodal curves calculated from both a fourth-degree polynomial, such as (19.2), and one containing the sixth power of the order parameter ξ.[241] Recent experimental evidence[242,243] appears to locate the spinodal temperature for an Al-7.5 at. % Zn alloy at about 130°C, which places it well outside the root-three construction. But, as mentioned above, experimental determinations of the spinodal rest essentially on nonequilibrium measurements and rather imprecise criteria for spinodal decomposition (Section 26), so that good agreement with model-dependent theoretical estimates is not to be expected.

As noted by Cahn,[170] "the word 'spinodal' has been somewhat of a mystery." On etymological grounds, one would expect the spinodal to be a locus of mathematical cusps or spinodes. Van der Waals[244] identified

---

[240] H.E. Cook and J.E. Hilliard, *Trans. Metall. Soc. AIME* **233**, 142 (1965).
[241] D. Amit, *Phys. Lett. A* **26**, 466 (1968).
[242] A. Junqua, J. Delafond, J. Mimault, and J. Grilhé, *Scr. Met.* **8**, 317 (1974).
[243] A. Naudon and J. Allain, *Scr. Metall.* **8**, 1105 (1974).
[244] J.D. van der Waals and P. Kohnstamm, "Lehrbuch der Thermodynamik," 1st ed., p. 133. Maas and van Suchtelen, Amsterdam, 1908.

cusps at the spinodal in an internal energy–entropy–composition diagram, but a simpler representation is provided by grand potential–chemical potential curves which are useful in calculating phase diagrams from CVM free energy models.[245] For a binary system, the grand potential per lattice point is, from Eq. (20.20),

$$g \equiv G/N = f - \mu\xi \qquad (24.11)$$

where f is the free energy per lattice point, ξ is the concentration difference $c_2 - c_1$, and μ is the half difference of chemical potentials. Grand and chemical potentials at ξ_0 are indicated on Fig. 35. If g is plotted vs μ and ξ varies from -1 to $+1$, a "swallowtail curve" is traced, as seen in perspective in Fig. 45.[245] If the temperature $\tau = T - T_c$ is allowed to vary, the full swallowtail surface is obtained, just as in Thom's catastrophe theory.[246] The sharp ridges of this surface are easily identified as the spinodal, also shown projected onto the μ-τ plane in Fig. 45. The cusplike nature of the spinodal is thus made very apparent. Note also that the self-intersection of the swallowtail surface (dot–dash curve) marks the equality of chemical potentials, hence the equilibrium states determined from the common tangent rule.

b. Binary Ordering

It has not always been recognized that a given binary system with fixed values of pair interaction parameters $V_1, V_2, \ldots$ can possess simultaneously a phase separation and various ordering spinodals. A necessary (but not sufficient) condition is that pair interactions beyond the first coordination shell be considered. This is best illustrated by means of an example: consider three pair interaction parameters V_1, V_2, V_3 with ratios $V_2/V_1 = -1.2$ and $V_3/V_1 = -0.8$ in an fcc solution. These are the coordinates of a point located inside the $\langle 000 \rangle$ or "clustering" region in Fig. 10a. Hence the lowest special-point minimum of the k-space potential $V(h)$ must be located at the BZ origin and the uppermost stability limit must be the phase separation spinodal. The proximity of this point to the $\langle 1\frac{1}{2}0 \rangle$ region indicates the possibility that an ordering spinodal at that BZ boundary point may also play a significant role. The stability analysis of this fcc solution can be pursued by use of Eq. (8.7) and the eigenvalue tables of de Fontaine.[49] Let us arbitrarily assign the

---

[245] R. Kikuchi and D. de Fontaine, *Scr. Metall.* **10**, 995 (1976).
[246] R. Thom, "Stabilité Structurelle et Morphogénèse." Benjamin, Reading, Massachusetts, 1973.

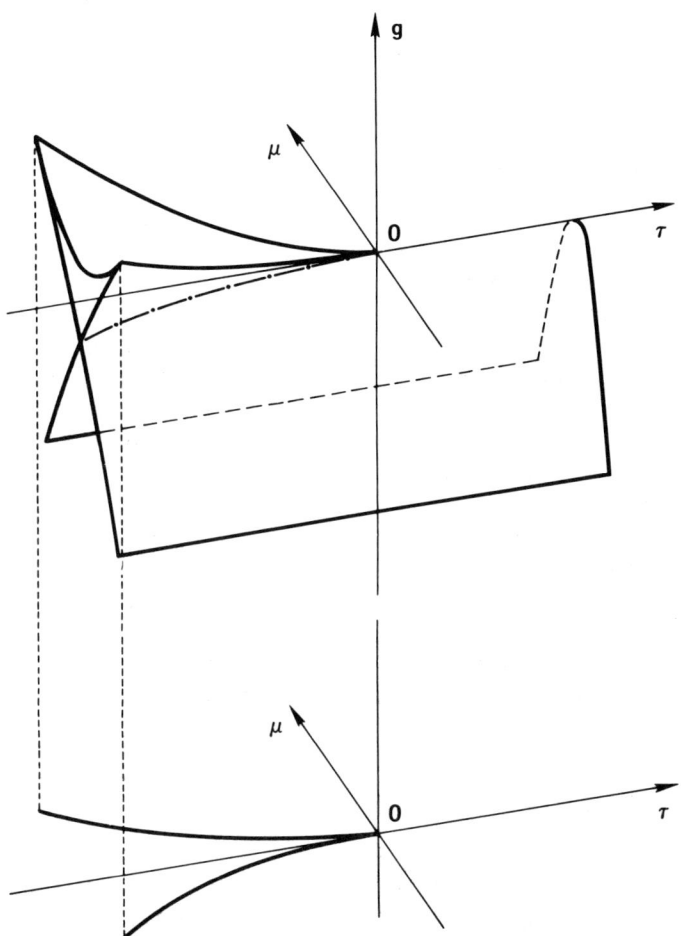

FIG. 45. Swallowtail surface of grand potential g plotted against chemical potential μ and temperature $\tau = T - T_c$ for phase-separating binary solution [R. Kikuchi and D. de Fontaine, *Sci. Metall.* **10**, 995 (1976)]. Ridge lines are spinodal loci; self-intersection line is locus of phase equilibrium.

value $\frac{5}{8}$ to V_1. Then we find

at $\langle 000 \rangle$: $V_0 = -9;$ $\lambda_1 = \lambda_2 = \lambda_3 = 25/4$ (24.12a)

at $\langle 1\frac{1}{2}0 \rangle$: $V_0 = -8;$ $\lambda_1 = \lambda_3 = 25/8,$ $\lambda_2 = 5/2$ (24.12b)

at $\langle 100 \rangle$: $V_0 = -3;$ $\lambda_1 = 19/4,$ $\lambda_2 = \lambda_3 = -5/2$ (24.12c)

at $\langle \frac{1}{2}\frac{1}{2}\frac{1}{2} \rangle$: $V_0 = 9/2;$ $\lambda_1 = 7/4,$ $\lambda_2 = \lambda_3 = -25/8,$ (24.12d) where V_0 is the value of $V(h)$ at the indicated point. Hence, at least according to the generalized BW model, Eq. (18.17), the highest (in T) special-point instability will be the $\langle 000 \rangle$, followed by the three ordering ones in the order indicated above.[247] However, local $V(h)$ minima are found only at $\langle 000 \rangle$ and $\langle 1\tfrac{1}{2}0 \rangle$, for which all three eigenvalues λ_i are positive. The two other ordering points are saddle points of $V(h)$ so that, as will be explained in Section 25, no SRO intensity peaks are expected at these two positions, no matter how far below the stability limits one ages the solution. Furthermore, from Eq. (24.12b), we have $\lambda_1 = \lambda_3 \cong \lambda_2$ so that according to Eq. (25.25), the $\langle 1\tfrac{1}{2}0 \rangle$ diffuse intensity peaks observed after decomposition of the solution below that instability limit should be roughly spherical in shape. All these conditions appear to be approximately met in a Cu-5 at. % Ti solution which can exhibit both $\langle 000 \rangle$ satellites and $\langle 1\tfrac{1}{2}0 \rangle$ SRO peaks depending upon where, with respect to these spinodals, the quenched solution is aged.[248] Normally, the uppermost spinodal will prevail since, at any lower temperature, the driving force for that particular decomposition will be greatest. However, the kinetics of ordering are faster than those of phase separation, leading to all manner of phase separation-ordering transformation modes whenever an ordering spinodal is situated just below the phase separation spinodal, as is apparently the case here and in such systems as Ni-Al, Ni-Si, Ni-Cu-Al, Ni-Cu-Si, and others, as evidenced by the early-stage decomposition observations of Manenc,[249] for example.

Another interesting example is that of fcc Ni-Mo solutions: by quenching from the disordered state over a wide range of compositions from about 8 to 33 at. % Mo, one finds in both X-ray[250] and electron[251,252] diffraction patterns roughly spherical diffuse intensity contours about $\langle 1\tfrac{1}{2}0 \rangle$ positions which do not coincide with any of the superlattice positions of the stable ordered phases. In this system, it thus appears possible to avoid, by rapid cooling, the nucleation of the equilibrium ordered structures and to obtain, instead, a spinodal ordering reaction occurring below the $\langle 1\tfrac{1}{2}0 \rangle$ stability limit. However, $\langle 1\tfrac{1}{2}0 \rangle$ waves by themselves cannot produce correct ordered structures at these average concentrations, so that the $\langle 1\tfrac{1}{2}0 \rangle$-modulated solid solution eventually evolves toward the appropriate ordered structures: stable $D1_a$ around the Ni_4Mo

---

[247] Actually the $\langle \tfrac{1}{2}\tfrac{1}{2}\tfrac{1}{2} \rangle$ instability temperature is negative.
[248] D.E. Laughlin and J.W. Cahn, *Acta Metall.* **23**, 329 (1975).
[249] J. Manenc, *Acta Metall.* **7**, 124 (1959).
[250] B. Chakravarti, E.A. Starke, C.J. Sparks, and R.O. Williams, *J. Phys. Chem. Solids* **35**, 1317 (1974).
[251] P.R. Okamoto and G. Thomas, *Acta Metall.* **19**, 325 (1971).
[252] S.K. Das, P.R. Okamoto, P.M.J. Fisher, and G. Thomas, *Acta Metall.* **21**, 913 (1973).

composition, and metastable Ni_2Mo (see Table VII for description of these structures). The DO_{22} structure itself must eventually give way to the incoherent Ni_3Mo orthorhombic structure.[180] According to the concentration wave picture, $\langle 1\tfrac{1}{2}0 \rangle$ spinodal ordering produces a structure with "average atoms" as indicated in Fig. 46: the effect of a $[1\tfrac{1}{2}0]$ concentration plane wave on the lattice is shown in [001] projection, and the resulting average three-dimensional structure is shown in perspective. This structure is seen to be in some way intermediate between the I4/mmm(DO_{22}) and $I4_1/amd$ illustrated in Fig. 29- a_5 and a_4, respectively (see also Table VII).

It is now possible to complete the classification of ordering mechanisms

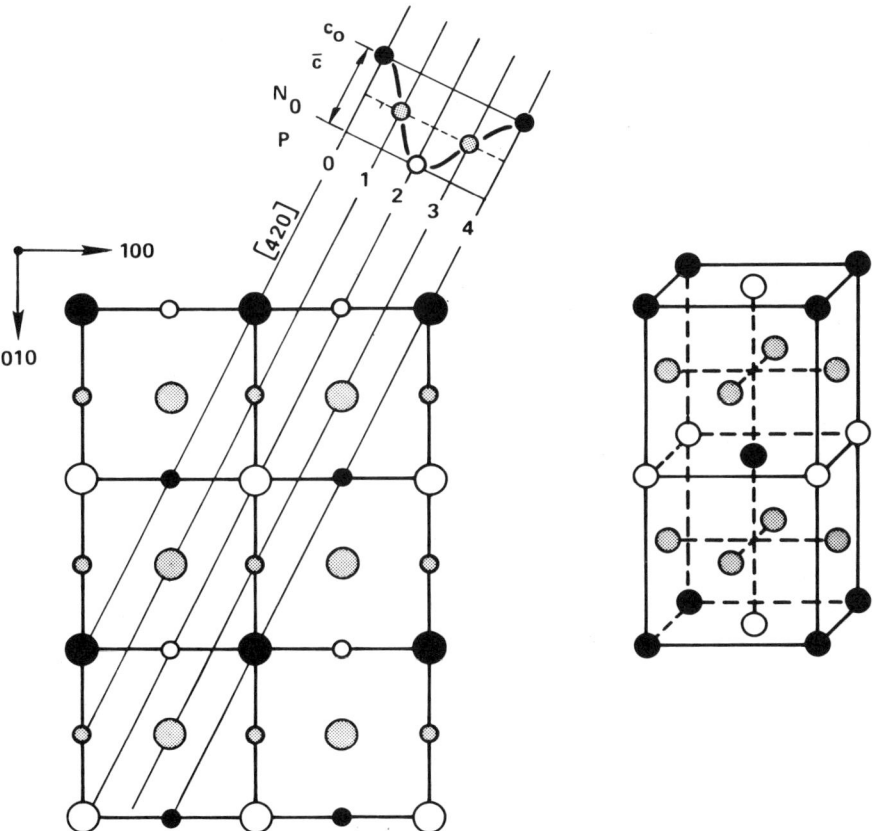

FIG. 46. Effect of $\langle 1\tfrac{1}{2}0 \rangle$ concentration plane wave on fcc lattice. Open circles: N-rich, filled circles: M-rich, shaded circles: intermedicate-composition "average atoms" [D. de Fontaine, *Acta Metall.* **23**, 553 (1975)].

given in Section 19 with the help of the Venn diagram of Fig. 22. If the equilibrium ordered structure belongs to set L, in other words if the ordering wave vectors verify all three Landau–Lifshitz criteria, then ordering may take place continuously, i.e., by continuous amplification of special-point concentrations waves, at all temperatures below the transition. If the equilibrium structure belongs to set K, continuous ordering can only occur at temperatures below the uppermost special-point ordering wave spinodal; above it, ordering must proceed by nucleation and growth. If the equilibrium structure belongs to set I, spinodal ordering is in principle possible by continuous amplification of the appropriate non-special-point ordering waves at low enough temperatures, but the uppermost special-point spinodal will dominate the decomposition, leading to a transient modulation such as the $\langle 1\frac{1}{2}0 \rangle$ partially ordered structure found on quenching Ni–Mo solid solutions, as described above.[49]

It is seen that the concepts of concentration waves, Landau rules and stability limits are invaluable in dealing with nonequilibrium transformations of unstable solid solutions. These concepts are intimately tied to mean-field or BW models which cannot provide accurate quantitative descriptions, or even, in the fcc case, qualitative ones. What is needed is stability analysis carried out in the framework of, say, the cluster variation method. This was recently attempted[199] in the Cu–Au case, and we close this section by briefly commenting on the results obtained thus far.

Clearly, the BW phase diagram of Fig. 38 is unsatisfactory from the instability point of view as well as falling far short of representing Cu–Au type phase equilibrium. The unique stability limit is the parabola (dashed curve)

$$k_B T_0 + \bar{c}(1 - \bar{c}) V(h) = 0 \tag{18.17}$$

a degenerate locus with $(h) = \langle 100 \rangle$ or $\langle 1\frac{1}{2}0 \rangle$, which is seen to lie *above* the stable AB ordered phase field. Also, this ordering spinodal touches the tops of both AB and A_3B (and AB_3) phase field at the 50% composition, indicating that both $L1_2$ and $L1_0$ phases appear by a second-order phase transformation at that composition, which is contrary to experiment and to the Landau–Lifshitz rules (Section 19).

These incongruities are absent from the stability locus calculated by the CV method (dashed line, Fig. 40). In the fcc CV calculation in the tetrahedron cluster approximation, there are three independent LRO parameters, one of which is the familiar

$$\xi_1 = x_A{}^\alpha - x_A{}^\beta, \tag{24.13}$$

i.e., the difference in A-component concentration on the two sublattices

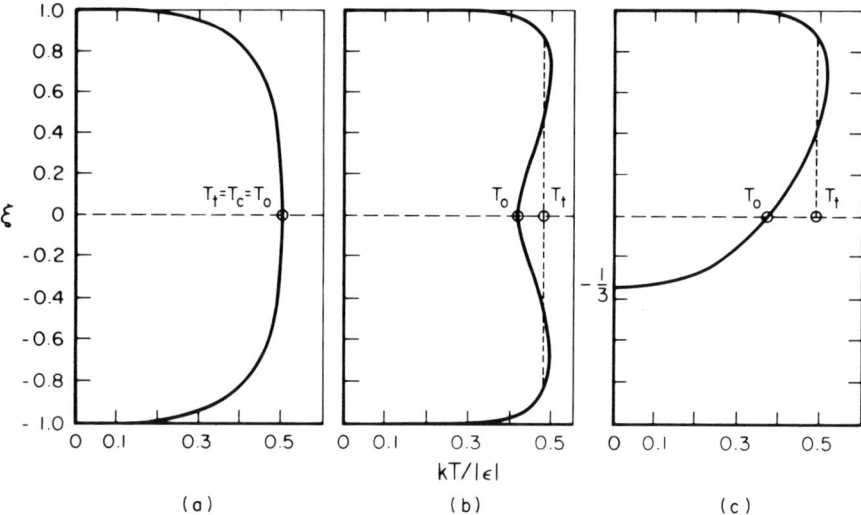

FIG. 47. LRO parameter ξ plotted against normalized temperature for (a) second-order transition, (b) symmetric first-order, (c) asymmetric first-order. Curves (b) and (c) are extensions to negative ξ of curves of Fig. 27. T_t are transition and T_0 instability temperatures.

α and β. The other two are related to tetrahedron concentrations. According to Eqs. (24.5) and (24.6), the stability locus is obtained by setting the determinant of second derivatives equal to zero, with all LRO parameters set equal to zero. This was carried out separately for Cu_3Au, $CuAu$, and $CuAu_3$ phases, and a unique continuous locus was obtained, as shown in Fig. 40. This locus must be none other than the $\langle 100 \rangle$ ordering spinodal. All phase transitions in this system are first-order, and this is correctly reflected in the fact that the ordering spinodal is everywhere below the ordering–disordering phase boundaries, or *ordus* and *disordus* lines.[253]

The relationship between instability and transition temperatures at the stoichiometric compositions is clarified by plotting the principal LRO parameter ξ_1, defined by Eq. (24.13) against reduced temperature. This is shown in Fig. 47 for (a) a second-order transition for an fcc miscibility gap system (symmetrical case) which is isomorphic to the Ising ferromagnet, (b) a first-order transition of CuAu type (symmetrical composition), and (c) a first-order transition of Cu_3Au type (asymmetric composition). Figures 47a and b are just the LRO diagrams of Figs. 27a and b

[253] Terms coined by P. Bardhan; J.B. Cohen, private communication.

extended to negative values of ξ_1. In all three cases, the point at which the LRO curve crosses the temperature axis is the instability temperature T_0. In the second-order case, T_0 coincides with the transition temperature T_t which is a real critical temperature T_c. Figures 47a and c are similar to familiar magnetization-temperature diagrams; Fig. 47b is particularly interesting since it shows a first-order transition in the 50% symmetrical case, which is forbidden in the BW approximation because of Eq. (18.18). Thus, here too, stability analysis in the CV framework shows marked improvement over that carried out in the BW or mean-field approximation.

c. Ternary Systems

In solid solutions containing more than two components of significant concentration it is necessary to introduce the concept of *polarization* of a normal concentration wave which was mentioned in connection with Eq. (24.7) and described in Section 3 with the help of Fig. 2. In Eq. (24.7), "normal" amplitudes $\bar{\Gamma}_l$ are defined by

$$\bar{\Gamma} = \mathbf{E}^{-1}\Gamma \tag{24.14}$$

where $\mathbf{E}$ is the orthogonal matrix which diagonalizes the free energy second-derivative matrix $\mathbf{\Phi}$. The eigenvectors $\mathbf{E}_l$ ($l = 1, \ldots, n$) of $\mathbf{\Phi}$ determine the *principal directions*, in composition space, of the free energy surface at the point $(T; \bar{c}_1, \ldots, \bar{c}_n)$ and wave vector $\mathbf{k}(h)$ considered.

Another useful concept is that of *asymptotic directions* obtained by equating to zero the quadratic form for ΔF:

$$\Gamma^*\mathbf{\Phi}\Gamma = 0. \tag{24.15}$$

For ternary systems, Eq. (24.15) represents degenerate conics (i.e., two intersecting straight lines), for quarternary systems, degenerate quadrics[3] (i.e., cones or intersecting planes), etc. The degenerate conics, quadrics, ... are real whenever the free energy matrix $\mathbf{\Phi}$ is indefinite. In any plane containing two principal directions (eigenvectors $\mathbf{E}_l$), the asymptotic directions are bisected by the principal directions.[3] Furthermore, along any stability surface, defined by the vanishing of an eigenvalue $\Lambda_l(h)$, the asymptotic directions collapse onto the included principal direction, which is the eigenvector corresponding to zero eigenvalue. This *instability direction* in composition space determines the polarization of the normal concentration wave of given wave vector which just becomes unstable at the phase diagram point $(T, \bar{c}_1, \ldots, \bar{c}_n)$. A complete stability analysis thus generally consists of the mapping onto the phase diagram of the uppermost stability surface along with indications of the associated principal directions. A few examples, taken from idealized ternary systems, will illustrate these concepts.

The equation for ternary stability surfaces in the generalized BW model is obtained by inserting the matrix elements (18.11) and (18.12) into the stability equation (18.14) with the result

$$(k_B T)^2 - 2H(h)(k_B T) - L(h)\bar{c}_1 \bar{c}_2 \bar{c}_3 = 0. \qquad (24.16)$$

This is just the general form of Eq. (23.2), valid for any wave vector $\mathbf{k}(h)$. The coefficients $H(h)$ and $L(h)$ are given by Eqs. (23.3) and (23.4) in which the pair parameters w_{ij} must be replaced by the k-space potentials $W_{ij}(h)$.

Meijering's classification can still be used in the more general case of h-dependent parameters: a given ternary system may be in one category with respect to a certain ordering wave vector (special point) and in a different category with respect to another. For given $\mathbf{k}(h)$, the number of stability surfaces encountered will depend on the number of positive roots $(k_B T)$ of Eq. (24.16): no instability in category IA, one instability in category IB, II, and III, and double instability in category IV with $L(h) < 0$. This can conveniently be represented graphically[254] by noting that by Eq. (23.4), $L = 0$ represents a double cone in W_{12}, W_{23}, W_{31} space (Fig. 48) so that L is positive outside the cone and negative inside. The cone is inscribed in a regular octahedron; the faces on which the cone emerges represent category IV on the positive octant of the W_{ij} coordinate system, category I on the negative octant. The other faces of the octahedron represent categories II and III.

Consider a hypothetical system for which a certain special-point k-vector imparts roughly equal positive values to all three $W_{ij}(h)$ functions. The system is thus represented by a point on the positive cone axis (case IVB, $L < 0$), and one may expect a double instability for that particular ordering wave. Isothermal sections of the phase diagram were shown in Fig. 41. At a certain temperature, the stability surfaces (or traces thereof in the isothermal section) will be as represented in Fig. 49,[167] where the definite or indefinite character of the free energy matrix has been indicated. The asymptotic directions are given by the explicit form of Eq. (24.14) valid for ternaries

$$\Gamma_1/\Gamma_2 = (-\Phi_{12} \pm \sqrt{-\Delta})/\Phi_{11} \qquad (24.17)$$

where, as in Eq. (18.14), Δ is the determinant of the symmetrical matrix Φ. Instability directions in concentration space are indicated in Fig. 49

[254] J.E. Morral and J.W. Cahn, private communication.

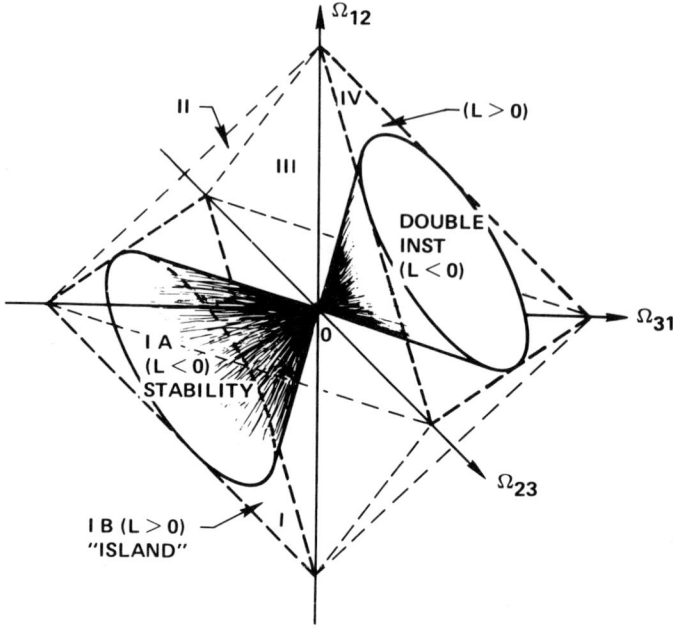

FIG. 48. Graphical representation of Meijering ternary instability categories. See also Table VIII.

by darkened areas delineated by the asymptotic directions for various points along CD in the indefinite region. The darkened full circle, which lies inside the second stability surface, indicates that the system is unstable to all possible concentration perturbations for the wave vector under consideration: the system is then "all-round unstable" in Meijering's terminology.[227] Along CD, the principal directions lie parallel and perpendicular to CD, and bisect the asymptotic directions, as mentioned above.

As another example, consider a bcc ternary solid solution for which the relevant ordering waves are the special-point k-vectors $\langle 100 \rangle$ and $\langle \frac{1}{2}\frac{1}{2}\frac{1}{2} \rangle$. Along a median vertical section of the ternary phase diagram, the stability surfaces may appear as shown in Fig. 50.[167] Note that this system belongs to category IA of Meijering (see Table VIII) with respect to the $\langle 000 \rangle$ clustering wave, to category IIIB with respect to the $\langle 100 \rangle$ ordering wave, and to category IVB ($L < 0$) with respect to the $\langle \frac{1}{2}\frac{1}{2}\frac{1}{2} \rangle$ and $\langle \frac{1}{2}\frac{1}{2}0 \rangle$ ordering waves. Hence, by analyzing the number of positive $(k_B T)$ roots of Eq. (24.16), one may predict, on the basis of the pair parameters used, complete $\langle 000 \rangle$ stability, one $\langle 100 \rangle$ instability surface, and two $\langle \frac{1}{2}\frac{1}{2}\frac{1}{2} \rangle$ and $\langle \frac{1}{2}\frac{1}{2}0 \rangle$ instability surfaces for this system. The $\langle 000 \rangle$ and $\langle \frac{1}{2}\frac{1}{2}0 \rangle$ surfaces have not been drawn, as the former has no positive-T portions, and the latter lie below their $\langle\frac{111}{222}\rangle$ counterparts. At point "1" in Fig. 50, the $\langle 100 \rangle$ wave is unstable with polarization along CD, giving rise to $L2_1$ ordering as shown in the upper portion of Fig. 2. At point "3" (or below) the $\langle\frac{111}{222}\rangle$ wave with polarization A–B is unstable, and ordering proceeds as shown in the lower portion of Fig. 2. The complete ordered structure, the so-called Heusler alloy structure, is shown in perspective in Fig. 2.

This type of ordering occurs in the vicinity of composition ABC_2 in bcc-based systems where A and B are Cu, Ag or Au, and C is Zn or Cd.[160] Another interesting example is that of the ternary system CuMnAl for which a spinodal phase separation occurs after the $\langle 100 \rangle$ and $\langle\frac{111}{222}\rangle$ ordering reactions have taken place.[225] In the fully ordered state, the structure can be described by four interpenetrating fcc sublattices: two of these are occupied by Cu atoms, one by Al atoms, and the fourth

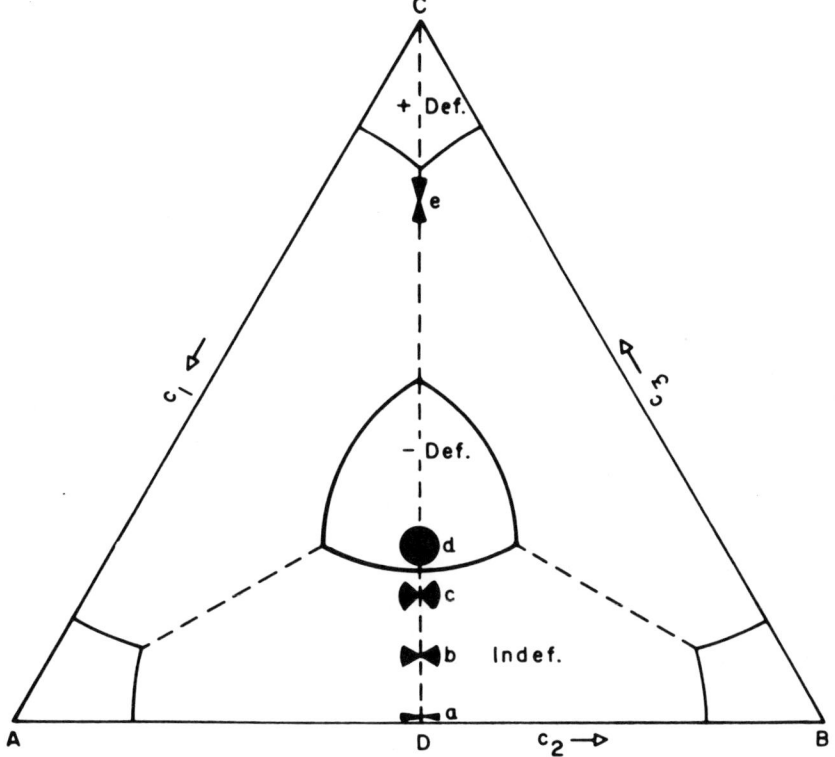

FIG. 49. Isothermal section through ternary spinodal surfaces for category IVB symmetric case [D. de Fontaine, *J. Phys. Chem. Solids* **34**, 1285 (1973)]. Decrease in free energy can occur for small composition fluctuations inside asymptotic directions represented by dark areas at selected points.

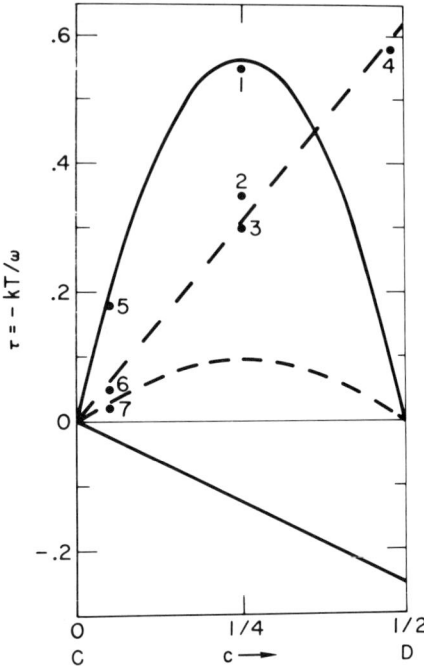

FIG. 50. Vertical section through ternary spinodals for ⟨100⟩ ordering (full line) and ⟨½½½⟩ ordering (dashed line) in bcc lattice [D. de Fontaine, *J. Phys. Chem. Solids* **34**, 1285 (1973)].

contains a distribution of Cu and Mn atoms, since the composition investigated was $AlMn_{0.5}Cu_{2.5}$, thereby containing an excess of Cu over the ABC_2 stoichiometry. At a temperature below that of the two successive ordering reactions, the fourth sublattice "spinodally decomposes" (see Section 26), i.e., a long-wavelength concentration wave of Cu–Mn polarization develops in the ⟨100⟩ space directions, creating alternating Cu-rich and Mn-rich regions. The remarkable electron micrograph obtained by Bouchard, reproduced in Fig. 51,[255] shows the modulated spinodal structure of about 150 Å wavelength and, superimposed, the ribbonlike contrast of antiphase boundaries produced during the ordering reactions.

The clustering system CuAgAu (category IIA) was discussed in Section 23, and the miscibility gap was shown in Fig. 43. The spinodal is indicated in Fig. 43b (dotted line) along with the eigenvector directions of the unstable mode. It is seen that, inside the spinodal, the eigenvectors lie

[255] M. Bouchard and G. Thomas, *Acta Metall.* **23**, 1485, (1975).

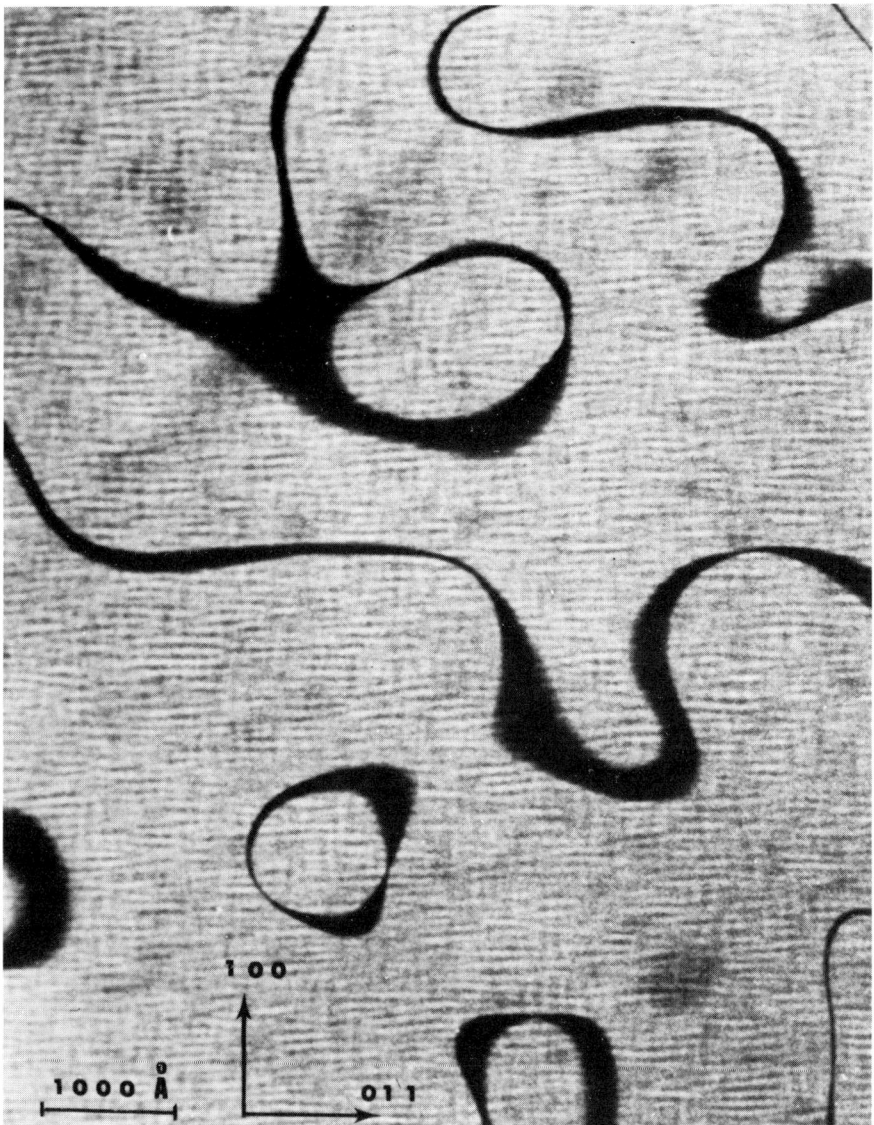

FIG. 51. (111) dark field micrograph of $Cu_{2.5}M_{0.5}Al$ aged 30 sec at 300°C showing ribbonlike ordering antiphase boundaries and phase separation spinodal modulations along $\langle 100 \rangle$ directions [M. Bouchard and G. Thomas, *Acta Metall.* **23**, 1485 (1975)].

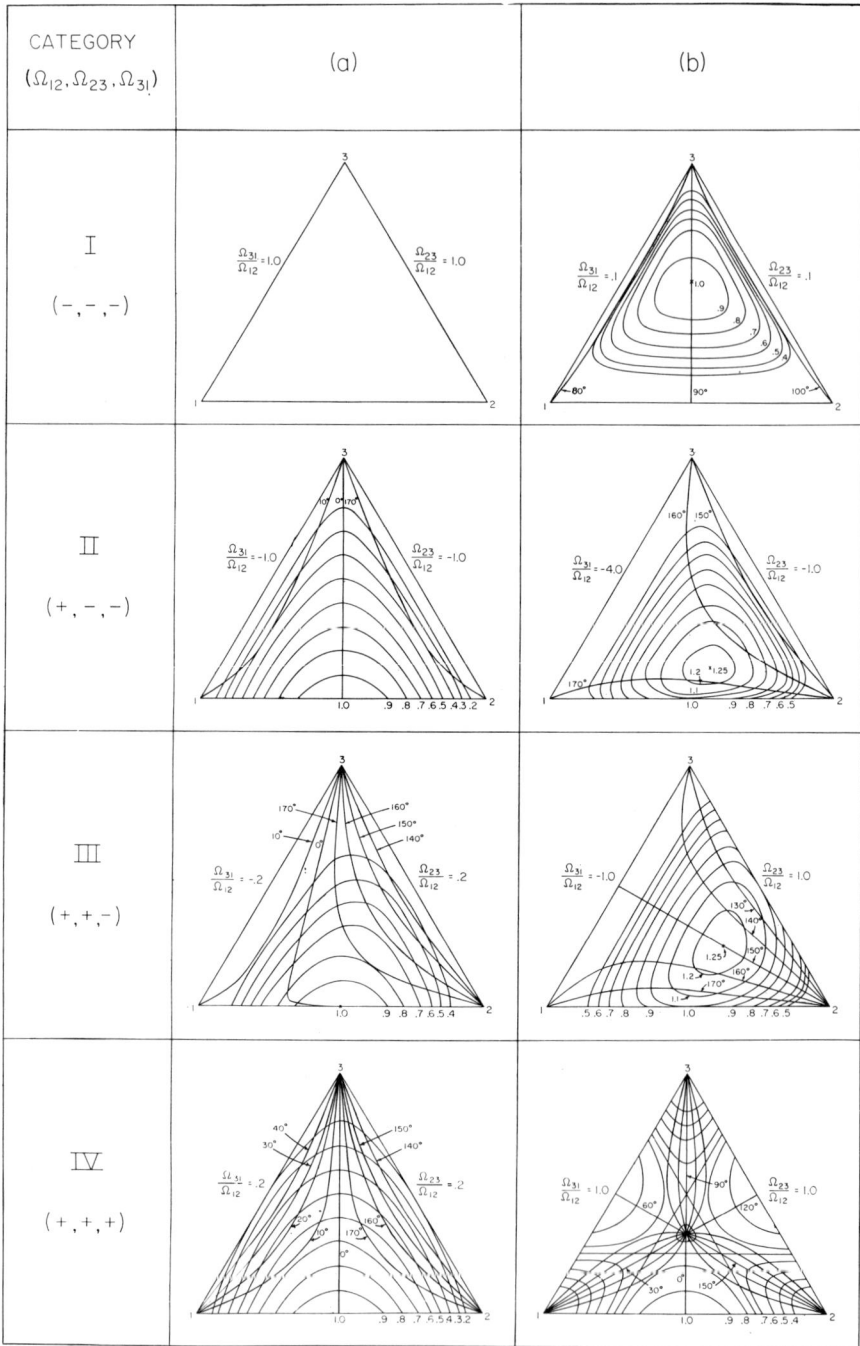

FIG. 52. Ternary spinodal and characteristic lines for each of the eight Meijering categories [J. E. Morral, *Acta Metall.* **20**, 1069 (1972)], for indicated values of Ω_{ij} interaction parameter ratios. Isotherms are labeled as T/T_c, principal direction orientations (as indicated by characteristic lines) are given in angles measured counterclockwise from the 1–2 binary side.

practically along the tie-lines, but swing away from these as the $\langle 000 \rangle$ stability surface is crossed, to become finally tangent to the MG line itself.

The above examples have shown that a great deal can be learned about a multicomponent system from stability analysis alone, a study which is much simpler to carry out than that of computing the complete phase diagram. For qualitative considerations, the BW formulation is often sufficient. A particularly convenient representation in the case of ternaries is that proposed by Morral[256] who plots the topmost instability surface by sets of contour lines on which *characteristic lines* are superimposed. The latter are lines along which the instability direction in composition space makes a constant angle with the sides of the Gibbs triangle. Examples of the eight Meijering categories are thus represented in Fig. 52.[256] Ratios of $W_{ij}(h)$ potentials used in the calculations are indicated. This procedure applies equally to ordering as to clustering reactions, as mentioned in connection with Eq. (24.16). The interested reader is referred to Morral's papers[256,257] for the use of a very elegant double-cone construction in analyzing stability of ternary BW models.

In principle, the techniques described above can be extended to quarternary and higher systems. The regular solution model was applied to quaternary systems by Meijering and Hardy[258,259] and can be generalized to quaternary ordering stability as discussed here for the ternary case. In particular, the conditions which cause a ternary clustering instability to be raised or lowered by the addition of a small quantity of a fourth component[258,259] can be applied as well to ordering instabilities. In quaternaries, asymptotic directions, defined by Eq. (24.15), form double cones which may degenerate into intersecting planes or straight lines.[3]

VII. Fluctuations and Kinetics

Thus far, atomic configurations have been regarded as static. For a more general description, however, the problem of fluctuations about an equilibrium state must be considered and also that of the rate of transformation from one state to another. The first of these two topics is considered in Section 25, while a special case of the second, namely the initial rate of change of an unstable solution, is described in Section 26. The general kinetic problem lies outside the scope of this article.

[256] J.E. Morral, *Acta Metall.* **20**, 1069 (1972).
[257] J.E. Morral, *Acta Metall.* **20**, 1061 (1972).
[258] H.K. Hardy and J.L. Meijering, *Philips Res. Rep.* **10**, 358 (1955).
[259] J.L. Meijering and H.K. Hardy, *Acta Metall.* **4**, 249 (1956).

25. Local Atomic Arrangements

We shall be primarily concerned with fluctuations in a disordered phase. Although, by definition, long-range order parameters vanish in the disordered state, certain atomic (or defect) configurations may be preferred locally over others; in other words, a certain amount of short-range order will be present at any finite temperature. As before, we have at our disposal two alternatives for treating the problem: (a) the wave method and (b) the cluster method.

a. Wave Method, General

Since variables used in the wave method are concentration amplitudes Γ_i, obtained by what has been termed microcanonical averaging (Section 2,b), only long-range order (LRO) can be described in this framework. Short-range order (SRO) can be introduced by the somewhat artificial procedure of considering locally ordered states as fluctuations on the unrealizable state of complete compositional uniformity. The probability that a given fluctuation be realized is then given by Eq. (16.1) with the free energy F of (16.3) being expressed in terms of concentration amplitudes:

$$\mathcal{P} = \mathcal{Z}^{-1} e^{-F\{\Gamma\}/k_B T}. \tag{25.1}$$

According to the approximation of Eq. (16.6), the partition function is written as

$$\mathcal{Z} = e^{-F_{eq}/k_B T} \tag{25.2}$$

where, in the present case, the equilibrium free energy F_{eq} in the disordered state is just the constant term $F^{(0)}$, since all variables in the Taylor's expansion (18.1) vanish. By combining Eqs. (25.1) and (25.2), we have

$$\mathcal{P} = e^{-\Delta F/k_B T} = e^{\Delta S/k_B} e^{-\Delta E/k_B T} \tag{25.3}$$

where the free energy difference is

$$\Delta F \equiv F - F_{eq} = F - F^{(0)} \cong F^{(2)}. \tag{25.4}$$

The latter approximation results from the neglect of third- and fourth-order terms in the concentration fluctuation amplitudes, which are assumed to be small. Hence, ΔF is given explicitly by Eq. (18.9) or (18.13) in matrix notation.

It is useful to define the "potential" $\Psi_k(h)$ conjugate to the variable $\Gamma_k(h)$ by

$$\Psi_k \equiv \frac{\partial \Delta F}{\partial \Gamma_k^*} = N \sum_{j=1}^{n} F_{kj} \Gamma_j. \tag{25.5}$$

Then the expectation value of the product of a variable times a potential at a particular h-point is given by[260]

$$\langle \Gamma_l^* \Psi_k \rangle \equiv \sum_{\{\Gamma\}} \Gamma_l^* \Psi_k e^{-\Delta F/k_B T} = \delta_{lk} k_B T, \qquad (25.6)$$

a very compact result. The expectation value of a product of amplitudes (at given h) is obtained by combining Eqs. (25.5) and (25.6):

$$\langle \Gamma_i^* \Gamma_j \rangle = (k_B T/N) F_{ij}^{-1}, \qquad (25.7)$$

where F_{ij}^{-1} denotes an element of the inverse matrix $\mathbf{F}^{-1}$. Alternatively, one may diagonalize the free energy quadratic as was done in Eq. (24.7) to obtain the expectation value of normal concentration modes

$$\langle |\bar{\Gamma}_i|^2 \rangle = k_B T/N\Lambda_i \qquad (25.8)$$

where $\Lambda_i(h)$ are the eigenvalues of $\mathbf{F}$ for the wave vector $\mathbf{k}(h)$ considered.

It is thus seen that SRO intensity is given by the expectation value of a product of LRO parameters: although the average values of the concentration amplitudes vanish identically in the disordered state, their convariance does not. Formula (25.7) or (25.8) shows that SRO intensity becomes unbounded at an instability limit. Actually, the SRO spectrum should approach a δ-function, giving rise to infinite-range correlations in direct space. Unfortunately, these simple formulas fail to produce a δ-function. This is because the superposition approximation (11.21) has been made implicitly without the use of the corrective factor $|S|^2$, hence the integrated intensity is not conserved and incorrectly increases without limit as the uppermost instability locus is approached from above. Below this locus the model predicts negative intensity, which is of course meaningless. Consequently, the mean-field "wave" treatment must be used with caution. Several modifications have been introduced in order to conserve integrated intensity (see below) but even with these alterations, the theory cannot be used very reliably in the immediate vicinity of a higher-order critical point. Above a first-order transition, anharmonic effects are expected to play a role, so that the cluster method appears to be preferable (see Section 25,d).

b. Binary Clustering

For binary solutions, Eq. (25.8) gives simply

$$\langle |\Gamma(h)|^2 \rangle = k_B T/NF''(h) \qquad (25.9)$$

---

[260] L.D. Landau and E.M. Lifshitz, "Statistical Physics," p.366. Addison-Wesley, Reading, Massachusetts, 1958.

where F'' is the k-space second derivative of the free energy with respect to concentration amplitude squared. In the BW approximation, F'' is given simply by Eq. (19.8). For phase separation systems, it is advantageous to use a long-wavelength expansion for F'' as was done in Section 9. For this purpose, one adds a bulk entropy contribution to the internal energy given by Eq. (9.5) to obtain, in the harmonic approximation,

$$\Delta F(\mathbf{k}) = F^{(2)} = \frac{V_N}{2} \sum_{\mathbf{k}} [f_0'' + 2\kappa k^2] |\Gamma(\mathbf{k})|^2 \qquad (25.10)$$

following the method used in deriving Eq. (9.10). In Eq. (25.10), f_0'' is the second derivative of the bulk free energy with respect to concentration squared, as in Eq. (24.8), the subscript indicating that the derivative must be evaluated at the average concentration. The gradient energy coefficient κ is given by Eq. (9.11), and V_N is the volume occupied by a crystal of N lattice sites. With these definitions, Eq. (25.10) is strictly valid only for fluids or for solids of cubic symmetry with negligible solute misfit. If misfit is present, then an elastic energy contribution must be added to the "incoherent" free energy, as shown by Cahn[217] and as indicated in Eq. (9.7). For cubic crystals in the long-wavelength limit, the required elastic energy contributions is, by Eq. (12.8) and (12.14),

$$E_e = V_N \eta^2 Y(\mathbf{k}) |\Gamma(k)|^2 \qquad (25.11)$$

where the effective modulus Y, depending only on the directional cosines of the wave vector $\mathbf{k}$, is defined by (12.14) but may be approximated by (12.17) or (12.18), as the case may be. By combining (18.15), (25.10), and (25.11), we finally get

$$F''(h) = \Omega_0 [f_0'' + 2\eta^2 Y(\mathbf{k}) + 2\kappa k^2], \qquad (25.12)$$

a formula first derived by Cahn.[89]

For binary fluid solutions the misfit parameter η vanishes and Eqs. (25.9) and (25.12) lead to the Landau formula[260]

$$\langle |\Gamma|^2 \rangle = \frac{k_B T}{V_N (f_0'' + 2\kappa k^2)} = \frac{k_B T}{2\kappa V_N} \frac{1}{K^2 + k^2} \qquad (25.13)$$

where

$$K^{-1} = \sqrt{2\kappa/f_0''} \qquad (25.14)$$

is a correlation length. The fluctuation spectrum given by the Landau formula has the Lorenzian shape $[1 + (k/K)^2]^{-1}$ about the origin, as shown in Fig. 53. The width of the peak at half-height is the inverse correlation length K.

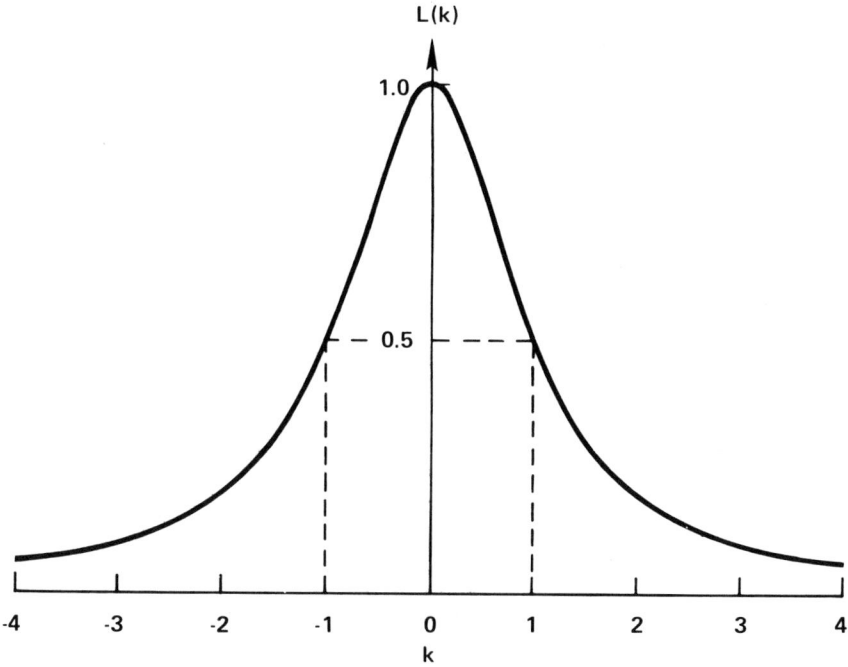

Fig. 53. Lorenzian profile $(1 + k^2)^{-1}$.

The Landau formula can be back-transformed into (isotropic) direct space to yield the pair correlation function as a function of radial distance r.[260]

$$q(r) \propto e^{-Kr}/r. \qquad (25.15)$$

This formula shows that, at an instability ($f_0'' = 0$), the correlation length becomes infinite and the correlation function behaves as $1/r$. At infinite temperature, the correlation length vanishes, as expected, and the correlation function behaves as a δ-function.

In solids, it is difficult to test clustering fluctuation theories because, as first pointed out by Cahn,[217] critical fluctuations must exist only above the coherent critical point, which, because of coherency strains, will always be located at a temperature lower than the top of the incoherent miscibility gap. To reveal large-amplitude fluctuations, one must perform scattering experiments just above the coherent critical point, i.e., inside the incoherent miscibility gap. It follows that incoherent (heterogeneous) precipitation may occur during the measurement, thereby tending to mask the effect one wishes to observe. Despite such difficulties, a suc-

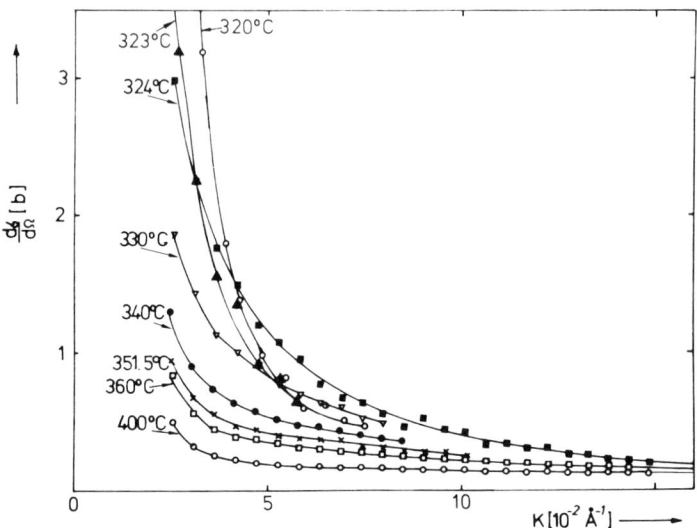

FIG. 54. Diffuse elastic neutron scattering cross section as a function of scattering vector at various temperatures above the coherent critical point in Al–Zn solid solution [D. Schwahn, Ph.D. Dissertation, Ruhr Universität, Bochum (1976); D. Schwahn and W. Schmatz, *Acta Metall.* **26**, 1571 (1978)].

cessful neutron diffraction experiment was performed at the Al–Zn critical composition by Schwahn and Schmatz.[261] Figure 54, taken from Schwahn's dissertation, shows a plot of diffuse, elastic, coherent neutron scattering cross sections as a function of the scattering vector $|k(h)|$ at various temperatures above the coherent critical point. It can be seen that there is no dramatic change in the nature of the fluctuation intensity at 351.5°C, the top of the incoherent miscibility gap (see Fig. 36). The scattering curves for the lower temperatures recorded do indicate the proximity of the true, or coherent, critical point which must be located between 324 and 323°C, in good agreement with the calculated coherent solvus shown in Fig. 36. Thus, the experiment of Schwahn seems to have provided the first direct verification of Cahn's fluctuation theory.

Al–Zn solutions at about 40 at. % Zn are very nearly elastically isotropic. In anisotropic systems, diffuse scattering above a critical point of unmixing should be highly anisotropic due to the presence of the $Y(\mathbf{k})$ term in the denominator of the modified Landau formula, obtained by inserting expression (25.12) for F'' into Eq. (25.9). In fact, diffuse scat-

---

[261] D. Schwahn, Ph.D. Dissertation, Ruhr Universität, Bochum (1976); D. Schwahn and W. Schmatz, *J. Phys. (Paris)* **38**, C7-411 (1977); D. Schwahn and W. Schmatz, *Acta Metall.* **26**, 1571 (1978).

tering should be confined to $\langle 100 \rangle$ streaks in the usual case of positive anisotropy [$A > 0$, see Eq. (11.13)], and to $\langle 111 \rangle$ streaks in the opposite case ($A < 0$). Although such streaks are often observed by electron diffraction, quantitative X-ray or neutron diffraction measurements have not been performed to date. Another interesting neutron diffuse scattering experiment is that performed by Mozer, Keating, and Moss[262] on $^{65}Cu_{0.435}$ $^{62}Ni_{0.565}$ by the so-called "null matrix" technique. More recent experiments on this system will be described below.

c. Binary Ordering

For ordering binaries, it is simplest to substitute (19.8) for F'' into Eq. (25.9). The result may be written as

$$\langle A(h) \rangle = \frac{C}{1 + q°V(h)/k_B T} \quad (25.16)$$

where A denotes the Fourier transform of the Warren–Cowley SRO parameters α, defined by Eq. (2.8). SRO intensities $A(h)$ are related to concentration wave amplitudes by

$$A = N|\Gamma|^2/q° \quad (25.17)$$

where $q°$, defined by Eq. (2.7), is here equal to the product $\bar{c}(1 - \bar{c})$. Formula (25.16) was derived independently by Krivoglaz[263] and by Clapp and Moss.[264] The constant C was introduced by the latter authors in order to take care of the normalization condition (3.9), written here as

$$\sum_h \langle A(h) \rangle = N, \quad (25.18)$$

which determines the value of C to be used at any given temperature above the uppermost instability. Other attempts at normalization include the use of a particular k-space entropy model,[265] or the introduction of a Lagrange multiplier in the denominator of (25.16) by analogy with the treatment of Bose–Einstein condensation.[266]

According to the Krivoglaz–Clapp–Moss (KCM) formula (25.16), the SRO intensity $A(h)$ will tend to peak at those points $\mathbf{k}(h)$ in reciprocal space where $V(h)$ takes on minimal values. In the vicinity of such min-

[262] B. Mozer, D.T. Keating, and S.C. Moss, *Phys. Rev.* **175**, 868 (1968).
[263] M.A. Krivoglaz, "Theory of X-ray and Thermal Neutron Scattering by Real Crystals." Plenum, New York, 1969.
[264] P.C. Clapp and S.C. Moss, *Phys. Rev.* **142**, 418 (1966).
[265] D.W. Hoffman, *Met. Trans.* **3**, 3231 (1972).
[266] D. de Fontaine, in "Critical Phenomena in Alloys, Magnets and Superconductors" (R.E. Mills, E. Ascher, and R.I. Jaffee, eds.), p. 277. McGraw-Hill, New York, 1971.

ima, the intensity peaks take on Lorenzian shape as shown readily by expanding the k-space potential $V(h)$ to second order in reciprocal space coordinates about the location $\mathbf{k}(h°)$ of the minima. Substituting (8.2) into (25.16) one thus obtains[49]

$$\langle A(H) \rangle = \frac{CT}{\Delta T}[1 + (q^0/2k_B\Delta T)\lambda_\alpha(h°)H_\alpha^2]^{-1} \qquad (25.19)$$

where $\Delta T = T - T_0$, T_0 being the instability temperature, and λ_α and H_α were defined in Eq. (8.2), summation over repeating Cartesian subscripts ($\alpha = 1, 2, 3$) being implied. By comparing (25.19) to (25.13), one sees that the predicted SRO intensity peaks have Lorenzian shape along any one of the principal directions of $V(h)$ at the minimum $\mathbf{k}(h)$. At ordering points, $V(h)$ need no longer have cubic symmetry, as in the $\langle 000 \rangle$ unmixing case, so that the width of the peak may differ along various k-space directions. One may then take the isointensity contour at half-peak height

$$\langle A(H) \rangle = CT/(2k_B\Delta T) \qquad (25.20)$$

as defining a characteristic ellipsoid for the short-range order intensity peak.[49]

$$(H_1/K_1)^2 + (H_2/K_2)^2 + (H_3/K_3)^2 = 1 \qquad (25.21)$$

where, as in Eq. (25.14), the semiaxes K_α are the inverse correlation lengths given explicitly by

$$K_\alpha = \sqrt{\frac{k_B(T - T_0)}{q_0\lambda_\alpha}} \qquad (25.22)$$

This equation has meaning only for $T > T_0$ and for $\lambda_\alpha > 0$. Formula (25.22) shows that the peak width broadens without limit along H_α as the associated eigenvalue λ_α [or the curvature of the $V(h)$ surface] along that principal value goes to zero, and also shows that the characteristic shape of the intensity peak, i.e., the eccentricity of the ellipsoid (25.21), depends only on the ratio of the eigenvalues λ_α but not on the temperature or average concentration. This property has been observed experimentally.[66]

The above analysis of peak shape is particularly useful when applied to the special points in k-space. At these points, the axes of the characteristic ellipsoid are directed along the principal directions given in Table IV, and the eigenvalues can be evaluated directly from Eq. (8.7), with the help of eigenvalue tables,[49] if the pair interaction parameters $v^{(s)}$ are known. Alternately, the values of these parameters may be estimated so

CONFIGURATIONAL THERMODYNAMICS OF SOLID SOLUTIONS   239

as to match the eccentricity of an experimentally determined diffuse intensity peak. One must then solve the (generally underdetermined) linear system

$$\sum_s v^{(s)}\lambda_\alpha^{(s)} = \Delta T/(q^0 K_\alpha^2). \qquad (25.23)$$

Equation (25.23) shows that the minimum number of pair interaction parameters needed for a fit at points of cubic, tetragonal (and trigonal), and orthorhombic symmetry is one, two, and three, respectively.

A simple example will illustrate the procedure: Moss and Clapp[66] chose $V_1 = 1$, $V_2 = -0.2$, and $V_s = 0$ ($s > 2$) on a relative scale, for their computation of theoretical intensity profiles according to Eq. (25.16) for the case of the (100) diffuse intensity maxima peaks in Cu_3Au. With these values, the eigenvalues calculated from Eq. (8.7) are found to be

$$\lambda_1 = 1\cdot 2 + 0.2\cdot 2 = 2.4$$

$$\lambda_2 = \lambda_3 = 0.2\cdot 2 = 0.4. \qquad (25.24)$$

The tetragonal wave vector group symmetry at the fcc special point (100) indicates that the diffraction spot is a circular disk with axis along the [100] direction. The eccentricity of the characteristic ellipse in an $h_1 h_2 0$ section is then, by Eqs. (25.22) and (25.24),

$$K_1/K_2 = \sqrt{\lambda_2/\lambda_1} = 0.408, \qquad (25.25)$$

which agrees well with the ratio of semiaxes of the elliptical intensity contours about the $\langle 100 \rangle$ superlattice positions seen in Fig. 55. This classical comparison between SRO X-ray intensity contours produced by Moss and Clapp from formula (25.16) illustrates quite well the usefulness of the latter formula.

Variants of the KCM formula have also been used. For instance, Walker and Keating[267] obtained a very good fit of their neutron diffraction data to the Zernike formula for the case of critical fluctuations above the higher-order transition of near-stoichiometric CnZn. Later, Als-Nielsen and Dietrich[268] fitted neutron diffraction data in the same system to the formula of Fisher and Burford.[269] More generally, Wilkins[67] has shown that the different formulas used may be written

$$\langle A(h) \rangle = \frac{[1 + G_1(T)V(h)]^{-1}}{\sum_h [1 + G_1(T)V(h)]^{-1}} \qquad (25.26)$$

[267] C.B. Walker and D.T. Keating, *Phys. Rev.* **130**, 1726 (1963).
[268] J. Als-Nielsen and O.W. Dietrich, *Phys. Rev.* **153**, 706 and 717 (1967).
[269] M.E. Fisher and R.J. Burford, *Phys. Rev.* **156**, 583 (1967).

which differs from Eq. (25.16) merely by the temperature dependence, here left unspecified, of the coefficient of the k-space potential. It is seen that the SRO intensity spectrum depends only on the function $G_1(T)V(h)$, i.e., on the quantities G_1V_1, G_1V_2, G_1V_3.... It follows that the ratios V_2/V_1, V_3/V_1... do not depend on the model chosen for the temperature dependence, and these ratios can thus be determined uniquely from experimental SRO intensity.

Wilkins[67] used this property of Eq. (25.26) to determine the pair parameter ratios to seventh coordination shell from the Cu_3Au X-ray data

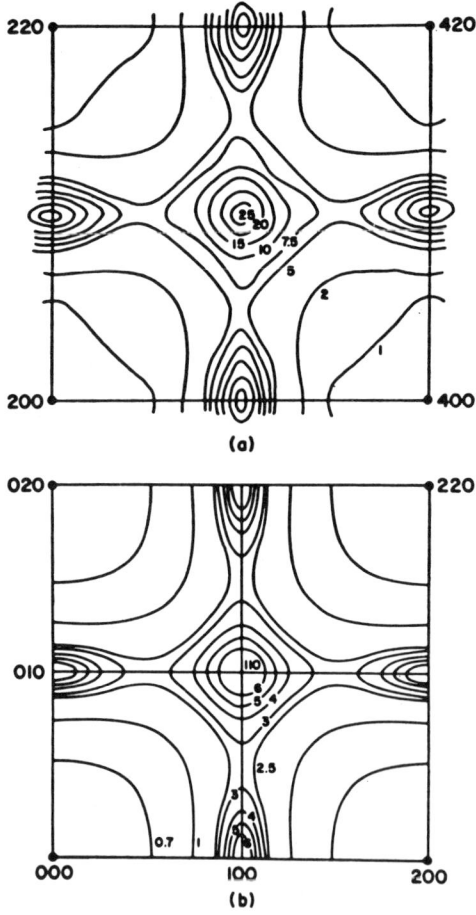

FIG. 55. Diffuse intensity contours in Cu_3Au in $(h_1 h_2 0)$ plane at 405°C; (a) experimental measurements, (b) contours calculated according to Eq. (25.16) with $V_2/V_1 = -0.2$, $V_i = 0$ for $i \geq 3$ [S. C. Moss and P. C. Clapp, *Phys. Rev.* **171**, 764 (1968)].

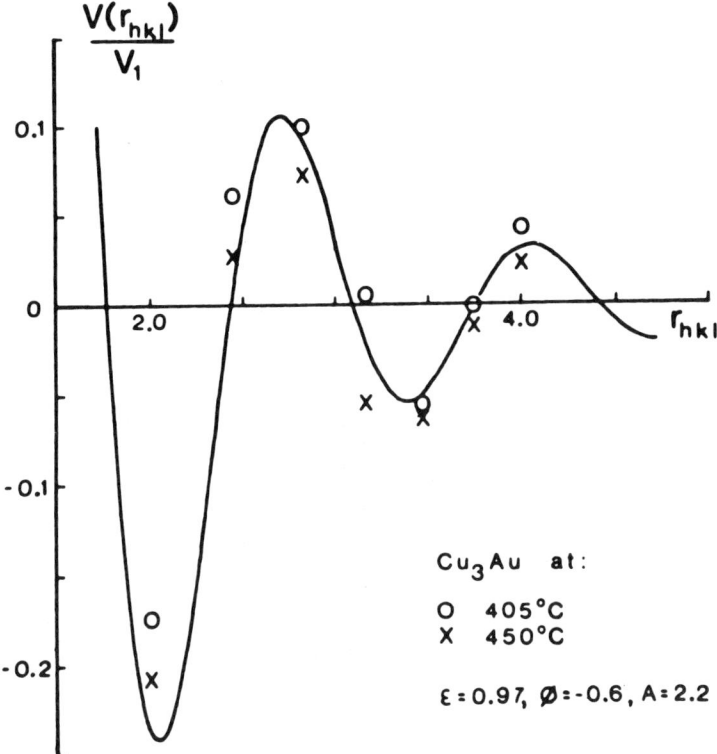

FIG. 56. Fit [S. W. Wilkins, *Phys. Rev. B* **2**, 3935 (1970)] of oscillating potential, Eq. (7.3) to pair interaction ratios V_i/V_1 deduced from experimental SRO parameters of Moss [S. C. Moss, *J. Appl. Phys.* **35**, 3547 (1964)] for Cu_3Au at two temperatures.

of Moss.[270] A plot of these ratios vs radial distance is shown in Fig. 56,[67] each set of points representing a different manner of handling the data. The continuous curve, representing the oscillatory potential $\omega(r)_{asy}$ of Eq. (7.3), was made to pass through the plotted points by fitting the parameters A (amplitude) and φ (phase shift) of this "Friedel oscillation" formula.

Clearly, measurements of SRO intensity can provide very important information concerning alloy potentials. The experiments are not easy to perform, however: very accurate diffracted intensity is required over large and peculiar-shaped volumes in reciprocal space for the purpose of removing from the total diffuse intensity that due to displacive effects, as discussed in Section 14. The interested reader is referred to articles

[270] S.C. Moss, *J. Appl. Phys.* **35**, 3547 (1964).

by Cohen and co-workers[13,132,271] for a treatment of experimental techniques, displacive corrections and interpretation of the data. Systems analyzed by these methods include, besides those already mentioned, $CuAu$,[272] Au_3Cu,[273] Cu-$14.5Al$,[274] $AuPd$,[275] Ni_4Mo,[250] Co_3Pt,[276] and various oxides.[277,278]

Recent work by Bardhan and Cohen[279] on that old war-horse, Cu_3Au, over a wide temperature range, has yielded some unexpected results. It appears that, although maximum diffuse intensity is always located at the $\langle 100 \rangle$ positions, $\langle 1\tfrac{1}{2}0 \rangle$ subsidiary maxima appear below 600°C, and $\langle \tfrac{1}{2}\tfrac{1}{2}\tfrac{1}{2} \rangle$ maxima appear above 850°C. Hence, all three special-point ordering wave vectors are evidenced in this system. Neutron diffraction has revealed other anomalies, from which it was concluded[280,281] that many-body (or anharmonic) effects play an important role just above the transition temperature in Cu_3Au. Likewise, CuNi alloys with compositions 20 to 80 at. % Ni, have recently been investigated in careful neutron diffraction experiments by Radelaar and co-workers.[282–282a] The results indicated that the diffuse SRO intensity data, for a range of compositions and temperatures, could not be interpreted on the basis of Eq. (25.16) by means of constant pair interaction parameters. Since the same anomalies were not detected in a corresponding Monte Carlo simulation (based on a pairwise interaction model),[282] the authors[282a] concluded that manybody interactions were playing a significant role in this system. A self-amplifying effect on the SRO parameters with decreasing temperature was also attributed to multiplet interactions.

A particularly illuminating way of plotting the data, first suggested by Cook,[283] is that shown in Fig. 57.[280] If fluctuations obeyed the KCM formula (25.16), I/I_{SRO} vs $(1 - T_c/T)$ should plot as a straight line

---

[271] J.B. Cohen, in "Phase Transformations" (H.I. Aaronson, ed.), Chapter 13, p. 561. Am. Soc. Metals, Metals Park, Ohio, 1970.
[272] B.W. Roberts, *Acta Metall.* **2**, 597 (1954).
[273] B.W. Batterman, *J. Appl. Phys.* **28**, 556 (1957).
[274] C.R. Houska and B.L. Averbach, *J. Appl. Phys.* **30**, 1525 (1959).
[275] W. Lin, J.E. Spruiell, and R.O. Williams, *J. Appl. Crystallogr.* **3**, 297 (1970).
[276] H. Berg, Jr. and J.B. Cohen, *Acta Metall.* **21**, 1579 (1973).
[277] M. Brunel and F. de Bergevin, *J. Phys. Chem. Solids* **30**, 2011 (1969).
[278] F. Koch and J.B. Cohen, *Acta Crystallogr., Sect. B* **25**, 275 (1969).
[279] P. Bardhan and J.B. Cohen, *Acta Crystallogr., Sect. A* **32**, 597 (1976).
[280] P. Bardhan, H. Chen, and J.B. Cohen, *Philos. Mag.* [8] **35**, 1653 (1977).
[281] H. Chen, J.B. Cohen, and R. Ghosh, *J. Phys. Chem. Solids* **38**, 855 (1977).
[282] J. Vrijen, C. van Dijk, E. W. van Royen, and S. Radelaar, *J. Phys. (Paris)* **38**, C7-341 (1977).
[282a] J. Vrijen, E.W. van Royen, D.W. Hoffman, and S. Radelaar, *J. Phys. (Paris)* **38**, C7-187 (1977).
[283] H.E. Cook, *Mater. Sci. Eng.* **25**, 127 (1976).

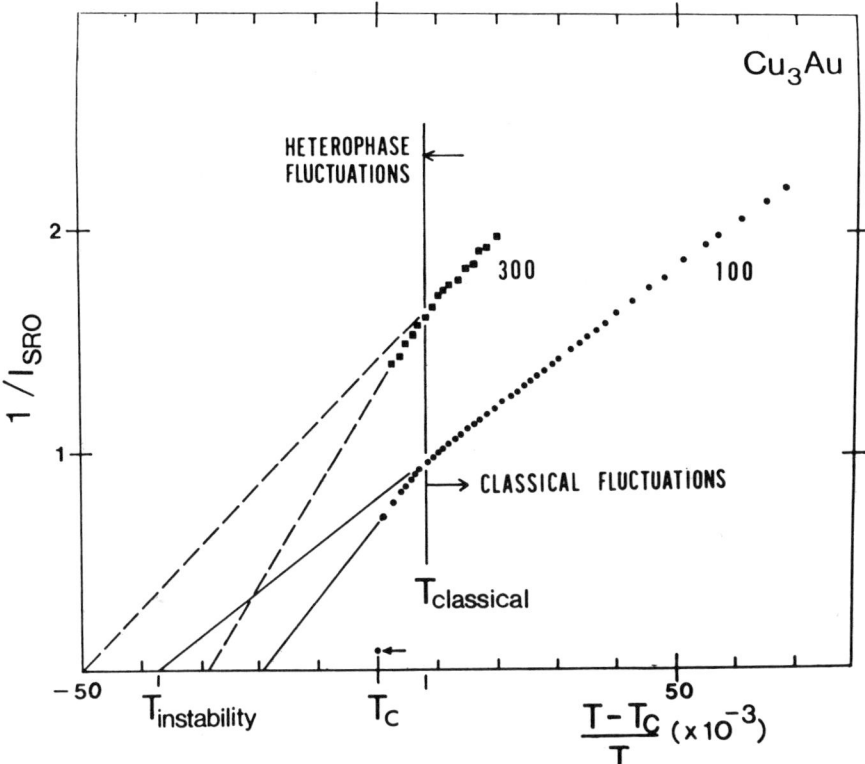

FIG. 57. Reciprocal of SRO intensity plotted vs reduced temperature for Cu_3Au at 100 and 300 superlattice positions [P. Bardhan, H. Chen, and J. B. Cohen, *Philos. Mag.* [8] **35**, 1653 (1977)]. Experimental points follow linear behavior predicted by Eq. (25.16) in "classical" region, extrapolating to instability temperature T_0.

extrapolating to the $\langle 100 \rangle$ instability temperature. At high enough temperature, this is indeed the case, but close to T_c (actually a first-order transition T_t), there is marked departure from the straight line. This "nonclassical" behavior of the fluctuations was interpreted[280] as evidence for the existence of "heterophase fluctuations" characterized by small metastable Cu_3Au clusters having relatively long lifetimes. This view is justified by referring to Fig. 47c: although above T_t the state of lowest free energy is that with vanishing long-range order parameter ξ, very close to T_t, a local minimum in the free energy function exists for values of ξ in the range 0.4 to 0.6 approximately. Indeed, in both CV and BW models, the LRO curve can be followed a short way into the disordered state, and small regions of the system can exist in a metastable state with finite LRO. It appears that the presence of small ordered clusters just above a first-order transition could best be studied by the CV method, but such an investigation has not yet been carried out systematically on large enough clusters. Some presently known results are described below (Section 25,d).

The KCM fluctuation formula (25.16) can of course also be applied to interstitial solutions, as was done by Khachaturyan and co-workers[150,284] for the case of carbon in martensite. The function $V(h)$ was calculated according to the methods outlined in Section 11, under the assumption that carbon-lattice interactions were of a purely elastic nature. Good agreement between calculated diffuse intensity contours and experimental electron diffraction patterns was obtained.[150]

d. Local Order: Cluster Method

Since expectation values $(y_{ij}, t_{ijk}, z_{ijkl}, \ldots)$ at equilibrium are calculated in the CV method, SRO information is obtained directly without the necessity of considering local order as fluctuations on an artificial state of complete compositional uniformity. Another advantage of the CVM over the wave method is that SRO can be obtained for the ordered as well as the disordered states, which is not permissible in the KCM formulation (25.16).

Thus far, the CVM has yielded information mostly on very short range order because of the practical difficulty of treating large clusters. Kikuchi and Sato[197] have plotted nearest-neighbor pair correlations as a function of temperature above both the first-order transition temperature in fcc (CuAu-type) and the higher-order transition in bcc (CuZn-type). Golosov et al.[233] have tabulated tetrahedron probabilities in disordered bcc and in the ordered structures B2 and D0$_3$, and, for fcc-based systems, in the disordered state at $\bar{c} = \frac{1}{2}$ and $\frac{1}{4}$, and in the ordered structures L1$_2$ and L1$_0$.

A more elaborate calculation was recently undertaken by Sanchez[196] in order to treat the "omega" displacive transformation by the CVM as a ternary order–disorder replacive transformation. The so-called ω-phase is a metastable structure which is produced upon rapid cooling of certain bcc (β) Ti, Zr, and Hf alloys. This structure results from formally subjecting the β-phase to a longitudinal static displacement wave of wave vector $\frac{2}{3} \langle 111 \rangle$,[285,286] which results in alternate {111} plane collapse as

---

[284] A.G. Khachaturyan and T.A. Onissimova, *Fix. Met. Metalloved.* **26**, 973 (1968).
[285] D. de Fontaine, *Acta Metall.* **18**, 275 (1970).
[286] D. de Fontaine, N.E. Paton, and J.C. Williams, *Acta Metall.* **19**, 1153 (1971).

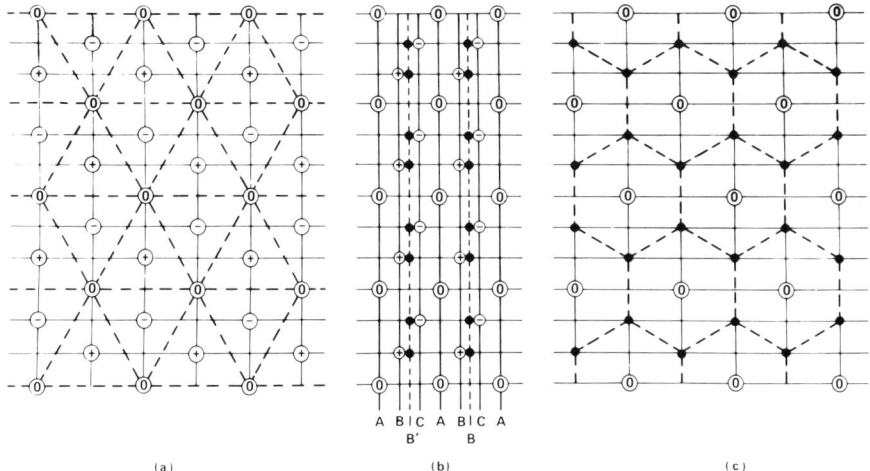

FIG. 58. Beta and omega stacking sequences: (a) projection of bcc (β) structure on (111) plane, (b) edge-on projection of (111) planes in β and ω (dotted) structures, (c) basal plane projection of ω structure [D. de Fontaine and O. Buck, *Philos. Mag.* [8] **27**, 967 (1973)].

indicated in Fig. 58[287]: on the left, a [111] projection of the bcc structure, on the right, the ω structure, with the plane collapse mechanism indicated in the center of the figure. Since, for a given ω variant ([111] direction), atomic displacements responsible for the $\beta \to \omega$ transition are: forward displacements by one half the (111) interplanar spacing, or backward by the same amount, or no displacement at all, the transformation can be modeled by assigning operators $+1$, -1, or 0 to each lattice point, as is done in the ternary replacive case. The smallest cluster that can be used successfully here is the eight-point bcc primitive unit cell with long diagonal along the chosen [111] direction. An "ω-cluster", i.e., the smallest unit of ω structure, is shown in Fig. 59 with atomic displacements indicated. Although the total number of clusters (not all distinct) was $3^8 = 6561$, equal to the number of simultaneous nonlinear equations to be solved, Sanchez[196] was able to compute cluster probabilities in the disordered state, by using a modification of the natural iteration method described in Section 20,b. The temperature dependence of the ω-cluster concentration is shown in Fig. 60,[196] and was used to explain anomalous diffusion in Ti, Zr, and Hf bcc metals and alloys. Now that methods have been developed to handle such large clusters, which contain first, second, third, and fourth neighbor pairs, one can expect increasing use of the

[287] D. de Fontaine and O. Buck, *Philos. Mag.* [8] **27**, 967 (1973).

ω CLUSTERS

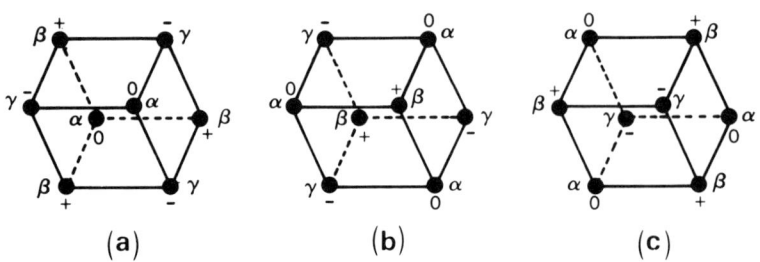

ACTIVATED COMPLEX

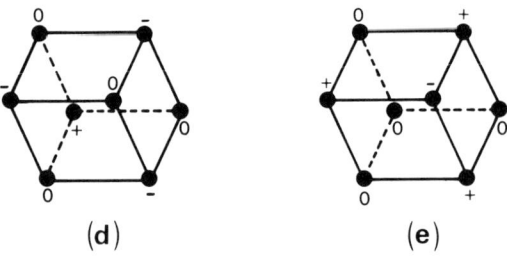

FIG. 59. bcc primitive unit cell and [111] atomic displacements arranged to form the three variants of the ω cluster (a,b,c) and the activated complex for diffusion (ω embryo, d and e) [J. M. Sanchez, Ph.D. Dissertation, School of Engineering and Applied Science, University of California, Los Angeles (1977); J. M. Sanchez and D. de Fontaine, *Acta Metall.* **26**, 1083 (1978)].

CVM to describe SRO in both disordered and ordered states. In other recent developments, Cook[84] was able to extend the KCM formation to fluctuations above a first-order transformation; in that case, the Fourier coefficients must be interpreted as representing large-amplitude quasi-static wave packets. This concept was applied by Cook to ω-like heterophase fluctuations[85] and then, in a return to second-order displacive transformations, to the problem of the central phonon peak in soft-mode transitions.[288]

A very interesting cluster method was developed by Clapp[9-11] for the purpose of determining cluster probabilities in a given solid solution from

---

[288] H.E. Cook, *Phys. Rev. B* **15**, 1477 (1977).

a knowledge of (experimentally determined) pair probabilities. Clapp pointed out that his method is, in a sense, the inverse of the CVM, and therefore called it the probability variation method (PVM). Although the basic equations were given an independent derivation in the original articles, it is instructive here to start from the CVM, as this clearly indicates the approximations which are made. One begins by postulating that the configurational internal energy depends exclusively on pair probabilities $y_{ij}(r)$, for arbitrary pair separations $x(r)$, but not on many-body interactions such as triplet (t_{ijk}), quadruplet (z_{ijkl}), etc. probabilities. Let the probabilities for the largest cluster used be denoted by $w_{ijklm...}$. One then writes the free energy according to the CVM as

$$F = E\{y_{ij}(r)\} - TS\{x_i, y_{ij}(r), t_{ijk}, z_{ijkl}, \ldots, w_{ijklm...}\}, \qquad (25.27)$$

as in Eq. (20.25), for example. In the present case, the average concentrations x_i are known (by chemical analysis, say) and so are the Warren–Cowley SRO parameters $\alpha_{ij}(r)$, determined by a Fourier back transform of suitably corrected SRO intensity, known from X-ray or neutron diffraction experiments. The SRO parameters in turn yield the pair probabilities $y_{ij}(r)$ by the linear equation

$$\alpha_{ij}(r) = \frac{y_{ij}(r) - x_i x_j}{x_i \delta_{ij} - x_i x_j} \qquad (25.28)$$

obtained from Eqs. (2.4), (2.7), and (2.8) rewritten in the CVM notation.

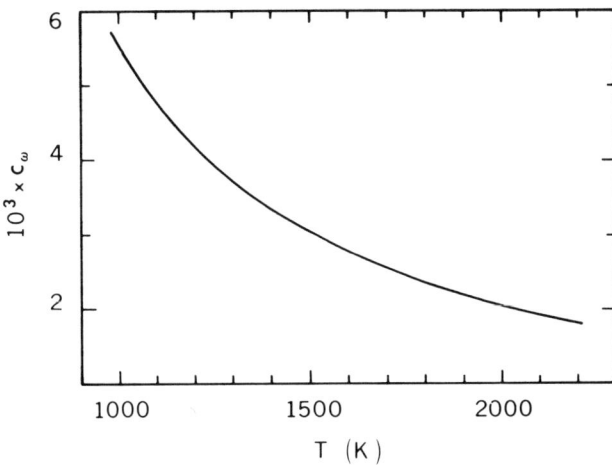

FIG. 60. Concentration of ω embryo (Fig. 59d and e) vs absolute temperature calculated by the CVM applied to the omega displacive transformation [J. M. Sanchez, Ph.D. Dissertation, School of Engineering and Applied Science, University of California, Los Angeles (1977); J. M. Sanchez and D. de Fontaine, *Acta Metall.* **26**, 1083 (1978)].

Just as was done in Section 20,b, the free energy (25.27) is minimized with respect to the higher cluster probabilities, but now Eqs. (20.22) and (20.23) (which must be rewritten in terms of summations of $w_{ijklm...}$ variables) are fixed constraints for which Lagrange multipliers must be introduced alongside that corresponding to Eq. (20.21). In performing the minimization of the free energy (25.27), subject to the imposed constraints, the internal energy drops out, since it has fixed values along with $y_{ij}(r)$, so that one is really seeking the values of the cluster variables (beyond pairs) which maximize the entropy S, the latter being given by standard CVM formulas which depend only on the lattice type and on the largest cluster retained.

The PVM thus determines uniquely all higher cluster probabilities, or triplet, quadruplet, etc. correlations up to a given higher member, from a knowledge of point and pair probabilities. The method is exact insofar as the internal energy depends exclusively on pair correlations and insofar as the CVM entropy approximates closely the true configurational entropy. Actually, the latter requirement is too strong; it is only necessary that the values t^*_{ijk}, z^*_{ijkl}, ..., $w^*_{ijklm...}$, which maximize the model entropy, lie very close to those which maximize the real configurational entropy. As for the internal energy assumption, its validity may be questioned in view of the importance of including many-body interactions, as was demonstrated in the derivation of the CuAu phase diagram (see Section 23,b). Nevertheless, the results obtained thus far by the PVM (see below) appear to be as accurate as the experimentally determined SRO parameters currently warrant.

A PVM calculation based directly on the method outlined above has not yet been performed. The reason was, probably, that the CVM entropy formulas can be difficult to derive whenever the clusters are large and, more practically, that the number of cluster probabilities become excessively large. Clapp avoided the first difficulty by using a simplified entropy formula, the PVM "entropy measure" or "ignorance function"

$$I[P_k] = - \sum_k P_k \ln P_k \qquad (25.29)$$

where the P_k are cluster probabilities denoted by $w_{ijklm...}$, above. The use of Eq. (25.29), which considerably overestimates the entropy, instead of a much more accurate CVM formula, was justified by Clapp by appealing to information theory[10]. Also the PVM calculations for multisite probabilities gave the correct result in those simple cases where the exact answer was known. The second difficulty remains: in order to form a good mental picture of what partial (local) order really looks like, a large enough cluster must be selected. Clapp[9] chose a site and all of its nearest neighbors, i.e., the centered 9-point cube for the bcc lattice, and the 13-point cuboctahedron shown in Fig. 33 for the fcc lattice. For the latter, the total number of $w_{ijklm...}$ variables is $2^{13} = 8192$, but the number of distinct P_k cluster variables for the 13-point figure populated by A and B atoms in "only" 288. The minimization equations in these variables can be solved by iteration,[9,289] but the problem of listing correctly, without omission or repetition, the distinct clusters and their respective weights is a difficult one, although a recently devised technique[15] should simplify matters considerably.

The PVM was used to compute triplet probabilities in Cu_3Au at 450°C,[10] with pair probabilities supplied by the X-ray diffraction data of Moss.[270] Clapp showed that the PVM values agreed closely in the triplet probabilities obtained from the same pair data by the Monte Carlo simulation method of Gehlen and Cohen[12] (see Section 5). Interestingly, the Kirkwood superposition approximation (5.1) gave triplet values which fell outside the allowed range, i.e., that which insures that all probabilities remain positive. Hence, at the very least, the PVM gives allowed triplet values, whereas the superposition approximation does not, and is therefore unacceptable.

Clapp[9] used the 9-point cubic cluster to analyze SRO in disordered bcc CnZn and the 13-point cuboctahedron.(Fig. 33) to analyze SRO in the disordered fcc solutions Cu_3Au, $CuAu$, $CuAu_3$, $CuAl$, $CuNi$, and $AuPd$, since reliable diffraction data was available for these systems. The "most enhanced clusters" were determined by the PVM, the enhancement factor for a given cluster being defined as its actual expectation value in the solution divided by its expected value in the completely disordered state. With the curious exception of $CuAu_3$, the most enhanced cluster was the one which represented the ordered phase occurring below the transition. This would tend to support the "microdomain" idea of local order although more work is needed, both experimental and theoretical, before a clear-cut distinction can be made between Krivoglaz-Clapp-Moss-type "classical" fluctuations on the one hand and "heterophase" fluctuations on the other. An interesting case was that of AuPd for which the most enhanced clusters indicate that although no ordering transition can be observed experimentally in this system, the hypothetical ground state may be the $I4_1/amd$ structure, (or an antiphase derivative thereof), the equiatomic member of the $\langle 1\frac{1}{2}0 \rangle$ ordering wave family (see Table VII). If this is indeed the structure that the system is tending toward, it would be the first example of the $I4_1/amd$ ground state

---

[289] R.D. De Ridder, *Physica* (*Utrecht*) **79A**, 217 (1975).

in alloys, although this structure is commonly observed in antiferromagnetic order.

Finally, let us mention the interesting "cluster" interpretation made by Sauvage and Parthé[74] of interstitial SRO in vanadium carbides, and a similar interpretation of Ni–Mo substitutional order made by De Ridder et al.[75]

e. Multicomponent Fluctuations

Whereas a considerable body of SRO intensity data exists for binary systems, there is practically none at all for ternaries. Undoubtedly the reason is that there are three linearly independent pair correlation functions in ternaries, instead of just one in binaries. There are six such functions in quaternaries, and in general $n(n + 1)/2$ in $(n + 1)$-component systems.[1] Hence, very difficult and costly experiments must be performed with distinct radiations, for example, with one X-ray source and one neutron source with samples enriched with various isotopes, as mentioned in Section 4,a. Another possibility would be to tune the continuous white X-ray spectrum available from synchrotron radiation to take advantage of anomalous dispersion.

Theoretically, the extension of the KCM formula to multicomponents poses no difficulty, however; it is merely required to write out the explicit form of Eq. (25.7), i.e., to perform an $n \times n$ matrix inversion. For example, the A–A (or 1-1) SRO intensity formula for a ternary system is[3]

$$\langle |\Gamma_1(h)|^2 \rangle = \frac{k_B T}{N} \frac{\bar{c}_1(1 - \bar{c}_1)(k_B T) - 2\bar{c}_1 \bar{c}_2 \bar{c}_3 W_{23}(h)}{(k_B T)^2 - 2H(h)(k_B T) - L(h)\bar{c}_1 \bar{c}_2 \bar{c}_3}, \quad (25.30)$$

with similar formulas for $\langle \Gamma_1^* \Gamma_2 \rangle$ and $\langle |\Gamma_2|^2 \rangle$. The denominator of Eq. (25.30) is just the left-hand side of Eq. (24.15), proportional to the determinant of the symmetric 2×2 free energy matrix $\mathbf{F}$, the numerator being proportional to the 1-1 minor.

Cook[84] has presented an interesting application of multicomponent fluctuation theory to martensitic transformations. Here, the "components" are transformation defects, defined in Section 11,a. To analyze the transformation, the $\mathbf{F}$ matrix is diagonalized, and the fluctuation formula is then that given in Eq. (25.8). The temperature at which the lowest eigenvalue $\Lambda_l(h)$ becomes negative gives the start temperature for the elastoplastic instability responsible for the martensitic product, whereas the corresponding eigenvector gives the relative mix of dislocation and transformation defects present in the normal mode.

26. KINETIC EFFECTS

The general kinetic problem is far too complex to be treated here, so that only a brief exposition of the setting up of a suitable mathematical model will be given (Section 26,a). In principle, differential equations for the time-rate of change of moments of a *master equation* can be derived, then decoupled and linearized to produce differential equations which can be readily solved by a "wave method."[290,291] In the present context however, it is simpler to derive these linear equations directly from the models presented thus far, thereby providing a time-dependent perturbation analysis for solid solutions of uniform concentration. This will be done in Sections 25,b-e. Finally, the kinetic extension of the CVM will be briefly described, the so-called path probability method (Section 25,f).

a. Master Equation

As was done in Section 2, the state of order at time t is considered to be given by a set of multiplet probabilities defined by

$$\langle \sigma_i(p + p_1)\sigma_j(p + p_2)\cdots\sigma_l(p + p_m) \rangle \qquad (26.1)$$

$$= \sum_{\{\sigma\}} \sigma_i(p + p_1)\sigma_j(p + p_2)\cdots\sigma_l(p + p_m)\mathcal{P}(\{\sigma\}, t).$$

In this equation, which is an extension of Eq. (2.1) to time-dependent problems, $\mathcal{P}(\{\sigma\}, t)$ is the probability that the system be found at time t in the configuration specified by the set of occupation operators $\{\sigma\}$. For time-independent problems, the probability would be that given by (25.1) or (25.3), for example. For time-dependent problems, the time-rate of change of the probability is required. A general expression for this was given by Kawasaki[292]:

$$\frac{\partial}{\partial t}\mathcal{P}(\{\sigma\}, t) = \frac{1}{2}\sum_{i,j=0}^{n}\sum_{p,r}\{W[\sigma_i(p), \sigma_j(p + r)]\mathcal{P}(\{\sigma\}^*, t) \qquad (26.2)$$

$$- W[\sigma_j(p + r), \sigma_i(p)]\mathcal{P}(\{\sigma\}, t)\},$$

where $W[\sigma_i(p), \sigma_j(p + r)]$ denotes the probability per unit time that σ_i on site (p) interchange with σ_j on site $(p + r)$ and where the symbol $\{\sigma\}^*$ denotes the configuration $\{\sigma\}$ with $\sigma_i(p)$ and $\sigma_j(p + r)$ interchanged. Equation (26.2), an extension of that first proposed by Glauber[293] for the

---

[290] H. Yamauchi, Ph.D. Dissertation, Northwestern University, Evanston, Illinois (1973).
[291] H. Yamauchi and D. de Fontaine, in "Order-Disorder Transformations in Alloys" (H. Warlimont, ed.), p. 148. Springer-Verlag, Berlin and New York, 1974.
[292] K. Kawasaki, *Phys. Rev.* **145**, 224 (1966).
[293] R.J. Glauber, *J. Math. Phys.* **4**, 294 (1963).

linear chain Ising model, expresses the fact that configuration $\{\sigma\}$ is destroyed by a $[\sigma_i(p), \sigma_j(p + r)]$ interchange, but is also created by the inverse exchange starting from configuration $\{\sigma\}^*$. The summation over (r) is generally taken to be over nearest neighbors of (p) in this direct exchange mode. The fact that atomic interchanges are actually mediated by vacancies does not invalidate the model, as Yamauchi[290] found that the introduction of a vacancy mechanism does not significantly alter the results of the calculations.

The interchange rate must be of the form

$$W[\sigma_i(p), \sigma_j(p + r)] = \theta[\sigma_i(p), \sigma_j(p + r)]e^{-\Delta E/2k_B T} \qquad (26.3)$$

where the preexponential factor θ represents the hypothetical interchange rate in an ideal solution, and the Boltzmann factor represents the fraction of successful interchange attempts in the actual solution, ΔE being the difference in internal energy before and after the contemplated interchange.

The time evolution of multiplet probabilities are obtained from the master equation by taking appropriate moments. The rate of change of single-site probabilities, $\partial \langle \sigma_i(p) \rangle / \partial t$, follows from Eq. (26.1) by multiplying both sides of (26.2) by $\sigma_i(p)$ and summing over all configurations $\{\sigma\}$, i.e., by taking the first moment of the master equation. There results a set of nN coupled nonlinear differential equations, which is completely intractable. One can then use the zeroth approximation to replace multisite probabilities by products of single-site probabilities. In order for these averages to differ from the trivial average composition, it is necessary to perform sublattice averaging, as explained in Section 2,b. Yamauchi[290,291] was able to show that all familiar models of ordering kinetics could thus be derived from the master equation by making appropriate simplifications. The models so derived included Cahn's linear equation for spinodal decomposition,[169] the k-space linear equations of Cook et al.,[38,294] the LRO models of Dienes[295] and Vineyard,[296] and subsequent modifications of Nowick and Weisenberg.[297] The linear k-space SRO intensity equation of Cook[298] could be obtained[290] by multiplying both sides of Eq. (26.2) by $\sigma_i(p)\sigma_j(p + r)$ and summing over all configurations $\{\sigma\}$, i.e., by taking the second moment of the master equation.

Langer also used the master equation approach to derive an equation

---

[294] H.E. Cook, D. de Fontaine, and J.E. Hilliard, *Acta Metall.* **17**, 765 (1969).
[295] G.J. Dienes, *Acta Metall.* **3**, 549 (1955).
[296] G.H. Vineyard, *Phys. Rev.* **102**, 981 (1956).
[297] A.S. Nowick and L.R. Weisenberg, *Acta Metall.* **6**, 260 (1958).
[298] H.E. Cook, *J. Phys. Chem. Solids* **30**, 2427 (1969).

for the kinetics of phase separation.[299-303] Langer used essentially the averaging procedure described by Eq. (11.20), leading to what he terms a "coarse-grained" description of a binary solution, with corresponding nonconvex "coarse-grained free energy," an example of which is shown in Fig. 21a. In this way, Langer was able to treat both nucleation kinetics and spinodal decomposition. Since the continuum approximation was made throughout, Langer's model cannot be used for order-disorder kinetics. Langer's theory clearly shows that nucleation and growth and spinodal decomposition are but two extreme cases of the general process of phase separation kinetics, as mentioned at the beginning of Section 24. The same conclusion can in fact be drawn from various one-,[219,304] two-,[305-309] and three-dimensional[208,310] computer simulations of decomposition kinetics.

In what follows, only a simple kinetic model derived by a perturbation method will be described. This allows one to perform a *kinetic stability analysis,* in a manner analogous to the static stability analysis presented in Section 24.

b. Multicomponent Perturbation Treatment

The microcanonical averaging technique frequently referred to in this article allows one to define a continuously varying "composition"

$$0 \leq c(p) \leq 1 \qquad (26.4)$$

at each lattice point (p), or, equivalently, on each sublattice. It is then permissible to define a "composition flux," or occupation probability flux $J_i(p; r)$, for species i from point $(p + r)$ to point (p). The simplest assumption to make is that this flux should be proportional to a linear combination of differences of chemical potentials at (p) and $(p + r)$ of

---

[299] J.S. Langer, *Ann. Phys. (N.Y.)* **65**, 53 (1971).
[300] J.S. Langer and M. Bar-on, *Ann. Phys. (N.Y.)* **78**, 421 (1973).
[301] J.S. Langer, *Acta Metall.* **21**, 1649 (1973).
[302] J.S. Langer, *Physica (Utrecht)* **73**, 61 (1974).
[303] J.S. Langer, M. Bar-on, and H.D. Miller, *Phys. Rev. A* **11**, 1417 (1975).
[304] D. de Fontaine, Ph.D. Dissertation, Northwestern University, Evanston, Illinois (1967).
[305] L.H. Shendalman and J.T. O'Toole, *J. Colloid Interface Sci.* **27**, 145 (1968).
[306] K. Binder and H. Müller-Krumbhaar, *Phys. Rev. B* **9**, 2328 (1974).
[307] K. Binder, *Z. Phys.* **267**, 313 (1974).
[308] A.B. Bortz, M. Kalos, J.L. Lebowitz, and M. Zendejas, *Phys. Rev. B* **10**, 535 (1974).
[309] M. Rao, M.H. Kalos, J. Marro, and J.L. Lebowitz, *Phys. Rev. B* **13**, 4328 (1976).
[310] J. Marro, A.B. Bortz, M. Kalos, and J.L. Lebowitz, *Phys. Rev. B* **12**, 2000 (1975).

all n independent concentration variables[3]:

$$J_i(p; r) = - \sum_{k=1}^{n} M_{ik} m(r)[\psi_k(p) - \psi_k(p + r)] \qquad (26.5)$$

where M_{ik} are elements of a symmetrical positive definite mobility matrix, where $m(r)$ is a geometrical factor allowing for different diffusion mechanisms for different neighbors, and where $\psi_k(p)$ is the potential at (p), i.e., the Fourier back transform of the potential defined by Eq. (25.5). In direct space, the potential is thus given by

$$\psi_k(p) = \sum_{j=1}^{n} \sum_{r} f_{kj}(r)\gamma_j(p + r) \qquad (26.6)$$

where the coefficient f_{kj} is just the quantity (in parentheses) being transformed in Eq. (18.10), i.e., a sum and difference of second derivatives of the free energy. Equation (26.5) reads "flux equals mobility times driving force."

The discrete counterpart of the continuity equation is here

$$\dot{\gamma}_i(p) = \sum_r J_i(p; r), \qquad (26.7)$$

where the dot indicates differentiation with respect to time t, the sum of fluxes to (p) from all contributing neighbors (r) representing the discrete-space divergence. The required linear kinetic equations are obtained by combining Eqs. (26.5) and (26.7). By Fourier transforming both sides of the resulting equations one obtains

$$\dot{\Gamma}_i(h, t) = -\beta(h) \sum_{k,j=1}^{n} M_{ik} F_{kj}(h) \Gamma_j(h, t) \qquad (26.8)$$

where F_{kj} are elements of the free energy matrix $\mathbf{F}$ used previously, Eq. (18.10), and where the positive geometrical coefficient $\beta(h)$ is given by[38]

$$\beta(h) = \sum_r m(r)[1 - \cos \mathbf{k}(h)\cdot\mathbf{x}(r)]. \qquad (26.9)$$

In matrix notation, Eqs. (26.8) have the form

$$\dot{\boldsymbol{\Gamma}} = \mathbf{A}\boldsymbol{\Gamma}, \qquad (26.10)$$

with

$$\mathbf{A} = -\beta \mathbf{D}, \qquad (26.11)$$

where the diffusion matrix

$$\mathbf{D} = \mathbf{MF} \qquad (26.12)$$

is the product of two symmetric matrices, one of which, $\mathbf{M}$, is positive definite. The positive-square-root matrix ($\mathbf{P}$) of $\mathbf{M}$ is therefore real, and we may write,[3] from Eq. (26.12)

$$D = P(PFP)P^{-1}. \tag{26.13}$$

It follows that D is similar to the symmetric matrix

$$S = PFP, \tag{26.14}$$

so that the eigenvalues $\Delta_i(h)$ of D are real along with those $\alpha_i(h)$ of A. The solution of Eq. (26.10) is then

$$\bar{\Gamma}_i(h, t) = \bar{\Gamma}_i(h, 0) \exp\left[\alpha_i(h)t\right]. \tag{26.15}$$

The normal diffusion modes $\bar{\Gamma}_i$ form the column matrix

$$\bar{\Gamma} = T^{-1}\Gamma \tag{26.16}$$

where T is the matrix which diagonalized D. We have

$$T = PR \tag{26.17}$$

where R is the orthogonal matrix which diagonalizes S. The diffusion matrix diagonalization operator T is not orthogonal, but the relation[311]

$$T^*M^{-1}T = I \tag{26.18}$$

shows that the column vectors obtained from T, denoted here as *kinetic principal directions*, are mutually orthogonal in a space whose metric is M^{-1}, the inverse of the mobility matrix.[3]

Furthermore, the relation

$$T^*FT = R^{-1}SR = \Delta \quad \text{(diagonal)} \tag{26.19}$$

shows that the free energy (F) and diffusion (D) matrices can be diagonalized simultaneously, so that positive, negative, and zero eigenvalues of these two matrices can be put in a one-to-one correspondence and ordered similarly. The eigenvalues of A are obtained from those of D by

$$\alpha_i(h) = -\beta(h)\Delta_i(h), \tag{26.20}$$

from Eq. (26.11). These α_i eigenvalues may be termed *amplification rates,* or *inverse relaxation times* of the corresponding concentration mode. According to the linear kinetic formulation presented here, a normal concentration wave of wave vector $k(h)$ will initially grow (decay) exponentially in time with positive (negative) rate α_i whenever the corresponding eigenvalue $\Lambda_i(h)$ of the free energy matrix is negative (positive). Hence, a "static" instability (system below a limit of metastability) implies a "kinetic" instability and conversely. It also follows from these

---

[311] The star denotes the hermitian conjugate matrix.

considerations that, although the effect of the mobility matrix **M** is to rotate the kinetic principal directions away from the static ones (eigenvalues of **F**), this rotation cannot cause a principal direction to cross an asymptotic direction. Furthermore, since asymptotic directions condense into a single line *at* a limit of metastability, static, and kinetic directions corresponding to that mode which just becomes unstable there must coincide.[3]

In certain cases of chemical systems removed far from equilibrium, a perturbation analysis similar to the one performed here may yield complex eigenvalues α_i, thereby providing a possible model for biological clocks, as shown by Prigogine and his collaborators.[312] Although, because **A** is similar to a symmetrical matrix, periodicity in time is not allowed here, space oscillations may be present, as will be shown below on various examples. Space oscillations are also predicted in Martin's perturbation analysis of the vacancy concentration in irradiated materials,[313] thereby providing a possible mechanism for void periodicity (see Fig. 20).

Cook[283,298] has emphasized the need of including a fluctuation source term in the diffusion equations to take interactions with the heat bath into account. This source term must be included in an *intensity* equation rather than in the *amplitude* equation (26.10). Yamauchi[290] and Langer[299] have given general derivations, and the former author has shown what simplifications are required in order to obtain Cook's equation. Here we outline a simple derivation of this equation, generalized to multicomponent systems[3]: multiply both sides of the normal mode equation

$$\dot{\bar{\Gamma}}_i(h, t) = \alpha_i(h)\bar{\Gamma}_i(h, t) \tag{26.21}$$

by the complex conjugate amplitude $\bar{\Gamma}_i^*(h, t)$ and add the resultant equations to their complex conjugates to obtain

$$\dot{\bar{Q}}_i(h, t) = 2\alpha_i(h)\bar{Q}_i(h, t) + X_i(h) \tag{26.22}$$

in which a time-independent fluctuation source term X_i has been added. The condensed notation

$$\bar{Q}_i = |\Gamma_i|^2 \tag{26.23}$$

has been used. The source term can easily be evaluated in the case of negative α_i since, for large t, steady state obtains so that

$$X_i(h) = -2\alpha_i(h)\bar{Q}_i(h, \infty). \tag{26.24}$$

---

[312] P. Glansdorff and I. Prigogine, "Thermodynamics of Structure, Stability and Fluctuations." Wiley, New York, 1971.

[313] G. Martin, *Philos. Mag.* [8] **32**, 615 (1975).

The expectation value of $\bar{Q}_i(h, \infty)$ can be taken as the equilibrium concentration fluctuation spectrum $\langle \bar{Q}_i \rangle$ given by Eq. (25.8):

$$\bar{Q}_i(h, \infty) = k_B T / N \Lambda_i(h) \tag{26.25}$$

where Λ_i are the eigenvalues of the free energy matrix **F**. Hence the Langevin equation (26.22) has solution

$$\bar{Q}_i(h, t) = [\bar{Q}_i(h, 0) - \bar{Q}_i(h, \infty)] \exp[2\alpha_i(h)t] + \bar{Q}_i(h, \infty). \tag{26.26}$$

Applications will now be given to various special cases.

c. Spinodal Phase Separation

In the unmixing case, BZ center instabilities are expected and only long wavelengths need be retained in the Fourier concentration spectrum. Then, for small $k = |\mathbf{k}|$, $\beta(h)$ in Eq. (26.9) goes into the wave vector squared k^2 in cubic crystals for which $m(r) = 1/a_0^2$ (a_0 being the lattice parameter). In this case, the linear diffusion equation (26.10) has solution

$$\Gamma(\mathbf{k}, t) = \Gamma(\mathbf{k}, 0) \exp[\alpha(\mathbf{k})t] \tag{26.27}$$

where, by Eq. (25.12), the amplification rate is given by

$$\alpha(\mathbf{k}) = -M\Omega_0 [f_0'' + 2\eta^2 Y(\mathbf{k}) + 2\kappa k^2] k^2, \tag{26.28}$$

a formula first given by Cahn.[89] Below the *coherent spinodal*

$$f_0'' + 2\eta^2 Y_{\min} = 0 \tag{26.29}$$

(where $Y_{\min}$ is the value of the effective modulus in the softest elastic direction), the solid solution is unstable to long-wavelength concentration fluctuation, so that $\alpha(\mathbf{k})$ for certain ranges of k values will be positive, resulting in growth with time of the corresponding concentration waves.

A typical plot of $\alpha(\mathbf{k})$ vs $\mathbf{k}$ for the unstable case is given in Fig. 61. According to this perturbation analysis, all waves of wave vector less than a *critical* one k_c, solution of

$$\alpha(k_c) = 0 \tag{26.30}$$

should grow in amplitude, all those with $k > k_c$ should decay, and maximum growth should occur for the *optimal wave* k_m solution of

$$d\alpha/dk = 0 \tag{26.31}$$

for a given direction **k**. Optimal and critical wave vector magnitudes are related by

$$k_c = k_m \sqrt{2}. \tag{26.32}$$

This kinetic model, known as *spinodal decomposition*,[89,169] predicts a resulting metastable structure consisting of concentration plane waves of optimal wavelength lying along the soft elastic directions. Since long-wavelength concentration waves are involved, this process may be termed *spinodal phase separation* to distinguish it from *spinodal ordering*, a similar process which involves ordering waves (Section 26,d). Cahn's predictions have been amply substantiated experimentally: see for example various review articles devoted to this subject[104,170,219] and note the fine $\langle 100 \rangle$ periodic striations visible in the micrograph of a spinodally decomposed ternary alloy shown in Fig. 51.

Quantitative verification of the linear diffusion equation has not fared as well. In practice, the amplitude-squared spectrum exhibits a maximum which does not grow exponentially in time, and which tends to shift toward the origin of k-space (longer wavelengths) as the decomposition progresses. The cause of this departure from the linear theory's predictions is found in the strongly nonlinear character of the actual diffusion equation, as derived for example from a master equation. Langer's equation, in particular, appears to reproduce experimentally determined intensity spectra quite well, as seen in Fig. 62,[301] where the inset shows a sequence of spectra measured by Rundman and Hilliard[314] by X-ray small-angle scattering in an Al-22 at. % Zn alloy aged below the spinodal. The interpretation of this sequence of curves is as follows: during the

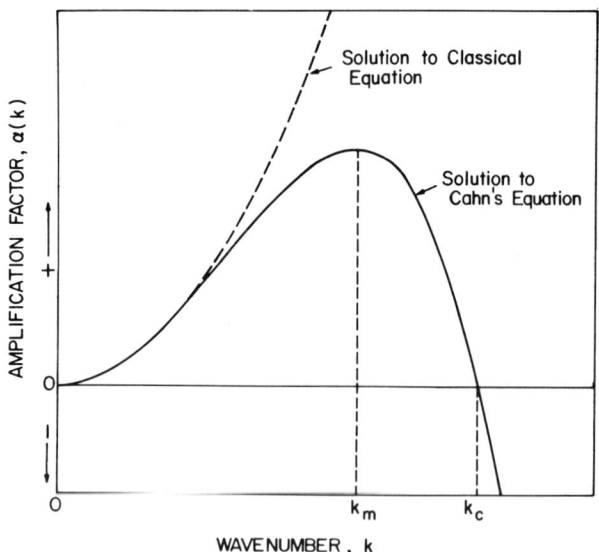

FIG. 61. Amplification rate $\alpha(k)$, Eq. (26.28), vs wave vector k; k_m and k_c are optimal and critical wave vectors, respectively.

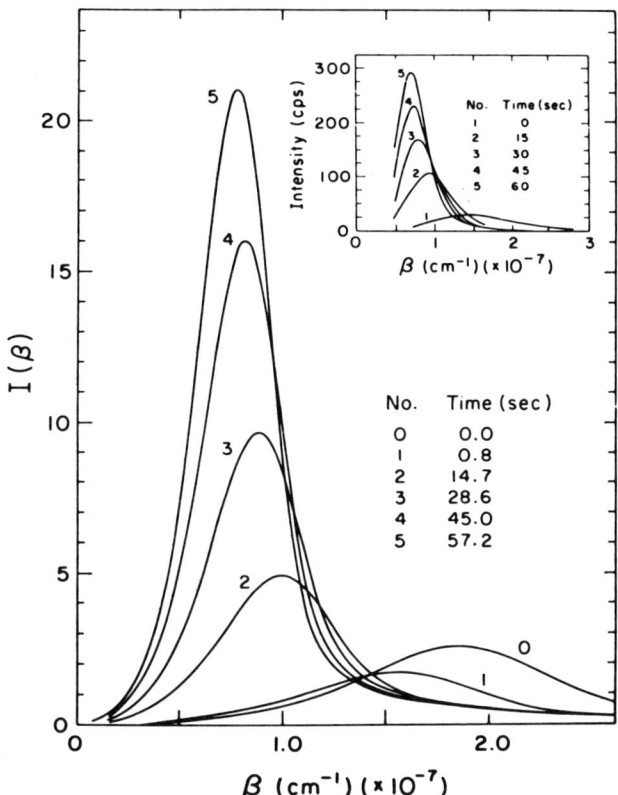

FIG. 62. Theoretical structure factor spectra $I(\beta)$ calculated by Langer [J. S. Langer, *Acta Metall.* **21**, 1649 (1973)] for Al-22 at. % Zn alloy aged at 150°C. Inset shows corresponding experimental data of Rundman and Hilliard [K. B. Rundman and J. E. Hilliard, *Acta Metall.* **15**, 1025 (1967)].

first annealing stages, the optimal wavelength is adjusted (increased) from that resulting from the quenching operation to that predicted by Cahn's formula. Then growth of the concentration modulations takes place with coarsening of the decomposition structure occurring simultaneously. This can be seen quite readily by solving the nonlinear diffusion equation numerically.[219,304] Early claims[314] that an exponential growth stage of the optimal wave could be observed were later retracted,[315] and recent results on phase separation of liquid critical solutions pressure-quenched into

---

[314] K.B. Rundman and J.E. Hilliard, *Acta Metall.* **15**, 1025 (1967).
[315] T.L. Bartel and K.B. Rundman, *Met. Trans.* **6A**, 1887 (1975); D.T. Lewandowski and K.B. Rundman, *ibid.* p. 1895.

the miscibility gap appear to exhibit no exponential growth stages.[316] Likewise, Langer's theory and Monte Carlo simulations[208] cannot confirm the existence of exponential growth of the optimal wave for any time stage or concentration range.

Lack of accurate quantitative agreement between the linear kinetic theory and experiment should not detract from the remarkable success of Cahn's spinodal decomposition theory which correctly predicts concentration modulations of wavelength close to those actually observed in the early stages of decomposition, in directions which are the elastically soft ones. Cahn's linear equation should not really be regarded as reliable kinetic model, the nonlinear terms are too dominant for that to be true, rather, the linear equation should be regarded as useful for analyzing the stability of a uniform solution by a time-dependent perturbation method. Note that, particularly for phase separating systems, it is necessary to perform such a time-dependent analysis if one wishes to predict the essential features of resulting metastable structures: a *time-independent* analysis only predicts the existence of a Brillouin zone center instability, a *time-dependent* analysis predicts a maximum in the intensity spectrum away from $\langle 000 \rangle$, at finite wavelengths.

The linear theory does work quite well for dissolution kinetics, i.e., for the time evolution of preexisting concentration fluctuations in stable or metastable solutions. Cook's intensity equation is particularly useful in this regard: as an example, consider a solution in which a rapidly quenched high-temperature fluctuation spectrum evolves to its equilibrium shape at a lower temperature (above the limit of metastability). Typical high- and low-temperature fluctuation profiles are shown in Fig. 63[298] (heavy lines), calculated according to Eq. (25.13), along with profiles at intermediate times during the annealing process (light lines) calculated according to Eq. (26.26). This sequence of curves is qualitatively very similar to that showing the time evolution of intensity spectra obtained by Monte Carlo simulation of a model binary alloy outside the spinodal.[310] The time evolution of diffuse intensity spectra for the Cu–Ni system also exhibits similar features, although the authors[282b] point out that a good quantitative fit to Eq. (26.26) is not possible. The equation derived by Yamauchi[290] appears to give better agreement.

d. *Spinodal Ordering*

Spinodal ordering is the BZ-boundary special-point kinetic instability corresponding to spinodal clustering which is a BZ-center kinetic instability. Much less work, both theoretical and experimental, has been

---

[316] N.-C. Wong and C.M. Knobler, *J. Chem. Phys.* **66**, 4707 (1977).

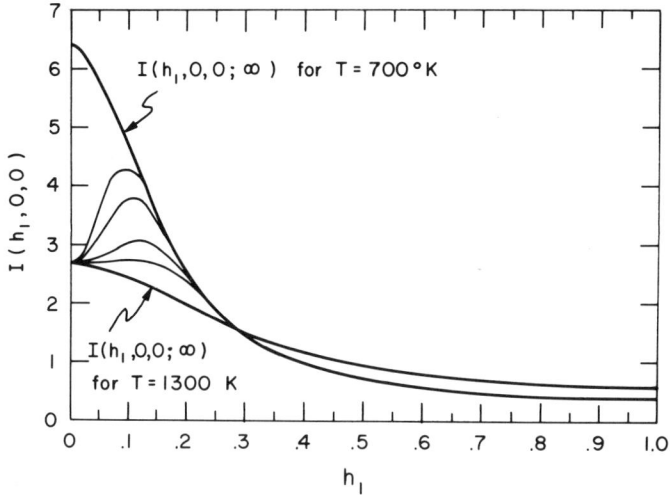

FIG. 63. Fluctuation intensity profiles calculated at 700 K and 1300 K according to Eq. (25.13) and diffuse intensity profiles at intermediate aging times calculated according to Eq. (26.26) [H. E. Cook, *J. Phys. Chem. Solids* **30**, 2427 (1969)].

done in the ordering case perhaps because the effects are less spectacular: because of the very short relaxation times, it is much more difficult to quench in an ordering metastable structure, and maximum kinetic amplification of ordering waves generally occurs *at* the special-point BZ-boundary points themselves rather than somewhat away from $\langle 000 \rangle$ as in spinodal segregation.

The concept of spinodal ordering was first introduced by transforming Cahn's linear, continuum diffusion equation into a "discrete" linear, differential-difference equation,[294] which turns out to be a special case of the general multicomponent equations (26.10). In the binary case, the amplification rate (26.20) has the form[38]

$$\alpha(h) = -M\beta(h)F''(h) \qquad (26.33)$$

with M a positive mobility, β given by Eq. (26.9), and F'' by Eq. (19.8). Typical plots of the amplification rate for $\mathbf{k}(h)$ along the $\langle 100 \rangle$ directions in a crystalline solution of cubic symmetry are shown in Fig. 64a for a temperature above the $\langle 100 \rangle$ limit of metastability and (Fig. 64b) below it (unstable case). The waves receiving maximum amplification in the latter case are just the $\langle 100 \rangle$ special-point ordering waves. Concentration waves which grow and those which decay are separated by a *critical surface* in k-space defined by

$$\alpha(h) = 0. \qquad (26.34)$$

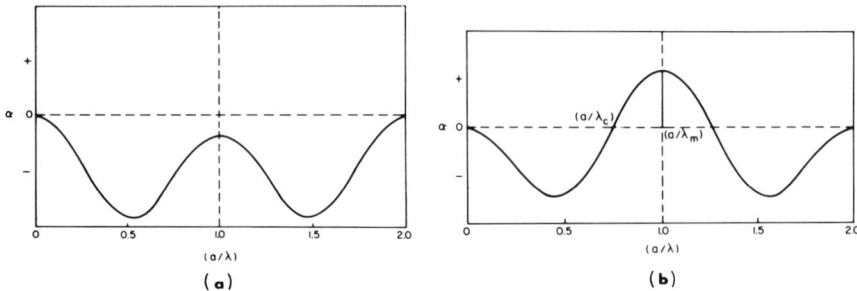

FIG. 64. Amplification rate curves $\alpha(h)$ vs $h = a/\lambda$ for ordering case in $\langle 100 \rangle$ direction in cubic crystals [H. E. Cook, D. de Fontaine, and J. E. Hilliard, *Acta Metall.* **17**, 765 (1969)]. (a) Above $\langle 100 \rangle$ ordering instability, (b) below $\langle 100 \rangle$ ordering instability.

According to Eq. (26.33), this defines two loci: the first is given by $\beta(h) = 0$ which guarantees the vanishing of $\alpha(h)$ at all reciprocal lattice points. The other locus is given by

$$F''(h) = 0 \tag{26.35}$$

and represents the true critical surface. These surfaces behave very much like Fermi surfaces in that, in the absence of strain effects, they are nearly spherical in cubic crystals for small $\mathbf{k}_c$ (spinodal segregation case), but progressively distort as $\mathbf{k}_c$ approaches the BZ boundaries. Examples of such surfaces, for bcc solutions are given in Fig. 65.

The only attempt at determining an amplification rate curve for an ordering system appears to have been that of Paulson.[317] The result of this study is shown in Fig. 66 for a Cu-16 at. % Au alloy thin film containing one-dimensional concentration modulations of the indicated wavelengths. The films were held at 225°C, i.e., above the ordering temperature, and the modulations annealed out with negative amplification rates which, despite experimental uncertainties, appear to follow a curve similar to the one predicted theoretically (Fig. 64a). More recently, Chen and Cohen[318] performed kinetic studies on several alloys (Cu_3Au, Cu-23 at. % Au, Cu-18.5 at. % Au and $CoPt_3$) above the transition temperatures. The results agreed fairly well with the linear kinetic theory in the form given by Eq. (26.26), with the amplification rate $\alpha(h)$ given by Eq. (26.33). It was concluded that first-neighbor interactions alone could not adequately account for the measured values of the function

---

[317] W.M. Paulson, Ph.D. Dissertation, Northwestern University, Evanston, Illinois (1972); W.M. Paulson and J.E. Hilliard, *J. Appl. Phys.* **48**, 2117 (1977).
[318] H. Chen, Ph.D. Dissertation, Northwestern University, Evanston, Illinois (1977); H Chen and J.B. Cohen, *J. Phys. (Paris)* **38**, C7-314 (1977).

$F''(h)$ which appears in the amplification rate (26.33), or its inverse, the relaxation time. This study again emphasizes that the important quantity to be considered is the $F''(h)$ harmonic coefficient or the k-space potential $V(h)$ since, from Eq. (19.8), the entropy part of F'' is a universal function of temperature and average concentration. Note that $F''(h)$ also appears in the general fluctuation formula (25.9), from which (25.16) may be derived, so that results from fluctuation and from kinetic studies may be compared directly. This is in fact how the authors proceeded,[318] the pair parameters $V_1, V_2, \ldots$ needed for the kinetic equation being taken from earlier SRO fluctuation measurements.[279]

There is very convincing qualitative evidence of spinodal ordering in the Ni–Mo system for average concentrations in the vicinity of Ni_4Mo and Ni_3Mo. This was already described in Section 24,b. The process leading to the initial enhancement of $\langle 1\frac{1}{2}0 \rangle$ diffuse intensity maxima upon quenching, and the absence of true superlattice maxima can be further

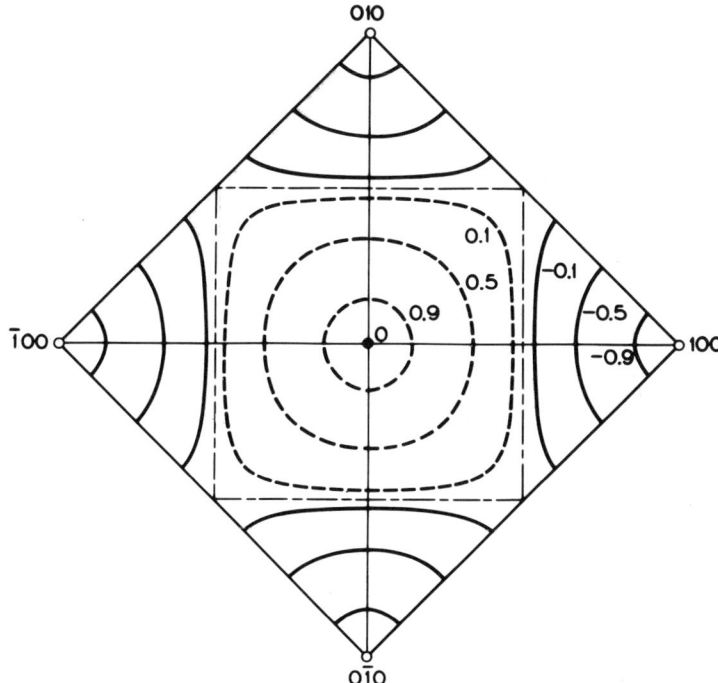

FIG. 65. Critical surfaces $\alpha(h) = 0$, Eq. (26.33), in bcc Brillouin zone. Numbers refer to reduced temperature T/T_0, positive for phase separation (dashed curves), negative for ordering (full curves) [H. E. Cook, D. de Fontaine, and J. E. Hilliard, *Acta Metall.* **17**, 765 (1969)].

elucidated by considering the amplification rate plot along $\langle h\frac{h}{2} 0\rangle$, Fig. 67. This curve was calculated from Eq. (26.33) after making some reasonable assumptions about pair interaction parameter ratios.[49] The $\alpha(h)$ curve shows that high-temperature fluctuations are, according to the perturbation theory, initially amplified preferentially at positions $\langle 1\frac{1}{2}0\rangle$ and $\langle 3\frac{3}{2}0\rangle$ at temperatures below the $\langle 1\frac{1}{2}0\rangle$ ordering spinodal. That diffuse intensity is initially absent from the $\langle 210\rangle$ positions (equivalent to $\langle 100\rangle$) can be explained by noting that $\alpha(h)$ has a local minimum there, as seen in Fig. 67. At later aging times, nonlinear terms in the diffusion equation must become important, so that wave-coupling produces the ordered structures of the equilibrium phase diagram.

e. Ternary Systems

Perturbation analysis of a bcc ternary model solution yields amplification rate curves typified by Fig. 68[3]: 68a exhibits double spinodal phase separation (despite appearances, there are two shallow positive α maxima at short wave vectors), 68b exhibits double spinodal ordering at special point $\langle 100\rangle$, 68c exhibits both spinodal phase separation and $\langle 100\rangle$-ordering, and 68d exhibits both $\langle 100\rangle$ and $\langle \frac{111}{222}\rangle$ ordering. For the latter case,

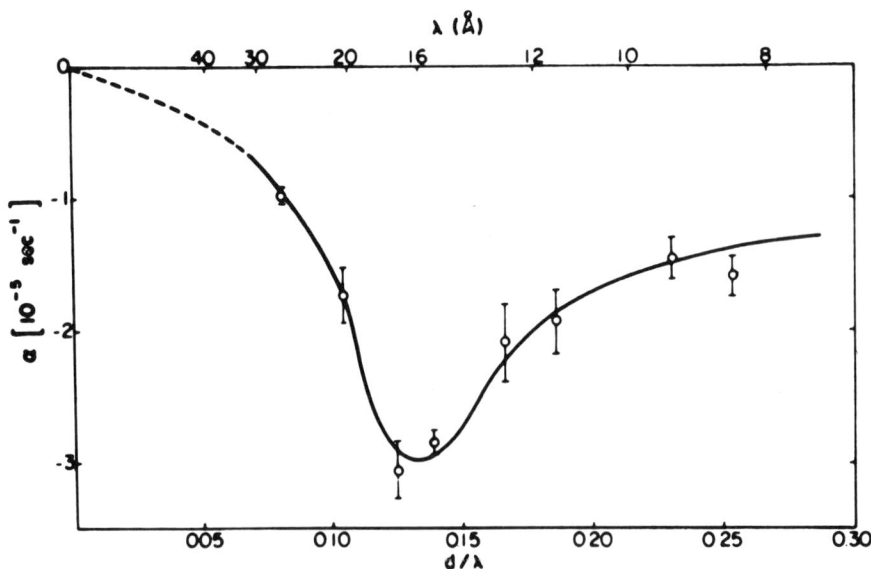

FIG. 66. Measured amplification rate $\alpha(h)$ vs $h/2 = d/\lambda$, obtained at 225°C in Cu-16 at. % Au along the [111] directions [W. M. Paulson, Ph.D. Dissertation, Northwestern University, Evanston, Illinois (1972); W. M. Paulson and J. E. Hilliard, *J. Appl. Phys.* **48**, 2117 (1977)].

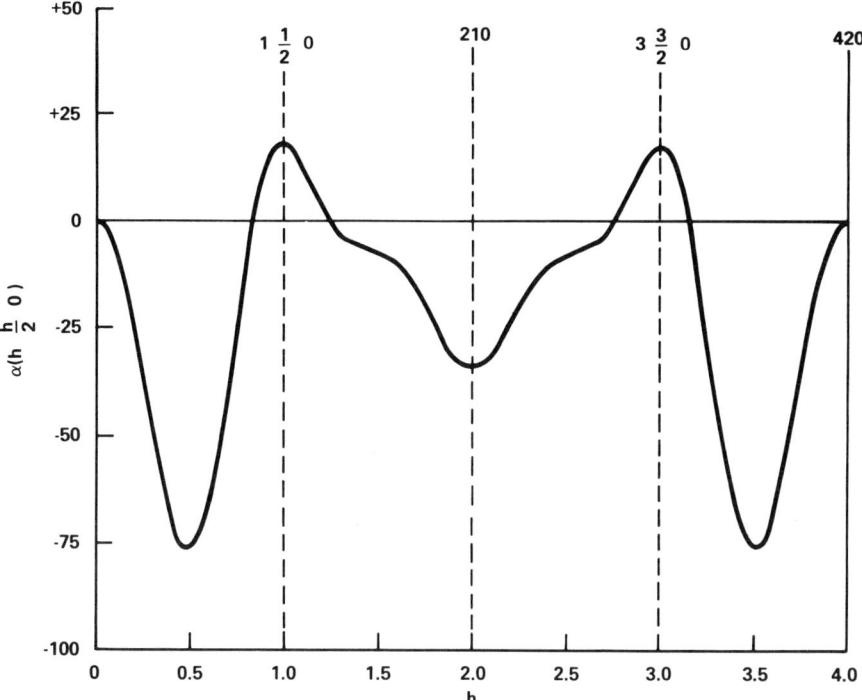

FIG. 67. Amplification rate $\alpha(h)$ along $[h\frac{h}{2}0]$ for system with $\langle 1\frac{1}{2}0\rangle$ instability [D. de Fontaine, *Acta Metall.* **23**, 553 (1975)].

the limits of metastability were shown in Fig. 50. As explained in Section 26,a, the polarizations of the unstable modes, or kinetic directions in concentration space, are given by the eigenvalues corresponding to positive α_i eigenvalues of the **A** matrix.

Spinodal phase separation is well documented in pseudobinary sections of certain ternaries, much as CuNiFe[319] and is also readily apparent in the CuMnAl micrograph, Fig. 51.[255] Quantitative measurement in complex unmixing–ordering cases is still lacking, however.

As was found to be true for binary solutions, the linear kinetic equation (26.12) describes fairly well the annealing out of composition modulations in stable solutions. In one such investigation,[320] alternate alloy layers in the completely miscible Ag–Au–Pd system were vapor-deposited on a mica substrate. The wavelength Λ of the composition modulations was

[319] E.P. Butler and G. Thomas, *Acta Metall.* **18**, 347 (1970).
[320] M. Murakami, D. de Fontaine, J.M. Sanchez, and J. Fodor, *Acta Metall.* **22**, 709 (1974).

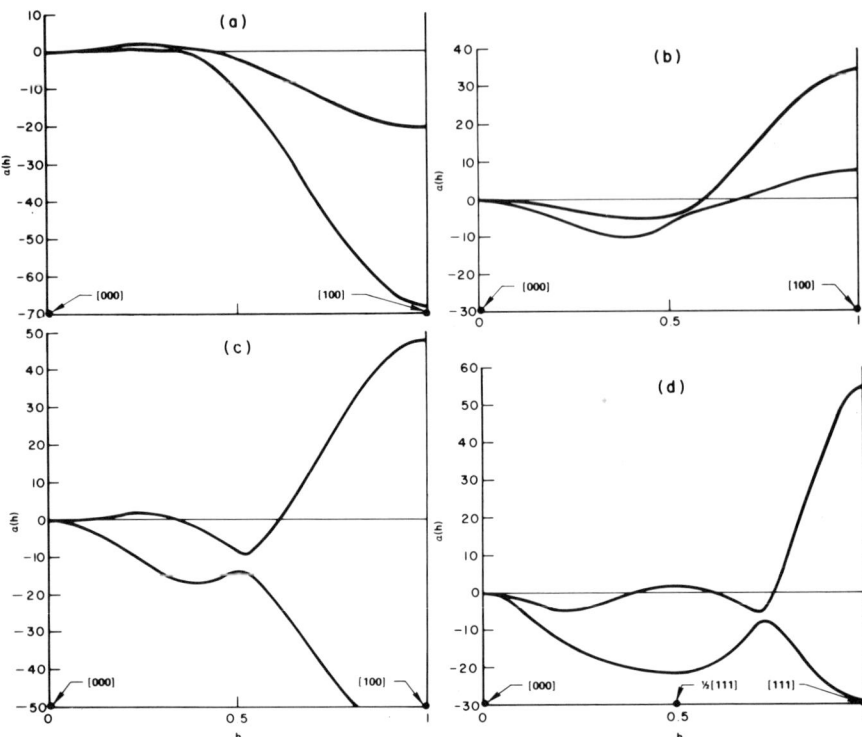

FIG. 68. Amplification rates $\alpha(h)$ in ternary systems for (a) double spinodal phase separation, (b) double spinodal ordering, (c) spinodal phase separation and ordering, (d) Heusler alloy ordering through combination of $\langle 100 \rangle$ and $\langle \frac{1}{2}\frac{1}{2}\frac{1}{2} \rangle$ instabilities [D. de Fontaine, J. Phys. Chem. Solids 34, 1285 (1973)].

of the order of 15 Å so that X-ray diffracted intensity satellites, due to the modulation, were clearly visible near the 111 Bragg reflections. The annealing out of the modulations at 300°C were followed by measuring the decay of normalized satellite intensity I/I_0 vs time. The logarithm of the intensity ratio against annealing time should plot as a straight line in binary systems,[321] according to linear kinetic theory. In ternaries, the satellite decay is characterized by two decay rates (negative amplification rates), α_1 and α_2, which could be back-calculated from the data shown in Figs. 69a and b. The open circles are calculated from Eq. (26.10). The two samples used had Au/(Au + Pd) layering (designated microcouple #1, Fig. 70) and Ag/(Au + Pd) layering (microcouple #2). From the data, it was possible to determine the eigenvector directions T_1 and T_2 of the

[321] H.E. Cook and J.E. Hilliard, J. Appl. Phys. 40, 2191 (1969).

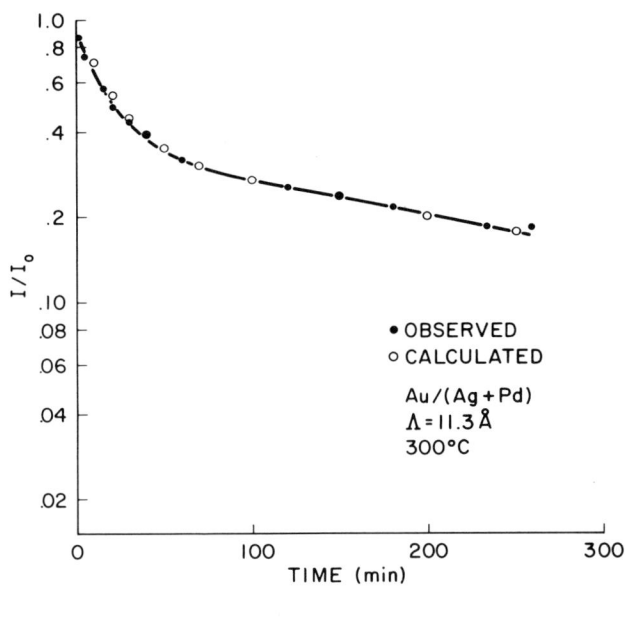

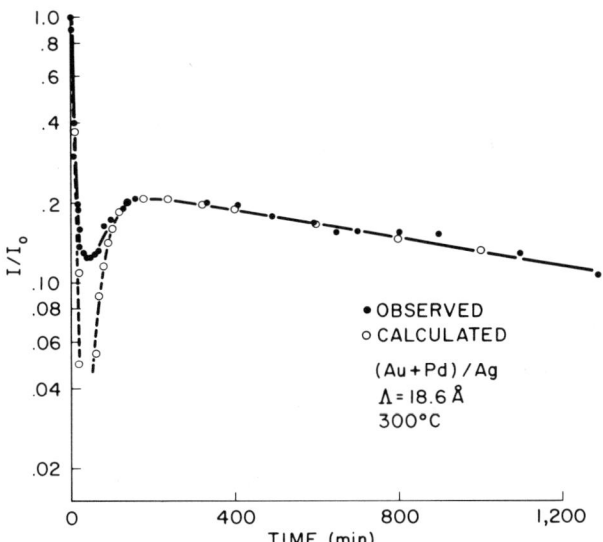

FIG. 69. Normalized satellite intensity decay as a function of time as observed and as back-calculated from Eqs. (26.10–26.15) for two types of Au–Ag–Pd layered thin films: (a) microcouple #1, (b) microcouple #2. Composition modulation wavelength designated by Λ [M. Murakami, D. de Fontaine, J. M. Sanchez, and J. Fodor, *Acta Metall.* **22**, 709 (1974)].

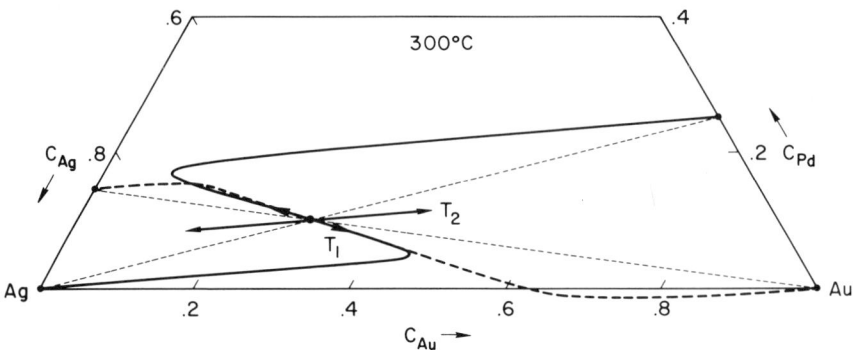

FIG. 70. Diffusion paths for microcouples #1 (dashed line) and #2 (full line) calculated by Eqs. (26.10–26.15) [M. Murakami, D. de Fontaine, J. M. Sanchez and J. Fodor, *Acta Metall.* **22**, 709 (1974)]. T_1 and T_2 are directions of minor and major eigenvectors, respectively.

diagonalization matrix **T**, Eq. (26.19), and hence it was possible to rotate back the diagonalized matrix **A** to obtain complete knowledge of system (26.10). Hence, the average compositions of the layers of both microcouples at all intermediate annealing times could be calculated according to linear kinetics and corresponding "tie-lines" could be plotted as shown in Fig. 70. The resulting diffusion paths start out in directions approximately parallel to the major eigenvector T_1 (corresponding to the largest negative amplification rate α_1) and end by being tangent to the minor eigenvector T_2. The path for microcouple #1 has a slight excursion outside the Gibbs triangle, where the composition modulation has large amplitude, but on the whole, the linear theory appears to give quite satisfactory results.

f. Path Probability Method

The path probability method (PPM), developed by Kikuchi,[322] is an extension in the time domain of the CVM. Corresponding to the cluster variables of Eq. (2.1), one now introduces a set of time-dependent *path variables* $\{A(t, t + \Delta t)\}$ which denote transitions from a given cluster type to another in time Δt. The probability that such a transition actually takes place is given by the probability function $\mathscr{P}[A(t, t + \Delta t)]$. As was the case for the "static" probability of Eq. (25.3), this "kinetic" probability can also be written as the product of two factors

$$\mathscr{P}\{A\} = \mathscr{P}_1\{A\}\mathscr{P}_2\{A\}, \qquad (26.36)$$

[322] R. Kikuchi, *Prog. Theor. Phys.* **S35**, 1 (1966).

the first factor being a combinatorial one related to the total number of ways that given atomic jumps may take place, and the second one being related to the rate at which an atom or defect can jump from one site to another for given temperature and local environment.

The second factor involves the usual Boltzman factors with activation energies given by the difference between the energy of the activated state and the energy of the initial state, the latter being usually obtained by bond counting.

The first factor $\mathcal{P}_1$ of the probability function (26.36) has caused considerable difficulties in the past. Vineyard[296] pointed out the unsatisfactory state of combinatorial theories existing at the time, and Makishima[323] called for a "generalized entropy of higher order which depends on time as well as space." It is precisely here that the PPM makes its most original contribution in the way suggested by Makishima: the lattice is now assumed to be populated by CVM-type clusters but with an added dimension, that of time. Hence each CVM point variable x_i is replaced by $X_{i \to k}$, the subscripts now ranging over all allowed transitions such as A → B, B → A, A → A, B → B. Likewise, the y_{ij} pair variables are replaced by the path variable $Y_{ij \to kl}$, there being as many of these as there are allowed transitions of n.n. pair configurations into one another. The combinatorial factor $\mathcal{P}_1$ is then the corresponding CVM combinatorial factor but with cluster variables replaced by path variables. For the pair approximation, for example, one simply rewrites Eq. (20.19) in terms of path variables.

The most probable kinetic path followed by the system is derived by maximizing the logarithm of the path probability function (26.36). Simultaneous differential equations are obtained which, in the pair approximation, take the form[322]

$$\frac{dy_{AV}}{dt} = \lim_{\Delta t \to 0} \frac{1}{\Delta t} \left(\sum_{i,j} Y_{ij \to AV} - \sum_{i,j} Y_{AV \to ij} \right) \quad (26.37)$$

for an [A-atom, vacancy] pair, for example. Equation (26.37) gives the intuitively correct result that the time rate of change of the pair A–V is equal to the sum of all path variables Y having AV as final state, minus the sum of all path variables having AV as initial state. Each Y variable itself is shown to be given by a product of Boltzmann factors and n.n. pair probabilities y_{ij}.[322] The resultant kinetic equations (26.37) constitute a set of coupled nonlinear differential equations of first order. There are corresponding equations for the point probabilities $x_i(t)$. These nonlinear differential equations must then be solved numerically. An advantage of

[323] S. Makishima, quoted by Kikuchi.[322]

BINARY SOLUTION STATE VARIABLES

A) POINT STATE VARIABLES

a) SUBSTITUTIONAL

b) INTERSTITIAL

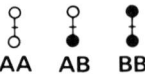

COMPLETE SET: 6 VARIABLES

B) PAIR STATE VARIABLES

a) REDUCED SET (EXAMPLES):

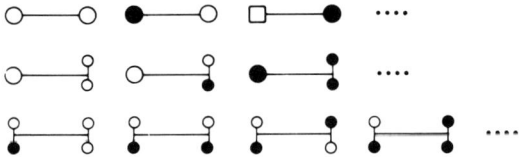

....
REDUCED SET: 22 DISTINCT PAIR VARIABLES

b) EXTENDED SET

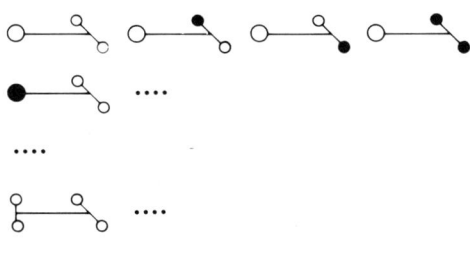

....

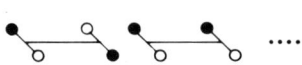

....

....

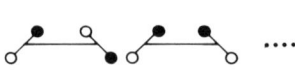

....

EXTENDED SET: 50 DISTINCT PAIR VARIABLES
COMPLETE SET: 72 DISTINCT PAIR VARIABLES

(a)

BINARY SOLUTION PATH VARIABLES

A) POINT PATH VARIABLES

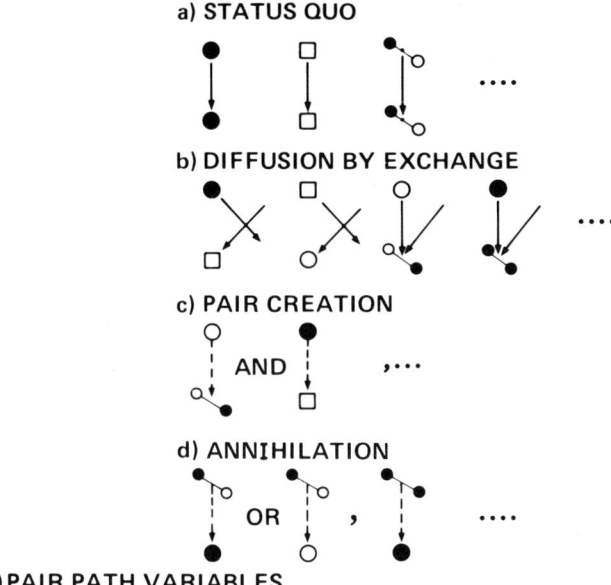

B) PAIR PATH VARIABLES

Diagrams are obtained simply by combining pair state diagrams and point path diagrams. Of particular interest are the following examples:

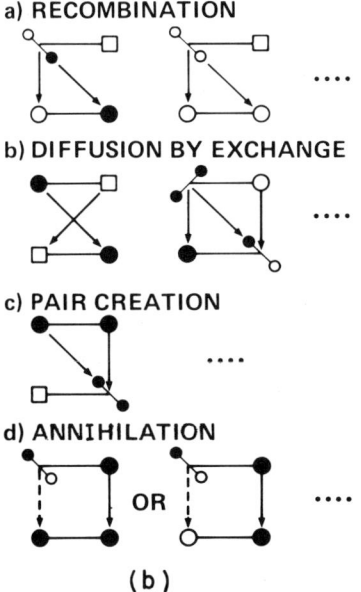

(b)

FIG. 71. Clusters required for kinetic study of vacancy and split interstitials diffusion in irradiated crystals in the pair approximation: (a) clusters for CVM, (b) clusters for PPM.

the PPM is that by construction, it necessarily converges at equilibrium to the state predicted by the CVM of the same order of approximation.

The PPM has been used successfully to treat, among other things, substitutional diffusion in ordered systems,[324] diffusion and superionic conductivity in solid electrolytes,[325] nucleation and growth in two dimensions,[326] and the kinetics of order–disorder transformations in bcc alloys.[327] In principle, fairly complex diffusional process can be investigated by this method, for example, vacancy and interstitial diffusion with defect pair creation and annihilation in irradiated crystals. Some of the state (cluster) variables and path variables required are indicated in diagrammatic form in Fig. 71. The distinction between reduced and extended sets lies in that the latter distinguishes all the possible orientations of $\langle 100 \rangle$ split interstitials (dumbbells) and the former does not. It is apparent that the introduction of the time dimension considerably increases the number of variables and hence the number of simultaneous nonlinear equations to be solved. This rapidly escalating complexity is one of the practical limitations of the method.

VIII. Conclusion

An attempt has been made in this review to bring together various topics related to concentrated solid solutions: description of the state of order, diffraction of radiation, electronic and elastic internal energy, configurational entropy, ground states and phase diagrams, static and kinetic instabilities, and fluctuations. Throughout, two modes of description proved to be useful: the wave method and the cluster method.

The former, the wave method, is particularly valuable when treating long-range, essentially delocalized phenomena. It leads to a particularly compact description whenever basic functionals are expressed in the harmonic approximation. Since the zeroth approximation is made, pair probabilities are decoupled to give products of "point" probabilities, this, in turn, implies a mean-field formulation, the point probabilities really representing sublattice averages, as in the Bragg–Williams model. Equivalently, the harmonic mean-field functional represents a pseudo two-body approach, Fourier-transformed into a single-body problem for-

---

[324] R. Kikuchi and H. Sato, *J. Chem Phys.* **51**, 161 (1969); **53**, 2702 (1970); **57**, 4962 (1972).
[325] H. Sato and R. Kikuchi, *J. Chem. Phys.* **55**, 677 (1971); R. Kikuchi and H. Sato, *ibid.* p. 702.
[326] R. Kikuchi, *J. Chem. Phys.* **47**, 1646 and 1653 (1967).
[327] H. Sato and R. Kikuchi, *Acta Metall.* **24**, 797 (1976).

mulated in terms of noninteracting concentration waves, the amplitudes of which may be considered as long-range order parameters. The wave method often gives a very useful qualitative description, particularly when used in conjunction with the appropriate symmetry rules, as in the Landau theory.

The cluster method is the former's antithesis: it is particularly suitable for long-range, localized phenomena, is essentially many-body, yields improved quantitative results, and the use of mean-field or decoupling schemes is delayed until after small neighborhoods of lattice points have been treated (almost) correctly. The cluster variation equations are hardly transparent, however, in the sense that actual results are obtained after fairly elaborate numerical computations. New algorithms are presently being developed, and their generalized use should make the CVM particularly attractive for a wide range of problems in the general field of phase transformations.

On the experimental front, increasingly accurate diffracted intensity data from defective or partially ordered structures are expected from sources such as X-ray, neutron, electron, and γ radiation. With the knowledge of pair correlation functions, one can then hope to determine such things as pair interaction parameters which serve as primary input in setting up internal energy functionals. Soon, it is hoped, very high resolution electron microscopy will allow us to "see" small ordered clusters (or microdomains) of minority components or defects, perhaps through the use of lattice imaging [328] or weak beam techniques treated as general spacial filtering methods. Increased resolution coupled with improvements in the CVM, say, lead to the exciting prospect of the experimenter's actually "seeing" the very elements which the theorist uses as building blocks in setting up his free energy functionals.

On the theoretical front, increasingly accurate model potentials are expected, along with applications of the coherent potential approximation or "cluster" calculations. In the meantime, phenomenological energy parameters should be used to describe a variety of order–disorder phase transitions, particularly in interstitial solutions, nonmetals and ternary solutions, the latter being virtually unexplored. Such approximate techniques as the CVM should be extended to include volume changes, lattice distortions, and vibrational entropy. The subject of nonequilibrium transformations and of steady-state or metastable structures held far from thermodynamic equilibrium are receiving due attention, and should continue to do so for years to come.

[328] R. Sinclair, *J. Phys. (Paris)* **38**, C7–453 (1977).

Acknowledgments

The author has benefited greatly from helpful conversations with many colleagues, particularly with Drs. S.M. Allen, J.B. Cohen, H.E. Cook, W.A. Harrison, E. Johnson, J. Kanamori, R. Kikuchi, M. Murakami, J.M. Sanchez, D. Schwahn, E. Seitz, and H. Yamauchi, who read parts of early drafts of this article and made valuable suggestions and/or who allowed previously unpublished material to be used here. The help of Drs. Sanchez and Yamauchi in correcting the proofs is also gratefully acknowledged. The author also wishes to express his gratitude to one of the Editors, Professor David Turnbull, for his constant encouragement and extraordinary patience during the preparation of this article. Research on elastic interactions was partially funded by the Energy Research and Development Administration, that on phase diagrams by the National Science Foundation, and that on the cluster variation method applied to ordering by the Army Research Office (Durham).

Inelastic Electron Scattering Spectroscopy

S. E. Schnatterly[*]

Joseph Henry Research Laboratories, Princeton University, Princeton, New Jersey

I. Introduction 275
II. Elastic Scattering 278
    1. Born Approximation and Wentzel Model 278
    2. Electron Diffraction and Radiation Damage 280
III. Inelastic Scattering 282
    3. Born Approximation Expression 282
    4. Wentzel Model and Total Cross Section 286
    5. Approximate Mean Free Paths 288
    6. Oscillator Strength Sum Rule 290
    7. Kramers-Kronig Relations and Multiple Scattering 292
    8. Localized States and Angular Momentum Selection Rules 294
    9. Local Field Effects 294
IV. Basic Spectrometer Description 296
    10. Energy Selectors 298
    11. Momentum Transfer Selection 303
    12. Detectors 304
V. Sample Preparation 305
    13. Thin Film Evaporation 306
    14. Sputtering 308
    15. Thinning Macroscopic Crystals 308
VI. Experimental Examples 309
    16. Plasmons in Simple Metals 309
    17. Core Spectra of Simple Metals 329
    18. Semiconductors and Insulators: Valence Excitations 340

I. Introduction

Spectroscopy, in one form or another, has played a key role in the evolution of modern physics—optical spectroscopy and atomic physics being the canonical example. In general, one studies the energy depend-

[*] Present address: Department of Physics, University of Virginia, Charlottesville, Virginia 22901.

ence of the emission or absorption of some kind of particle or field by a selected sample material. This energy dependence reveals dynamical information about the sample. The most successful kinds of spectroscopy are those in which the particle and sample interact only weakly, since then the dynamical information is characteristic of the unperturbed sample.

In condensed matter physics where the sample is extended in space, it is useful to be able to study not only the energy dependence but also the spatial dependence of the emission or absorption rates. The relevant length scale is in the range between the size of an atom and the sample size; or in Fourier transform language, the important wave vectors are in the approximate range 0–1 Å^{-1}. The important energy range depends on the problem being studied.

Figure 1 shows a comparison between four spectroscopies of solids revealing the range of energies and wave vectors accessible to each. The diagonal line represents optical absorption. It is impossible to independently vary the energy and momentum of a photon, since Maxwell's equations fix one once the other is known. Therefore any excitation in a solid that is created by absorption of a photon must lie on the line shown. This means it is impossible, for example, to create an excitation with a 1 Å wavelength and a 10 eV energy, since 10 eV photons have a wavelength of 1000 Å (and a photon of wavelength 1 Å has an energy of 10 keV).

To move off this fixed line, a scattering experiment must be performed. A beam of particles is incident on the sample and scatters through some angle, losing a certain energy. If multiple scattering is not a problem, then the energy lost and momentum transferred in the sample can be ascribed to the creation of a single excitation. By varying the scattering angle and the energy loss, an entire region in energy–momentum space can be studied.

Figure 1 shows that in the low-energy range inelastic neutron scattering reigns supreme. Slow neutrons from a reactor have been used for many years to study an enormous number of excitations in the thermal energy range. Neutrons cannot presently be used to study higher energy excitations because the flux from a reactor drops rapidly above about 0.1 eV.

Recently inelastic X-ray scattering (Compton, or X-ray Raman scattering) has been developed as an experimental tool. This technique is presently limited in its ability to study low-energy excitations by an energy resolution of approximately 2 eV. With further development, including the advent of high-energy synchrotron storage rings as intense X-ray sources, this number may change. X-ray scattering is well suited to high-momentum transfer experiments since the scattering cross section increases with momentum transfer.

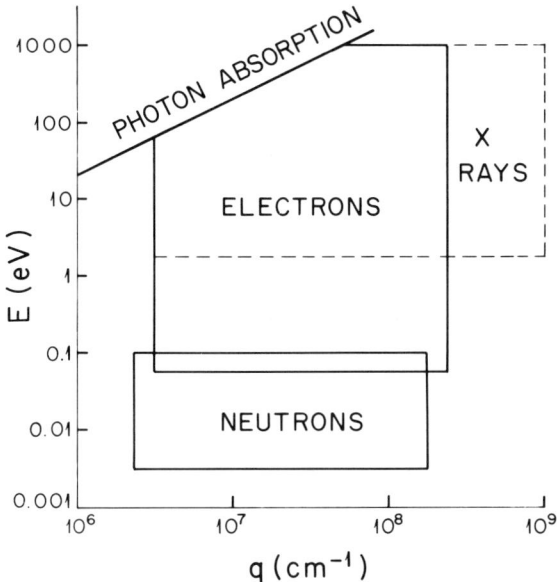

FIG. 1. Regions in E, q space accessible to different experimental probes presently available.

Inelastic electron scattering is nicely complementary to all of the above techniques since the range in energy-momentum space which can readily be covered overlaps each of their regions. The principal difficulty with electrons is that the scattering cross section is larger than one would like, making it necessary to use higher beam energies than is convenient. These difficulties can be overcome, however, as will be described in this article; so electron scattering stands as one of the most versatile of all spectroscopies for the study of condensed matter.

Figure 2 shows, as an example, the measured excitation spectrum of Al from 1-250 eV. In order of increasing energy the features in the spectrum are: an interband transition at 1.5 eV; the surface plasmon at 7 eV; the bulk plasmon at 15 eV; double, triple, and quadruple plasmons at 30, 45, and 60 eV; and the $L_{II,III}$ soft X-ray threshold at 72.5 eV with its associated structure to higher energies. With one instrument, all the elementary excitations in Al from the near infrared to the soft X-ray region can be studied.

Previous reviews of inelastic electron scattering studies include those of Raether[1] and Daniels et al.[2] The present review extends many of the

---

[1] H. Raether, *Springer Tracts Mod. Phys.* **38**, 84 (1965).
[2] J. Daniels, C. Van Festenberg, H. Raether, and K. Zeppenfeld, *Springer Tracts Mod. Phys.* **54**, 78 (1969).

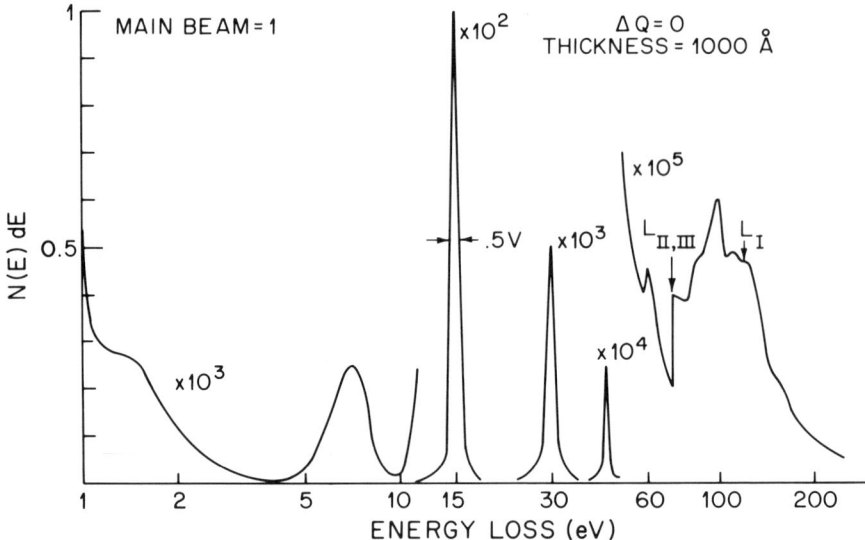

FIG. 2. Excitation spectrum of Al from 1–250 eV.

results previously described and in addition presents new applications of inelastic electron scattering.

II. Elastic Scattering

1. Born Approximation and Wentzel Model

First let us consider the elastic scattering of a fast electron from a single atom. If we assume the Born approximation is adequate, we immediately have for the elastic scattering cross section,[3]

$$\frac{d\sigma}{d\Omega} = \left(\frac{m}{2\pi\hbar^2}\right)^2 |V(q)|^2 \tag{1.1}$$

with

$$V(q) = \int d^3r \, V(r) e^{i\mathbf{q}\cdot\mathbf{r}}, \tag{1.2}$$

where $V(r)$ is the total electrostatic potential of the atom and $\mathbf{q} = \mathbf{k}_0 - \mathbf{k}$ is the momentum transfer in the collision. This expression is traditionally rewritten with the substitution

$$V(q) = -(4\pi e^2/q^2)[Z - F(q)], \tag{1.3}$$

---

[3] L. D. Landau and E. M. Lifshitz, "Quantum Mechanics, Non-Relativistic Theory," Sect. 110. Pergamon, Oxford, 1958.

where

$$F(q) = \int d^3r e^{i q \cdot r} n(r) = \sum_i e^{i q \cdot r_i} \quad (1.4)$$

and $n(r)$ is the electron density of the atom and i labels the electrons.[3a] $F(q)$ is the X-ray form factor of the atom, so that the X-ray scattering is proportional to $[F(q)]^2$. Electron scattering differs from X-ray scattering because electrons scatter from the total electrostatic potential of the atom, while X rays scatter from the electronic density. The rewritten form for the cross section is

$$\frac{d\sigma}{d\Omega} = \frac{4}{a_0^2 q^4} |Z - F(q)|^2, \quad (1.5)$$

where a_0 is the Bohr radius. This is simply the Rutherford cross section for scattering from a point charge multiplied by $|Z - F(q)|^2$.

It is possible to obtain approximate elastic cross sections for all atoms using the form factor of Wentzel[4]

$$F(q) = Z/[1 + (qR)^2] \quad (1.6)$$

which corresponds to an atomic potential of the form

$$V(r) = (Ze/r)e^{-r/R} \quad (1.7)$$

with R being an effective radius of the atom. This leads to a cross section:

$$\frac{d\sigma}{d\Omega} = \frac{4}{a_0^2} \frac{Z^2}{(q^2 + 1/R^2)^2}. \quad (1.8)$$

Thus elastic scattering is independent of q for $qR \ll 1$ and varies as $1/q^4$ for $qR \gg 1$. This gives for a total elastic scattering cross section,

$$\sigma_e = 4\pi Z^2 R^2 / k^2 a_0^2. \quad (1.9)$$

The only relativistic correction that must be applied to this result as well as the above cross sections is the mass change which enters in a_0. The corrected result is

$$\sigma_e = 4\pi Z^2 R^2 \gamma^2 / k^2 a_0. \quad (1.10)$$

[3a] To obtain Eq. (1.3), use Green's second identity

$$\int d^3r (\phi \nabla^2 \psi - \psi \nabla^2 \phi) = \int\int \left(\phi \frac{\partial \psi}{\partial n} - \psi \frac{\partial \phi}{\partial n}\right) \cdot d\mathbf{S}$$

with $\phi = e^{iq \cdot r}$ and $\psi = V(r)$. This leads to

$$\int \nabla^2 V(r) e^{iq \cdot r} d^3r = -q^2 V(q)$$

Then Poisson's equation, $\nabla^2 V(r) = 4\pi e^2 [Z\delta(r) - n(r)]$ leads directly to Eq. (1.3).

[4] G. Wentzel, Z. Phys. **40**, 590 (1927).

This result can be extended to all atoms by assuming a suitable form for $R(Z)$. The form

$$R = a_0 Z^{-1/3}$$

has the advantage that it agrees with the Thomas–Fermi model in the exponent of Z, but the coefficient is more accurate, Thomas–Fermi values being too large.[5]

Assuming this Z dependence for R, we have

$$\sigma_e = (4\pi\gamma^2/k^2)Z^{4/3}. \tag{1.11}$$

Note that the total elastic scattering cross section increases more slowly than the Z^2 result for X rays, due to the partial cancellation between nuclear and core electron potentials.

2. Electron Diffraction and Radiation Damage

Now let $V(r)$ represent the potential due to an atom in a solid, i.e., not an atomic potential but the actual total potential in the neighborhood of an atom. Divide up the solid into nonoverlapping polygons surrounding the atoms. Indicate the position of a unit cell by R_α and atoms in a unit cell by r_i. Then the total potential of the solid can be written

$$V(r) = \sum_{i\alpha} V_i(r - R_\alpha - r_i). \tag{2.1}$$

Adding up all scattered beams leads to

$$\frac{d\sigma}{d\Omega} = |\sum_\alpha e^{iq\cdot R_\alpha}|^2 |\sum_i [Z - F_i(q)]e^{iq\cdot r_i}|^2. \tag{2.2}$$

The first term restricts allowed scattering wave vectors to values such that $q = G$, a reciprocal lattice vector. The second term determines the intensities of the various allowed scattering peaks and is called the structure factor for the lattice.

Electron diffraction can, then as is well known, produce the same kind of structural information as X-ray diffraction. There are some differences however. First, since for high-energy electrons q is very nearly perpendicular to k_0, the incident beam wave vector, not all reciprocal lattice vectors will be accessible. Only a two-dimensional net of points in the plane perpendicular to k_0 represent allowed scattered beams.

Second, electrons scatter much more strongly than X rays. Approximate total mean free paths for 10 keV X rays and 300 keV electrons are

---

[5] von F. Lenz, *Z. Naturforsch.*, Teil A **9**, 185 (1954).

0.25 cm and 10^{-5} cm, respectively. This is primarily a disadvantage for electrons, since it means that extremely thin single crystals must be grown. If this can be accomplished using epitaxial techniques, then there is the advantage that only a very small amount of the material of interest is needed, so that very rare materials can more easily be studied.

Third, there is the problem of radiation damage. Both electrons and X rays damage certain materials readily. This is especially true of crystals of large organic molecules such as amino acids. The radiation damage problem is a subtle and complex one, not fully understood even for simpler materials such as alkali halides. Nevertheless a few words can be said about it. Electrons lose energy gradually, the average energy loss in a single inelastic collision being typically 30 eV. X Rays lose their energy all at once by exciting a photoelectron. This electron then deposits its energy in a relatively small volume of the crystal, due to its very short mean free path. Thus the distribution of damage is expected to be completely different in the two cases. If we ignore this difference, however, and simply take the ratio of elastic to inelastic total cross sections as a figure of merit for diffraction, we can easily make a comparison. For 10 keV X rays the ratio of elastic to inelastic mass absorption coefficients for low-Z materials is typically[6]

$$\frac{\mu_{ray}}{\mu_{abs}} \simeq \frac{0.2 \text{ cm}^2/\text{gm}}{4 \text{ cm}^2/\text{gm}} = 0.05. \qquad (2.3)$$

Using the Wentzel model of atomic potentials, the corresponding ratio for electrons is approximately (see Section 4)

$$\sigma_e/\sigma_i \simeq Z/22, \qquad (2.4)$$

where Z is the atomic number of the target atom. For typical organic matter this ratio is approximately 0.5. From this simple point of view electrons are a factor of 10 better than X rays for diffraction purposes. If we include the fact that electrons typically deposit much less energy in the sample in an inelastic scattering event than do X rays, i.e., 30 eV per inelastic collision as opposed to 10 keV, then electrons appear to be better than X rays by a factor of 3000.

In summary, electron diffraction appears to be superior to X-ray diffraction as a means of studying easily damaged materials, but use of electrons necessitates developing techniques for growing extremely thin crystals.

[6] "American Institute of Physics Handbook," Chapter 8e. McGraw-Hill, New York.

III. Inelastic Electron Scattering

3. BORN APPROXIMATION EXPRESSION

High beam energy means the energy loss and momentum transfer are small compared with energy and momentum of the beam, so the Born approximation should be valid. This means that the interaction between beam and sample is averaged over the unperturbed beam and sample. This is very important since the result gives information about excitations from the *unperturbed* ground state of the sample. No model of the sample wave function is needed. Experimentally in the case of single atom inelastic scattering, the Born approximation is known to be accurate for inelastic scattering from He, for electron energies as low as 250 eV. Higher Z materials require higher beam energies. In solids as will be discussed below, the problem is not so much one of the validity of the Born approximation as the presence of multiple scattering due to the high electron density in the sample.

The Hamiltonian of the system (electron and sample) can be written

$$H = H_0 + \frac{p^2}{2m} + \sum_i \frac{e^2}{|r - r_i|}, \tag{3.1}$$

where H_0 is the Hamiltonian of the unperturbed sample, $p^2/2m$ the kinetic energy operator of the fast electron, and r and r_i the position operator of the fast electron and a sample electron labeled by i. Let λ label the eigenfunctions of the sample. Then the unperturbed system has eigenfunctions of the form

$$|\lambda, k\rangle = \phi_\lambda(r_1 \cdots r_N)(1/\sqrt{V})e^{ik_0 \cdot r}, \tag{3.2}$$

where $\phi_\lambda(r_1 \cdots r_N)$ is the many-body wave function for the sample in state λ with energy E_λ, and V is the volume of a box chosen for normalization of the fast-electron plane wave.

Let $\mathbf{q} = \mathbf{k}_0 - \mathbf{k}$, where k_0 is the momentum of the fast electron before scattering, and k after scattering. Decompose q into components parallel and perpendicular to k. Then

$$q^2 = q_\parallel^2 + q_\perp^2. \tag{3.3}$$

For small angles (i.e., $|q| \ll |k_0|$), $q_\perp = k_0\theta$, where θ is the scattering angle. By energy conservation, the energy loss is

$$2m(\Delta E/\hbar^2) = k_0^2 - k^2 = k_0^2 - (\mathbf{k}_0 - \mathbf{q})^2 = 2\mathbf{k}_0 \cdot \mathbf{q} - q^2. \tag{3.4}$$

For $q \ll k$,

$$(2m/\hbar^2)E = 2\mathbf{k}_0 \cdot \mathbf{q} = 2k_0 q_\parallel. \tag{3.5}$$

So

$$q_\parallel = \frac{2m}{\hbar^2}\frac{\Delta E}{2k_0} \equiv k_0 \theta_E. \qquad (3.6)$$

Then

$$q^2 = k_0^2(\theta^2 + \theta_E^2). \qquad (3.7)$$

Fermi's golden rule can be written[7]

$$W_{\lambda_0 \to \lambda}(\omega) = \frac{2\pi}{\hbar^2}|\langle \lambda k| \sum_i \frac{e^2}{|r - r_i|}|\lambda_0 k_0\rangle|^2 \delta\left(\frac{E_{\lambda_0} - E_\lambda}{\hbar} - \omega\right). \qquad (3.8)$$

Since Van Hove's work the matrix element has usually been rewritten as[8]

$$\langle \lambda k|\sum_i \frac{e^2}{|r-r_i|}|\lambda_0 k_0\rangle = \frac{1}{V}\langle \lambda|\sum_i \int d^3r\, e^{iq\cdot r}\frac{e^2}{|r-r_i|}|\lambda_0\rangle$$

$$= \frac{1}{V}\langle \lambda|\sum_i e^{iq\cdot r_i}\int d^3r\, e^{iq\cdot(r-r_i)}\frac{e^2}{|r-r_i|}|\lambda_0\rangle$$

$$= \frac{1}{V}\frac{4\pi e^2}{q^2}\langle \lambda|\sum_i e^{iq\cdot r_i}|\lambda_0\rangle. \qquad (3.9)$$

So

$$W_{\lambda_0 \to \lambda}(\omega) = \frac{2\pi}{\hbar^2}\left(\frac{4\pi e^2}{q^2}\right)^2 \langle \lambda_0|\frac{1}{V}\sum_i e^{-iq\cdot r_i}|\lambda\rangle$$

$$\langle \lambda|\frac{1}{V}\sum_i e^{iq\cdot r_i}|\lambda_0\rangle \delta\left(\frac{E_{\lambda_0} - E_\lambda}{\hbar} - \omega\right). \qquad (3.10)$$

Now use

$$\delta(\omega) = \frac{1}{2\pi}\int_{-\infty}^{\infty} \exp(i\omega t)\, dt \qquad (3.11)$$

$$W_{\lambda_0\to\lambda}(\omega) = \frac{1}{\hbar^2}\left(\frac{4\pi e^2}{q^2}\right)^2 \int dt \exp(-i\omega t)\langle \lambda_0|\frac{1}{V}\sum_i \exp(-iq\cdot r_i)|\lambda\rangle$$

$$\exp\left[\frac{i(E_{\lambda_0} - E_\lambda)}{\hbar}t\right]\langle \lambda|\frac{1}{V}\sum_i \exp(iq\cdot r_i)|\lambda_0\rangle$$

---

[7] L. D. Landau and E. M. Lifshitz, "Quantum Mechanics, Non-Relativistic Theory." Pergamon, Oxford, 1958.

$$= \frac{1}{\hbar^2} \left(\frac{4\pi e^2}{q^2}\right)^2 \int dt \exp(-i\omega t) \langle \lambda_0 | \frac{1}{V} \sum_i \exp\left(\frac{iE_{\lambda_0}}{\hbar} t\right)$$
$$\exp(-iq\cdot r_i) \exp\left(\frac{-iE_\lambda}{\hbar}\right) |\lambda\rangle\langle\lambda| \frac{1}{V} \sum_i \exp(iq\cdot r_i) |\lambda_0\rangle.$$
(3.12)

The two energy exponentials can be replaced with exponentials of the Hamiltonian, and together they form the time translation operator. Then the transition rate can be written

$$W_{\lambda_0 \to \lambda}(\omega) = \frac{1}{\hbar^2} \left(\frac{4\pi e^2}{q^2}\right)^2 \int dt \, e^{-i\omega t}$$
$$\langle \lambda_0 | \frac{1}{V} \sum_i e^{-iq\cdot r_i(t)} |\lambda\rangle\langle\lambda| \frac{1}{V} \sum_i e^{iq\cdot r_i(0)} |\lambda_0\rangle \quad (3.13)$$

What is measured is not $W_{\lambda_0 \to \lambda}$ but the total transition rate summed over all final states λ, with the δ-function eliminating terms that do not conserve energy:

$$W_{\lambda_0}(\omega) = \frac{1}{\hbar^2} \left(\frac{4\pi e^2}{q^2}\right)^2 \int dt \, e^{-i\omega t} \langle \lambda_0 | \frac{1}{V} \sum_{ij} e^{-iq\cdot r_i(t)} e^{iq\cdot r_i(0)} |\lambda_0\rangle$$
$$\equiv \frac{2\pi}{\hbar} \left(\frac{4\pi e^2}{q^2}\right)^2 \frac{N}{V^2} S(q, \omega),$$
(3.14)

where the dynamic structure factor has been introduced:

$$S(q, \omega) = \frac{1}{2\pi \hbar N} \int dt \, e^{-i\omega t} \langle \lambda_0 | \sum_{ij} e^{-iq\cdot r_i(t)} e^{iq\cdot r_j(0)} |\lambda_0\rangle$$
$$= \frac{1}{2\pi \hbar N} \int dt \, e^{-i\omega t} \langle \lambda_0 | n_q(t) n_{-q}(0) |\lambda_0\rangle,$$
(3.15)

where $n_q(t)$ is the Fourier transform of the density of electrons at time t. To convert the transition rate to a differential cross section per atom, we must divide W by the incident flux $\hbar k/Vm$, the number of atoms in the sample N, and multiply by the number of states available for the fast electron to lose energy between $\hbar\omega$ and $\hbar(\omega + d\omega)$. The density of states in k space is

$$\rho(k) \, d^3k = [V/(2\pi)^3] k^2 \, dk. \quad (3.16)$$

So in energy,

$$\rho(E) \, dE \, d\Omega = \frac{V}{(2\pi)^3} \frac{m}{\hbar} k \, d\omega. \quad (3.17)$$

And we find

$$\frac{d^2\sigma}{d\omega\,d\Omega} = \frac{4\hbar}{a_0^2 q^4} S(q, \omega), \qquad (3.18)$$

where a_0 is the Bohr radius and we have assumed $|k_0| \simeq |k|$. This equation is the meeting ground between experiment and theory. On the left-hand side we have counts per second in a certain solid angle; on the right-hand side is a particular correlation function—a property of the unperturbed sample material—which a complete theory of the material can provide. Multiplying $S(q, \omega)$ is a simple well-known function—some constants and the Fourier transform of the Coulomb potential of a point charge. This separation into an amplitude term multiplying a structure term is characteristic of experiments which can be analyzed with the Born approximation, as was pointed out by Van Hove 20 years ago.[8] This is an example of the fluctuation–dissipation theorem: The energy dissipation in the sample is directly proportional to the density fluctuations which can take place.

Another form for the scattering cross section can be obtained by relating $S(q, \omega)$ to the longitudinal dielectric response function of the sample[9]:

$$S(q, \omega) = \frac{q^2 V}{4\pi^2 e^2 N} \operatorname{Im}\left[\frac{-1}{\epsilon(q, \omega)}\right]. \qquad (3.19)$$

Aside from retardation effects which cause new modes of excitation to appear, the only important relativistic correction is due to the mass change in a_0 causing the result to be multiplied by γ^2. Making this correction we have the equivalent forms

$$\frac{d^2\sigma}{d\Omega\,d\omega} = \frac{\hbar\gamma^2}{(\pi e a_0)^2 n} \frac{1}{k_0^2} \frac{1}{\theta^2 + \theta_E^2} \operatorname{Im}\left[\frac{-1}{\epsilon(q, \omega)}\right] \qquad (3.20)$$

$$= \frac{\hbar\gamma^2}{(\pi e a_0)^2 n} \frac{1}{q^2} \operatorname{Im}\left[\frac{-1}{\epsilon(q, \omega)}\right] \qquad (3.21)$$

$$= \frac{\hbar\gamma^2}{2\pi^2 a_0 n} \frac{1}{E} \frac{1}{\theta^2 + \theta_E^2} \operatorname{Im}\left[\frac{-1}{\epsilon(q, \omega)}\right], \qquad (3.22)$$

where n is the number density for electrons, k_0 the beam wave vector, and E the beam energy. Since $\epsilon(q, \omega)$ has roughly the same magnitude for different q values, this result shows that the cross section falls off approximately as $1/q^2 = 1/(q_\parallel^2 + q_\perp^2)$.

[8] L. Van Hove, *Phys. Rev.* **95**, 249 (1954).
[9] See, for example, P. M. Platzman and P. A. Wolff, "Waves and Interactions in Solid State Plasmas," Chapter 2. Academic Press, New York, 1972.

4. WENTZEL MODEL AND TOTAL CROSS SECTION

In order to obtain a total inelastic scattering cross section for comparison with the total elastic cross section, we must integrate over all energies. Beginning with Eq. (3.15), we have for a single atom,

$$\int S(q, \omega) \, d\omega = \frac{1}{2\pi\hbar} \int dt \, d\omega \, e^{-i\omega t} \langle 0 | \sum_{ij} e^{-i q \cdot r_i(t)} e^{i q \cdot r_j(0)} | 0 \rangle$$

$$= \frac{1}{\hbar} \int dt \, \delta(t) \langle 0 | \sum_{ij} e^{-i q \cdot r_i(t)} e^{i q \cdot r_j(0)} | 0 \rangle \quad (4.1)$$

$$= \frac{1}{\hbar} \langle 0 | \sum_{ij} e^{-i q \cdot r_i(0)} e^{i q \cdot r_j(0)} | 0 \rangle.$$

So the total cross section (elastic plus inelastic) is

$$\sigma_T = \int \frac{d^2\sigma}{d\omega \, d\Omega} \, d\omega$$

$$= \frac{4}{a_0^2 q^4} \langle 0 | \sum_{ij} e^{-i q \cdot r_i(0)} e^{i q \cdot r_j(0)} | 0 \rangle. \quad (4.2)$$

To include all scattering, elastic as well as inelastic, the sums must be extended to include nuclei as well as electrons. Explicitly carrying out the sums over nuclei, we have

$$\sigma_T = \frac{4}{a_0^2 q^4} \langle 0 | \sum_{ij} e^{-i q \cdot r_i} e^{i q \cdot r_j} + Z^2$$

$$- Z(\sum_i e^{-i q \cdot r_i} + \sum_j e^{i q \cdot r_j}) | 0 \rangle$$

$$= \frac{4}{a_0^2 q^4} \{Z^2 - Z[F(q) + F^*(q)] \quad (4.3)$$

$$+ \langle 0 | \sum_{ij} e^{-i q \cdot r_i} e^{i q \cdot r_j} | 0 \rangle \},$$

where the sums are over electronic coordinates only. This bears a resemblance to the elastic cross section which is proportional to

$$|\langle 0 | Z - \sum_i e^{i q \cdot r_i} | 0 \rangle|^2 = Z^2 - Z[F(q) + F^*(q)] + |F(q)|^2. \quad (4.4)$$

So we have

$$\sigma_i = \sigma_T - \sigma_e = \frac{4}{a_0^2 q^4} (\langle 0 | \sum_{ij} e^{-i q \cdot r_i} e^{i q \cdot r_j} | 0 \rangle - |\langle 0 | \sum_i e^{i q \cdot r_i} | 0 \rangle|^2). \quad (4.5)$$

That is, the total inelastic scattering is proportional to the dispersion of the operator $\sum_i e^{iq \cdot r_i}$. The simplest way to evaluate the dispersion is to assume a Hartree-type product of one electron states $|i\rangle$ for the total ground-state wave function $|0\rangle$.

Then the first term in the expression for the dispersion can be rewritten

$$\langle 0 | Z + \sum_j \sum_{i \pm j} e^{iq \cdot (r_j - r_i)} | 0 \rangle \tag{4.6}$$

$$= (Z + \sum_j \langle j | e^{iq \cdot r_j} | j \rangle \langle i | \sum_{i \pm j} e^{-iq \cdot r_j} | i \rangle) \tag{4.7}$$

$$= Z - \sum_i |\langle 0 | e^{iq \cdot r_i} | 0 \rangle|^2 + |\langle 0 | \sum_i e^{iq \cdot r_i} | 0 \rangle|^2 \tag{4.8}$$

$$= (Z - \sum_i |f_i|^2 + |F(q)|^2), \tag{4.9}$$

where f_i is the structure factor for the ith electron. We then have for the total inelastic cross section

$$\sigma_i = \frac{4}{a_0^2 q^4} (Z - \sum_i |f_i|^2), \tag{4.10}$$

a result first obtained by Wentzel[4] and later used by Lenz.[5] To obtain approximate results as a function of atomic number, we shall use Eq. (1.6) for f, which yields

$$\sigma_i = \frac{4Z}{a_0^2 q^4} \left[\frac{(qR)^2}{1 + (qR)^2} \right]; \tag{4.11}$$

which shows that for $qR \ll 1$, σ_i varies as $1/q^2$ as previously mentioned, and for $qR \gg 1$, σ_i approaches the Rutherford (or Compton) limit, varying as $1/q^4$.

Integrating σ_i over all q, one obtains the total inelastic cross section

$$\sigma_i = \frac{8\pi R^2 Z}{k^2 a_0^2} \ln \left(\frac{1}{q_\parallel^2 R^2} \right), \tag{4.12}$$

where $q_\parallel$ is the longitudinal momentum transfer associated with the average energy loss.

This result should be multiplied by γ^2 to make it relativistically correct. Using $R = a_0 Z^{-1/3}$,

$$\sigma_i = \frac{8\pi \gamma^2 Z^{1/3}}{k^2} \ln \left(\frac{Z^{2/3}}{q_\parallel^2 a_0^2} \right). \tag{4.13}$$

Figure 3 shows total inelastic and elastic cross sections as a function of momentum transfer q. Note that in the small q range inelastic scattering dominates, while at large q the converse is true. In this case, $Z = 14$, which is typical, the crossover occurs at $q \approx 1.5$ Å^{-1}.

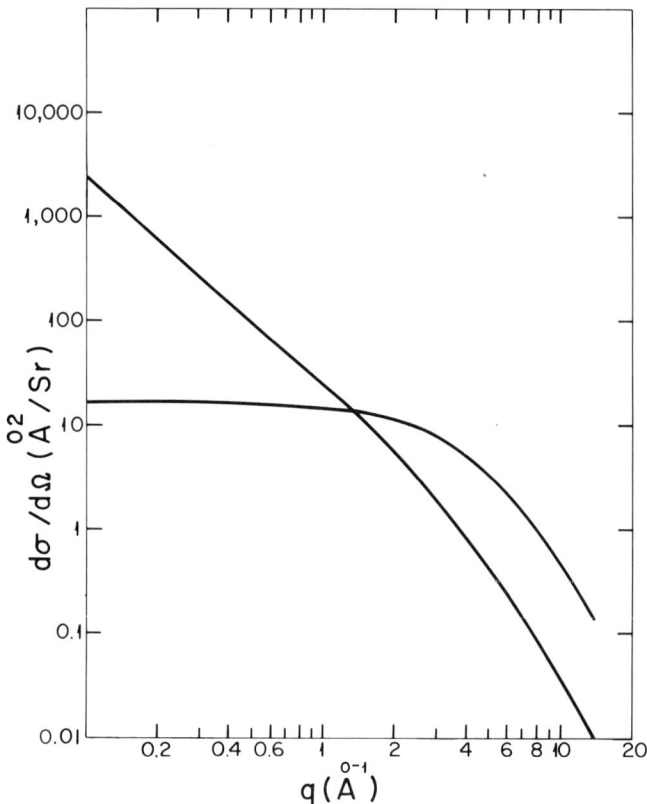

FIG. 3. Total elastic and inelastic cross sections as a function of q.

5. Approximate Mean Free Paths

One of the major difficulties with electron scattering experiments is that the target–probe interaction is stronger than one would like, so the fast electrons are likely to scatter more than once in passing through the sample. For this reason, it is important to be able to estimate approximate mean free paths for both elastic and inelastic scattering in various samples. The above-derived total cross sections can be used, but one needs an atomic density to obtain a mean free path. The easiest way around this problem is to define a mass absorption coefficient

$$\mu = \rho l = \frac{\rho}{n\sigma} = \frac{A}{N_0 \sigma} \quad \text{g/cm}^2, \tag{5.1}$$

where l is a mean free path, n the atomic density, σ the total cross section, A the atomic mass, and N_0 Avogadro's number. Figure 4 shows elastic, inelastic, and total mass absorption coefficients as a function of Z using the above-derived cross sections.

Mean free path estimates can be used in the following ways: Suppose the probability of creating a given excitation is given by tp, where t is the sample thickness. Then, ignoring multiple scattering, the counting rate for observing the excitation is proportional to

$$N = tpe^{-t/l_T}. \tag{5.2}$$

Differentiating this and setting it equal to zero results in $t = l_T$ for maximum counting rate. This condition applies also to maximizing the signal-to-noise ratio when there is a large background underlying the feature of interest due to other excitation processes. This condition, along with Fig. 4, is one way of deciding in advance on the desired thickness for a given sample.

Maximum counting rate is not the only criterion of importance in deciding on sample thickness, however. In high q experiments one is often bothered by multiple elastic and inelastic scattering, referred to loosely as "thermal diffuse" scattering. This is because, as explained earlier, elastic scattering is roughly q independent out to $q \simeq 1/R$, while inelastic scattering falls of as $1/q^2$ in this range. Therefore, at some value of q sufficiently large, the probability of elastic scattering with momentum

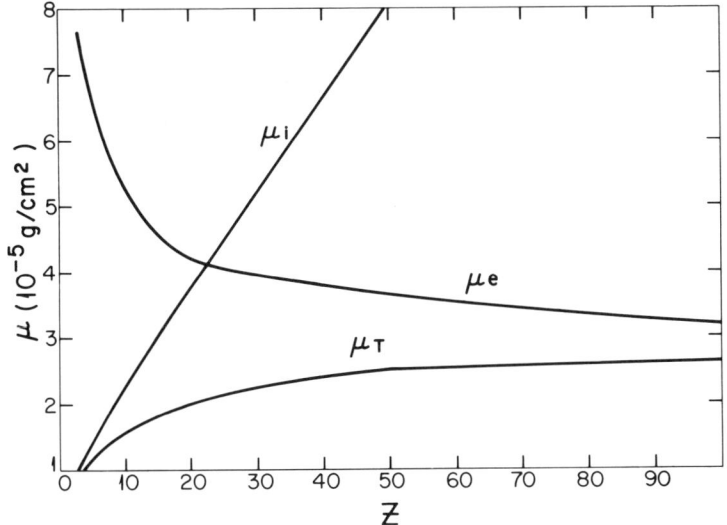

FIG. 4. Elastic, inelastic, and total mass absorption coefficients as a function of atomic number Z.

transfer q followed by inelastic scattering with $q \simeq 0$ may become comparable to a single inelastic scattering with momentum transfer q.

To make this explicit, assume a feature of interest has a counting rate proportional to $(t/l_i)(q_c/q)^2 f$, where q_c is a characteristic upper limit wave vector used in defining l_i, and f is the fraction of the total inelastic scattering corresponding to the feature of interest. Define the thermal diffuse mean free path as $l_{TD} = Bl_e$, where B depends on the crystallinity of the sample and the temperature. Then the total inelastic counting rate is proportional to

$$N(q) = f\frac{t}{l_i}\left(\frac{q_c}{q}\right)^2 e^{-t/l_T} + f\frac{t}{l_i}\ln\left(\frac{\delta q}{q_\parallel}\right)\frac{t}{Bl_e} e^{-t/l_T}, \quad (5.3)$$

where the first term represents single inelastic scattering and the second thermal diffuse scattering. δq is the momentum resolution of the spectrometer and $q_\parallel$ is defined in Eq. (3.3). The logarithm comes from averaging the inelastic scattering rate over the momentum resolution function, assumed to be a step function, centered on $q = 0$. Then the ratio of thermal diffuse to direct scattering is equal to

$$R(q) = \left(\frac{q}{q_c}\right)^2 \ln\left(\frac{\delta q}{q_\parallel}\right)\frac{t}{Bl_e}. \quad (5.4)$$

To keep the thermal diffuse ratio down to an acceptable level at a given momentum transfer, the sample thickness may need to be less than that which produces the optimum counting rate. Lowering the temperature and improving the crystallinity will also help. So, for instance, if it is found that the thermal diffuse ratio in Mg at room temperature is 15% at $q = 1.8$ Å^{-1}, Fig. 4 can be used to estimate how thick any other sample must be made to achieve a given thermal diffuse ratio at a specified momentum transfer.

6. Oscillator Strength Sum Rule

The important oscillator strength sum rule can be derived as follows: Multiply the definition of $S(q, \omega)$ by $e^{i\omega t'}$, and integrate

$$\int d\omega\, e^{i\omega t} S(q, \omega) = \frac{1}{N\hbar}\langle\lambda|\sum_{ij} e^{-iq\cdot r_i(t)} e^{iq\cdot r_j(0)}|\lambda\rangle. \quad (6.1)$$

Differentiate with respect to time:

$$\int d\omega\, e^{i\omega t}\omega S(q, \omega) = \frac{-1}{N\hbar^2}\langle\lambda|\sum_{ij} [e^{-iq\cdot r_i(t)}, H]e^{iq\cdot r_j(0)}|\lambda\rangle. \quad (6.2)$$

Set $t = 0$

$$\int d\omega \, \omega S(q, \omega) = \frac{-1}{N\hbar^2} \langle \lambda | \sum_{ij} [e^{-iq\cdot r_i(0)}, H] e^{iq\cdot r_j(0)} | \lambda \rangle. \quad (6.3)$$

The remarkable thing about this result is that the potential term in the Hamiltonian contributes nothing to the commutator, so the result is independent of the complexities of the many-body problem as long as the interaction between the particles in the system depends only on their positions. The commutator can be evaluated straightforwardly using

$$[e^{-iq\cdot r_i}, P_j^2] = [e^{-iq\cdot r_i}, P_j]P_j + P_j[e^{-iq\cdot r_i}, P_j] \quad (6.4)$$

and

$$[e^{-iq\cdot r_i}, P_j] = \hbar q \, e^{-iq\cdot r_i} \delta_{ij}. \quad (6.5)$$

This leads to

$$[e^{-iq\cdot r_i}, P_i^2] = e^{-iq\cdot r_i}(2\hbar q \cdot P_i - \hbar^2 q^2). \quad (6.6)$$

So

$$\langle \lambda | \sum_{ij} [e^{-iq\cdot r_i}, H] e^{iq\cdot r_j} | \lambda \rangle = \frac{-1}{2m} \sum_i \hbar^2 q^2 = -\frac{\hbar^2 q^2}{2m} N \quad (6.7)$$

and

$$\int d\omega \, S(q, \omega)\omega = \frac{q^2}{2m}. \quad (6.8)$$

Showing once again that the overall strength of the inelastic scattering varies as $1/q^2$. Using the relation between $S(q, \omega)$ and $\text{Im}[-1/\epsilon(q, \omega)]$ yields the final result

$$\int d\omega \, \text{Im}\left(\frac{-1}{\epsilon(q, \omega)}\right) \omega = \frac{\pi}{2} \omega_p^2. \quad (6.9)$$

Notes:

1. This rule is quite general—independent of $V(r_1 \cdots r_n)$.
2. It applies to *all* electrons in the system: $1/N$ of the sum does not necessarily belong to each electron. So for example, a filled d band with 10 electrons per atom will not necessarily contribute 10 electrons per atom worth to the sum. Outer electrons have more than their share, inner electrons less, because the oscillator strength for transitions to states of lower energy is negative, and transitions to these states are prevented by the Pauli principle.

3. This rule is useful theoretically in checking model calculations and in insuring that simple models work reasonably well. For example, the Penn model of the dielectric response function of semiconductors lumps all the absorption at a single frequency, and satisfies the oscillator strength sum rule.[10] With the frequency of the absorption appropriately chosen, this provides an extremely simple model of semiconductor screening which is adequate for many purposes.

4. Sum rules are useful to experimentalists in carrying out a Kramers-Kronig analysis of optical or inelastic electron scattering measurements. (See Section 7.)

7. Kramers-Kronig Relations and Multiple Scattering

As shown above the inelastic electron scattering cross section provides a direct measure of $\text{Im}[-1/\epsilon(q, \omega)]$. In order to obtain the longitudinal response function $\epsilon_1(q, \omega) + i\epsilon_2(q, \omega)$ we must use a Kramers-Kronig analysis.[11] This procedure is very similar to obtaining the transverse response function from measurements of the reflectivity of a solid surface. As $q \to 0$, these two response functions become identical. In practice, however, there are two differences between these two experimental methods. First, a broader energy range is more easily covered using electron scattering than in the case of optics, so the endpoint corrections are less of a problem. Second, multiple scattering is a serious problem in electron scattering work when absolute values of the cross section are desired. Multiple scattering generally results in the observed spectrum being too large at intermediate energies. This can spoil the usefulness of the sum rules if it is not corrected for.

Several schemes have been described for taking multiple scattering into account. Most have included only double scattering, since this is often the dominant correction. The shape of the double scattering contribution to the spectrum can be approximated by a convolution of the measured spectrum with itself. In doing so, two errors are made. First, the measured spectrum includes double scattering, which when convoluted will result in errors. This can be corrected for by subtracting the estimated double scattering from the measured spectrum before convoluting. In order to get started, it is necessary to assume that a certain range of the spectrum at low energies is free of double scattering. This is rigorously true in insulators, for which the lowest possible energy at which double scattering will appear is twice the band gap. In metals it is

---

[10] D. R. Penn, *Phys. Rev.* **128**, 2093 (1962).
[11] See, for example, D. Pines and P. Nozières, "The Theory of Quantum Liquids," Vol. 1. Benjamin, New York, 1966.

usually true that double scattering is weak below twice the plasma energy. Recently a scheme has been developed in which this procedure is modified to include triple scattering at the same time.[12]

The second error being made is the neglect of the momentum transfer dependence of the spectral shape. In a measurement made at momentum transfer q, the only restriction on each of the momenta involved in the observed double scattering is $q_1 + q_2 = q$. Since the shape of the spectrum generally depends on q, what is needed is a three-dimensional convolution of data, two momentum dimensions and one energy dimension. Most schemes used to date approximate the effect of the momentum convolutions by assuming that certain values of q_1 and q_2 dominate, such as, $q_1 \approx q_2 \approx q/2$.

Besides the shape of the multiple scattering contribution, one must obtain its amplitude. If the absolute cross section is measured and the sample thickness known, this can be done directly. Such an approach is difficult, and impossible if the sample thickness is nonuniform. It is still possible to make progress, however, by using the oscillator strength sum rule and one other piece of information.

Let $f(E)$ be the measured spectrum after multiple scattering has been subtracted as described above. Then the true single and double inelastic cross sections, S_1 and S_2 are

$$S_1 = Af(E) \tag{7.1}$$

$$S_2 = Bf(E) x f(E), \tag{7.2}$$

where A and B are unknown constants and x represents a convolution. Two equations must be provided to evaluate the two unknowns. One excellent candidate is the oscillator strength sum rule. This requires estimating the effective number of electrons appropriate for the energy range covered, but this is not too difficult since a considerable range is available. Usually only deep core excitations need be left out and their contribution is not great, as explained above. A second candidate, nicely complementary to the first is the value of $\text{Re}[(1/\epsilon(q,0)) - 1]$. This number complements the sum rule since its value is dominated by low-energy excitations, while the sum rule emphasizes high energies. In addition, one can often obtain this value from other sources. For $q \approx 0$ all that is needed is the optical index of refraction in insulators and semiconductors, while for metals, $\text{Re}[(1/\epsilon(0,0)) - 1] = -1$. At sufficiently high q values, a model of local field effects is needed to describe the variation of $\epsilon(q,0)$

---

[12] J. R. Fields, Ph.D. Thesis, Princeton University, Princeton, New Jersey (1975); and to be published.

with q. This procedure has been tested successfully for small q using LiF (see Section 18c).

8. Localized States and Angular Momentum Selection Rules

Consider a transition from an initial state λ_0, which is tightly bound and localized about a particular atom in the solid, to any kind of final state λ. For example λ_0 may be an atomic core state in any type of solid, or it may be a localized valence state in a molecular crystal. It is clear from Eq. (3.9) that the operator which causes this transition is $e^{iq\cdot r}$ where r is the position operator of the electron undergoing the transition, and q the momentum transfer. Suppose $1/q$ is large compared with the spacial extent of the localized state λ. Then, although the radial integral extends over all space, the extreme localization of λ_0 cuts it off at a small value of r and we may expand the exponential before integration:

$$e^{iq\cdot r} = 1 + iq\cdot r - \tfrac{1}{2}(q\cdot r)^2 + \ldots . \tag{8.1}$$

The first term contributes nothing since λ_0 and λ are orthogonal to each other. The second term is proportional to $Y_1(\theta,\phi)$ and causes electric dipole transitions. Therefore at small q the transitions excited by the fast electron will be the same as those seen in optical absorption measurements. The third term can be written

$$-\tfrac{1}{2}(q\cdot r)^2 = -\tfrac{1}{2}q^2 z^2 = -\tfrac{1}{6}q^2 r^2 - \tfrac{1}{6}(3z^2 - r^2), \tag{8.2}$$

so it contains terms proportional to both $Y_0(\theta,\phi)$ and $Y_2(\theta,\varphi)$ and therefore causes monopole and quadrupole transitions. Such transitions are, of course, regarded as "forbidden" in optical spectroscopy. Higher terms in the expansion of $e^{iq\cdot r}$ cause higher order multipole transitions.

Therefore we see that by scanning the energy loss spectrum at different values of q, different multipole transitions can be observed. This kind of experiment is well known among atomic physicists and chemists using inelastic electron scattering to study individual atoms in the gaseous state. In that case much lower beam energies can be used, so the energy resolution problem is not so severe. Only recently have high beam energies been combined with high energy resolution to allow similar studies to be carried out in solids. Some examples will be given later.

9. Local Field Effects

The dielectric response function of a solid is usually derived by assuming that the only frequencies and wave vectors present in the microscopic electric field are those present in the driving term—in our case the displacement due to the fast electron. Neglecting frequencies other than the applied ones corresponds to ignoring any nonlinearities in the response of the medium and is a good approximation for all the systems which will be described here. Ignoring other wave vectors, on the other hand, corresponds to ignoring the atomic nature of matter. This might be a good approximation when discussing plasmons in simple metals but need not be accurate in the case of insulators where the polarizable units of the material, the ions, are located on the points of a lattice. Then the electric field averaged over a unit cell (the macroscopic field) need not be the same as the field evaluated at the location of a particular ion.

The simplest way to include the effect of local fields on the dielectric response function $\epsilon(0,\omega)$ is to use the Lorenz–Lorenz formula

$$\epsilon(0,\omega) = 1 + \frac{4\pi n\alpha}{1 - (4\pi/3)n\alpha}, \tag{9.1}$$

where α is the atomic polarizability of the atoms of the material (here assumed all identical) and n the number density of the atoms. This formula is based on the classical point dipole approximation. Calculations have also been carried out on a more microscopic basis.[13]

Recently, Nagel and Witten[14] have used the point dipole approximation to study the effects of local field corrections to the complete response function $\epsilon(q,\omega)$. That is, they included the effect of variations in wave vector as well as frequency. Their model should be accurate when the important polarizable units of the solid are highly localized. This includes not only insulators such as alkali halides and molecular solids but also any material in a frequency range such that core states dominate $\epsilon(q,\omega)$.

Using this approach, the total microscopic field at the position of an ion is calculated by adding to the applied field the fields due to all the induced dipoles in the vicinity. We must include the fact that the phases of these dipoles vary with position due to the nonzero value of q. Thus the total field driving a given ion at a particular frequency varies with q due to the phase variation of the surrounding dipolar fields. Since the induced fields depend on the magnitude of the atomic polarizability, $\alpha(\omega)$, the total field is also strongly frequency dependent. Therefore, in order to understand quantitatively the shape of $\epsilon(q,\omega)$ it is important in general to include local field effects. Nagel and Witten[14] find

$$\epsilon(q,\omega) = 1 + \frac{4\pi n\alpha(\omega)}{1 - B(q)n\alpha(\omega)}, \tag{9.2}$$

where $B(q)$ is a function which they evaluate numerically, and n and α have the same meaning as in Eq. (9.1). Figure 5 shows examples of $B(q)$

[13] S. L. Adler, *Phys. Rev.* **126**, 413 (1962).
[14] S. R. Nagel and T. A. Witten, Jr., *Phys. Rev. B* **11**, 1623 (1975).

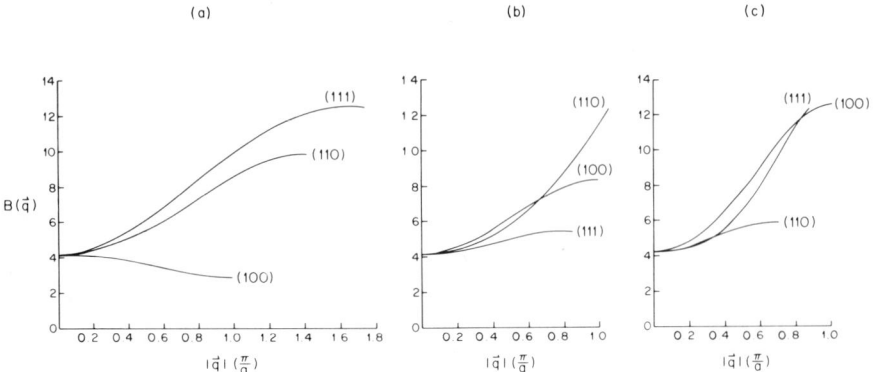

FIG. 5. Local field factor $B(q)$ for the three cubic crystal structures: (a) simple cubic, (b) face-centered cubic, (c) body-centered cubic [S. R. Nagel and T. A. Witten, Jr., *Phys. Rev.* **B11**, 1623 (1975)].

for the three cubic crystal structures. Figure 6 shows the prediction of the changes in $\epsilon_2(q,\omega)$, due solely to local field effects, on changing q from 0 to π/a in the (111) and (110) directions in CsCl.

Similar calculations have been carried out for covalent materials, but the resulting response functions have only been evaluated for $q = 0$ to compare with optical measurements.[15] At the time of this writing these works disagree with each other, so there appears to be a need for further effort in this area.

IV. Basic Spectrometer Description

An electron scattering spectrometer suitable for carrying out the kinds of experiments described here consists of seven major elements: (1) an electron gun that produces a beam of well-defined size and angular divergence, (2) when high-energy resolution is needed, a monochromator capable of reducing the energy spread in the beam below that produced by the cathode, (3) a power supply and accelerating optics to raise the kinetic energy of the beam to the desired value, (4) a sample holder that supports the thin film sample in the beam, (5) an aperture for selecting the angle of scattering, (6) an energy analyzer for selecting electrons that have lost a definite amount of energy in the sample, and (7) a detector that can measure and record the flux of electrons passed by the system.

---

[15] J. A. Van Vechten and R. M. Martin, *Phys. Rev. Lett.* **28**, 446 (1972); W. R. Hauke and L. J. Sham, *ibid.* **33**, 582 (1974); S. G. Louie, J. R. Chelikowsky, and M. L. Cohen, *ibid.* **34**, 155, (1975).

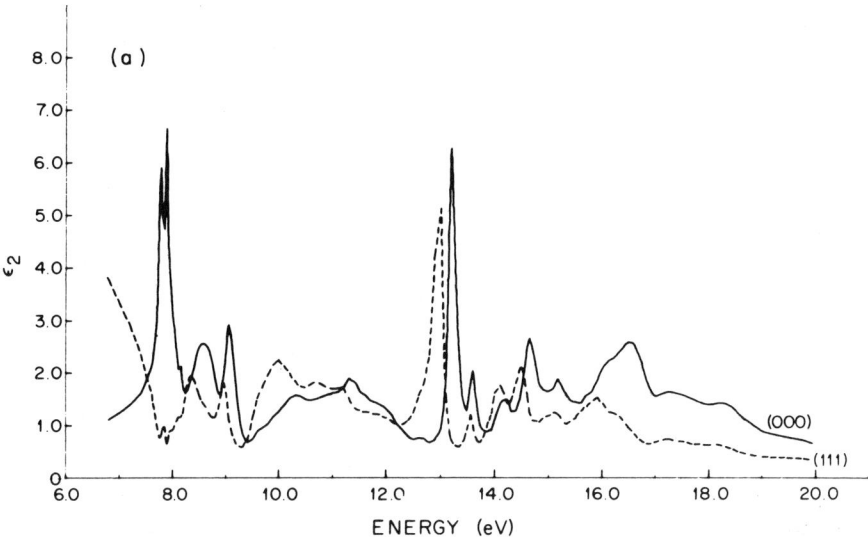

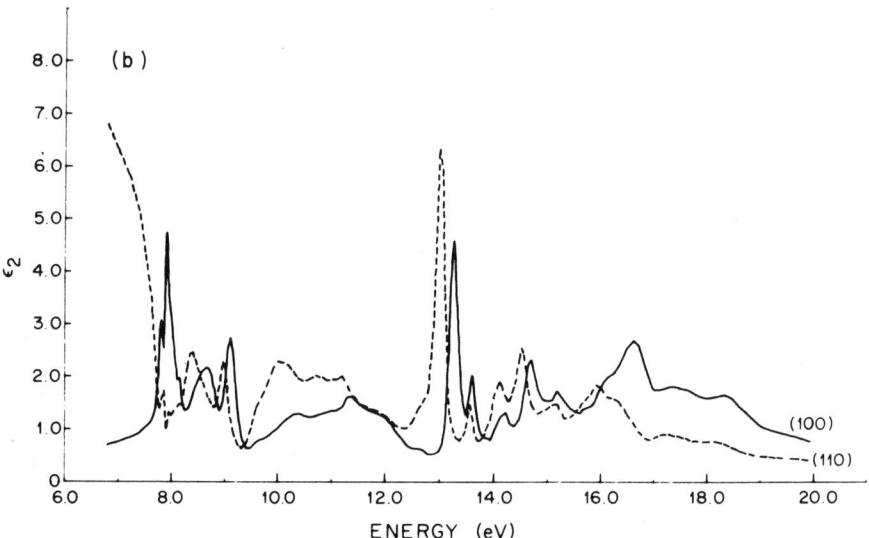

FIG. 6. Local field contributions to the change in spectral shape of ϵ_2 with q for CsCl in (a) the (111) direction and (b) the (110) direction [S. R. Nagel and T. A. Witten, Jr., *Phys. Rev.* **B11**, 1623 (1975)].

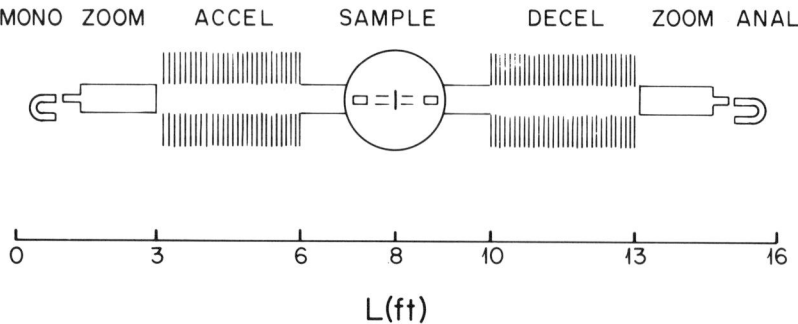

FIG. 7. Inelastic electron scattering machine schematic.

A schematic of the electron-optical elements of such a spectrometer are shown in Fig. 7.[16]

It is highly desirable that the sample chamber and the vacuum chamber housing the most sensitive electron optics be capable of baked ultrahigh vacuum to reduce reaction rates of active samples and stabilize the work functions of the exposed electron-optical elements. The entire system should be magnetically shielded for stability. In addition it is very useful to have data taking controlled by a mini-computer so the most tedious aspects of doing an experiment are left to machines. The computer can rapidly carry out repeated scans over the same energy range averaging out slow intensity drifts in the beam.

Each of the above major elements will be briefly described.

10. Energy Selectors

The dispersion of any energy selector can be approximately written

$$\Delta E/E = C(\Delta x/R), \qquad (10.1)$$

where E is the energy of the beam, ΔE the energy width transmitted by the selector, Δx a slit width or aperture diameter, R a characteristic path length for electrons in the selector, and C a constant of order unity. Energy selectors are used both as monochromators, to reduce the energy spread of the beam below that of the cathode, and as analyzers, to select electrons that have lost a specified amount by scattering from the sample.

To achieve a small energy width, one must either use a small ratio of Δx to R or reduce the kinetic energy of the beam as it passes through the selector. Both methods have been used successively. The principal dif-

---

[16] P. C. Gibbons, J. J. Ritsko, and S. E. Schnatterly, *Rev. Sci. Instrum.* **46**, 1546 (1975).

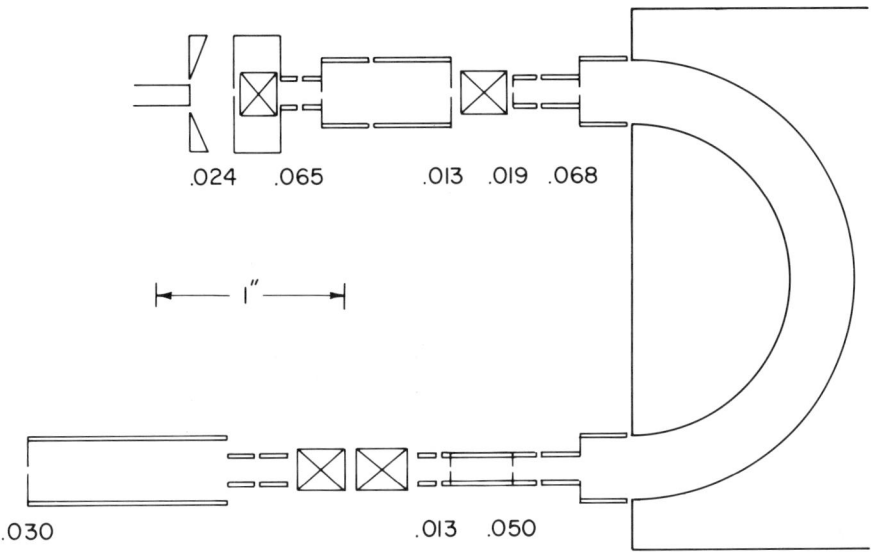

Fig. 8. Electron monochromator schematic.

ficulty with low energies is that space-charge effects become important when the selector is used as a monochromator, and are difficult to handle theoretically. The principal problem with extremely small apertures is that great demands are put on the stability of the fields in the selector. Only one paper exists in the literature which was written with the intent of choosing between these alternatives and designing an optimum electron monochromator. This was done for the case of a hemispherical electrostatic energy selector.[17] Further work on this important problem should be carried out using other types of selectors. A brief description of some representative energy selectors follows.

a. Hemispherical Electrostatic Deflector

Figure 8 shows an abbreviated schematic of a very carefully designed hemispherical energy selector.[17] This device is designed to operate at very low beam energies (a few eV), so additional acceleration and deceleration must be incorporated to produce beam energies at the sample high enough that the Born approximation is accurate. Kuyatt and Simpson[17] carry out the most complete calculation to date of the design parameters and performance characteristics of a hemispherical electron

---

[17] C. E. Kuyatt and J. A. Simpson, *Rev. Sci. Instrum.* **38**, 103 (1967); J. A. Simpson, *ibid.* **35**, 1698 (1964).

monochromator including all known intrinsic effects. The two problems most difficult to handle are space-charge spreading and the anomalous energy spread.[18]

Space-charge defocusing in the hemispheres was handled using a simple model in which the hemispherical path is straightened out into a tube and the focusing effect of the hemispheres replaced by two thin lenses in the tube. Then the effect of space charge is taken to be the same as that of a negative lens at the center of the tube. The anomalous energy spread is the increase in energy width of the beam over that of the cathode due to the Coulomb repulsion between the electrons as they drift around the hemispheres. This increase was measured and an empirical equation that fits the results was used along with the other electron-optical equations to calculate an optimum energy in the hemispheres for a given resolution. The result is

$$E = 116 \Delta E_k R^2 \Delta E , \qquad (10.2)$$

where ΔE_k is the energy spread due to the cathode, ΔE the final monochromatized energy spread, and R the radius of the orbit in the hemispheres in centimeters. This predicts beam energies between 2 and 10 eV for energy resolutions between 10 and 100 meV. Different modeling procedures and different beam characteristics undoubtedly would change the value of the coefficient in Eq. (10.2) but would presumably not change the functional dependence of E on the other parameters. The most important conclusion of this work is that there is an optimum beam energy for a given energy resolution, and that this energy is rather low.

The corresponding problem for energy analyzers is quite different since the current densities are characteristically much less, so space-charge and anomalous energy spreading should not be a problem. This problem has not received much attention. Most workers have simply built an analyzer that matches the monochromator being used, expecting it to work at least as well. This position is not fully justified, however, since the aberrations, which are reduced or completely eliminated in the space-charge-limited monochromator, may be large enough to cause trouble in the analyzer.

The Kuyatt–Simpson monochromator has produced a beam of 70 nA current with an energy spread of 40 meV. The monochromatized output current varies as the 5/2 power of the energy width.

b. Einzel Lens near Cutoff: The Mollenstedt Analyzer

Everyone who uses an einzel lens soon becomes aware of the peculiar way it operates when the potential of the central electrode is near that of

---

[18] H. Boersch, *Z. Phys.* **151**, 519 (1958).

the cathode. This is the region near cutoff for which the electrons are decelerated to very low kinetic energy at the center of the lens and which Grivet refers to as "transgaussian." [19] In this region the focal properties of the lens vary rapidly with the kinetic energy of the electrons. There are two different ways in which this high chromatic aberration has been used for the purpose of selecting the energy of an electron beam. Möllenstedt [20] showed that an off-axis beam approaching the lens from one side is focused at a point on the other side whose position is a rapid function of the beam energy near cutoff. In fact the focal position oscillates rapidly from above to below axis as the cutoff energy is approached. Historically this was the first energy selector to provide high resolution in energy loss experiments. An energy resolution of 40 meV at a beam energy of 20 keV has been achieved.[21]

A second method, in which the beam is on axis, makes use of the fact that the focal strength of the lens varies rapidly near cutoff. In this region the reciprocal focal length, $1/f$, passes rapidly through zero so that a small diaphragm placed on axis far from the lens will transmit predominantly electrons whose energy is very close to cutoff. An energy resolution of 30 meV has been achieved with a beam energy of 30 keV and a current of $10^{-11} A$.[22]

The remarkable feature of this type of energy selector is that its resolving power, $E/\Delta E$, is so high. It achieves this high resolution by decelerating the electrons to very low kinetic energy in the center of the lens where the transverse electric fields produce highly energy-dependent trajectories.

c. Wien Filter

In elementary physics courses it is usually taught that crossed electric and magnetic fields can serve as a velocity selector. This device ordinarily is not double focusing but has nevertheless produced beams of narrow energy spread and usable currents. Boersch, Geiger, and Stickel [23] carried out an analysis of this type of selector used as a monochromator. They include the effects of anomalous energy spread but not space-charge effects. They report inelastic scattering measurements on gaseous argon with an energy resolution of 40 meV, but do not mention the beam current.

Legler [24] has described how to modify the Wien filter to make it double

---

[19] P. Grivet, "Electron Optics," 2nd ed. Pergamon, Oxford, 1972.
[20] G. Möllenstedt, *Optik* **5,** 499 (1949).
[21] F. Leonhard, *Z. Naturforsch., Teil A* **9,** 727 (1954).
[22] W. Harth, *Phys. Lett.* **13,** 133 (1964).
[23] H. Boersch, J. Geiger, and W. Stickel, *Z. Phys.* **180,** 415 (1964).
[24] W. Legler, *Z. Phys.* **171,** 424 (1963).

focusing. This is accomplished by curving the electrodes which create the electric field so the electrons pass between two cylindrically shaped electrodes with the central ray parallel to the axis but displaced from it. The dispersion is given by Eq. (10.1) with $C = 1$ and R equal to the radius of curvature of the cylindrical electrodes. Legler achieves a beam of 1 nA current, 100 eV energy with an energy spread of 60 meV.

d. Cylindrical Selector

A magnetic cylindrical electron energy analyzer has recently been described by Crewe, Isaacson, and Johnson.[25] They have used it to analyze the energy of electrons passing through a thin sample in a remarkably successful electron microscope. Since the current in the analyzer is at most 0.1 nA, space charge and anomalous energy spreading were ignored. Calculations were carried out giving the shape of the curved pole pieces necessary to correct all median plane aberrations and at the same time result in a double-focusing instrument. The analyzer uses a 25 keV beam energy, a radius of curvature of 6 cm, and an ouput slit width of 2.5×10^{-5} cm. An energy resolution of 45 meV has been achieved which agrees well with the calculated value using the parameters of the analyzer.

Cylindrical electrostatic energy selectors are among the simplest to construct and have been widely used. Recently, it has been discovered that the weak tails of the energy pass function can be substantially reduced by grooving the cylindrical capacitors near their centers.[26]

e. Coaxial Cylindrical Selector

This electrostatic energy selector uses electrodes that are coaxial cylinders of different radii and different potentials. The source of electrons is on the axis of the cylinders, and the solid angle of transmission is a conical shell defined by a gap in the inner cylinder. Electrons drift from the source through the gap where they are repelled by the field between the cylinders back toward the axis. They pass through a second gap in the inner cylinder and are focused on axis at a point whose position depends on the kinetic energy of the electrons. The dispersion is given by Eq. (10.1) with Δx being a small displacement of the focal position along the axis of the cylinders, R the radius of the inner cylinder, and C a constant that depends on the off-axis angle of the central ray of acceptance. The minimum value of C is approximately 0.2 for an off-axis angle of 30°.[27]

---

[25] A. V. Crewe, M. Isaacson, and D. Johnson, *Rev. Sci. Instrum.* **42**, 411 (1971).
[26] H. Froitzhein, H. Ibach, and S. Lehwald, *Rev. Sci. Instrum.* **46**, 1325 (1975).
[27] H. Z. Sar-El, *Rev. Sci. Instrum.* **38**, 1210 (1967).

One difficulty in using such a selector as a monochromator or analyzer is the large solid angle which is transmitted. This may seem paradoxial since a large solid angle means a high beam current. The difficulty is that the other electron-optical lens elements in the complete system must be compatible with this large solid angle and this may result in problems with aberrations.

11. Momentum Transfer Selection

As the beam passes through the sample, the electron trajectories are straight lines that are nearly parallel. Electrons that pass through a given point on the sample appear to have come through an aperture some distance away. This aperture is called the pupil. The size and position of the pupil determine the angular divergence of the beam at the sample position. To select electrons that have been scattered through a given angle in the sample, an image of the pupil is formed downstream of the sample, and an aperture is placed in the plane of that image. The aperture should be of an appropriate size so as to transmit all electrons in the unscattered beam. Then the electron beam near the sample position is given a transverse deflection with a set of deflector plates. Now only electrons that receive an equal and opposite deflection in the sample will be transmitted by the angle selecting aperture. Because the momentum transfer in the sample is nearly perpendicular to the incident momentum, the magnitude of the momentum transfer is decoupled from that of the energy loss in the sample, so energy and momentum can be independently selected. The most common mode of operation is a constant q scan in which the momentum transfer is fixed and the energy loss scanned. Because of the small angle of scattering this is much simpler to accomplish than in the case of inelastic neutron scattering, where typically large scattering angles are used.

The angular resolution of the spectrometer is determined by the angular spread in the beam at the sample position: Electrons must be scattered through an angle larger than this to be distinguished from unscattered electrons. It is highly desirable that the angular resolution be variable. This is because some experiments require higher resolution than others and, even for the same incident beam current, the counting rate in a given experiment is proportional to the solid angle of acceptance of the spectrometer. Thus to avoid unnecessarily low counting rates when high momentum resolution is not needed, it is useful to be able to vary the momentum transfer resolution, and hence the pupil size, at will.

Variable angular resolution could be accomplished by replacing apertures both before and after the sample, but this would be cumbersome since it entails opening the vacuum system. A simpler way is to include zoom lenses between the sample and the monochromator and analyzer.

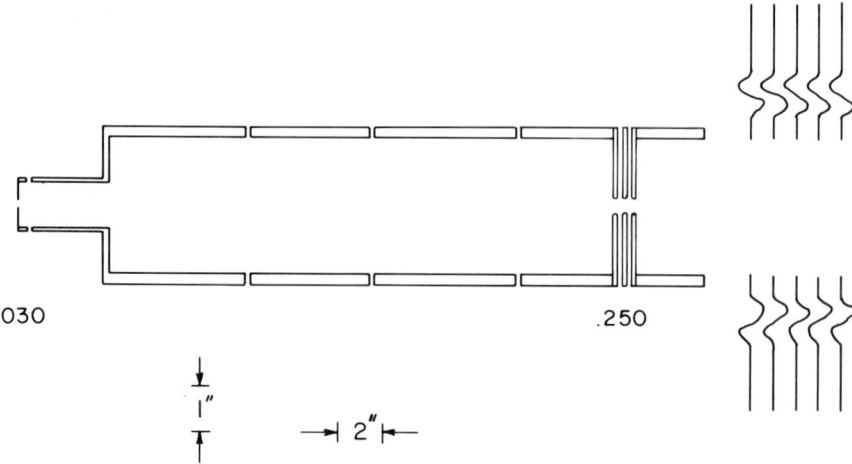

Fig. 9. Zoom lens schematic.

A zoom lens is a combination of lenses that allows the magnification to be varied while keeping the focal position fixed. Figure 9 shows a schematic of an electron-optical zoom lens.[16] It consists of a series of cylinder lenses any one of which can be used to focus the beam onto a given image plane. Different lens positions give different magnifications. The einzel lens at the output end which is near the desired image plane allows the position of the pupil to be adjusted without much affecting the image.

12. Detectors

a. Electronic Detection

A good detector must be capable of covering a great dynamic range since the currents detected may vary from less than one count per second for a scattering process with very small cross section to perhaps 1 μA (10^{13} counts/sec) in the unscattered beam. To cover this range it is necessary to make a transition at some point from counting individual electrons to recording the average beam current. This produces no real difficulty since using a preamplifier, amplifier, discriminator, and scalar capable of 100 MHz counting rate and an electron multiplier with no more than 10 nsec pulse width, the counting errors due to dead time at 10^6 counts/sec are about 1%, which is usually not important. At the same counting rate, the electron multiplier operated in the current mode at a gain of 10^6 produces a current of 100 nA which is easily detected with a picoammeter whose output can be simultaneously viewed on an XY recorder and stored in the memory of a computer or on tape. Thus there is plenty of overlap between the two modes of operation. In addition the gain of the electron multiplier can be turned down when detecting the unscattered beam directly, and a picoammeter used to record the result.

Conventional electron multipliers appear to be the most versatile and useful of presently available detectors of this type. Channeltrons lack the dynamic range at the high end, and a Faraday cup at the low end. This does not imply that new developments are not needed or anticipated. Experiments of this type are often statistics limited. Therefore some type of multichannel electronic detector would be highly desirable. Instead of placing a narrow aperture at the output of the analyzer, a broad aperture of many resolution widths would be used, and the beam imaged onto a position-sensitive detector which would then record the current in a range of energy losses simultaneously. Even more efficient would be a system that focuses momentum transfer along one axis and energy loss along a perpendicular axis so an entire spectrum could be recorded at once. Possible detectors of this type include charge-coupled devices and image orthicons. (See also Section 12,b.)

b. Film

The detector used in several early energy loss spectrometers was film, and it is still used today. The competition between film and electronic detection will probably continue for some time, just as it has in Raman scattering work, where the detector requirements are similar. The disadvantages of film are that it is nonlinear and that the result of a measurement cannot be immediately seen by the experimenters. The advantage is that it is a position-sensitive detector and so can be used to record a whole spectrum at once, by focusing different energies at different points on the film. In addition, output optics has been used which focuses momentum transfer along one axis and energy loss along a perpendicular axis so as to record directly the dispersion curve of an excitation.[28] The fact that this can be done simply and with present technology makes film an attractive detector, especially for samples that suffer from radiation damage, such as organic solids.

V. Sample Preparation

Ultimately, to measure the spectrum of a material, one encounters the problem of sample preparation. The ideal sample would be a single crystal a few hundred angstroms thick and several square millimeters in area.

[28] H. Watanabe, *J. Phys. Soc. Jpn.* **11**, 112 (1956).

It would be self-supporting and have atomically clean surfaces. Such a sample has not yet been prepared. Various ways of approximating this ideal are described below.

13. Thin Film Evaporation

Polycrystalline films of many materials in the thickness range desired can readily be made by evaporation onto a substrate in vacuum. The only problem then is what to do with the substrate. There are two solutions—use a thin substrate that does not scatter electrons much, or somehow remove the film from the substrate. Some recipes follow.

a. Solvent Removal of Substrate

Victawet 35B [29] is a water-soluble wax which can be applied as a thin film onto a microscope slide. The sample is then evaporated onto the slide and later floated off carefully in a bowl of water, then picked up on a sample holder with an aperture for the beam, or an electron microscope grid. The only tricky part is applying the Victawet to the slide. If the film is too thin, the sample will not float off readily; and if it is too thick, it may be rough. One method that works well is to place a few drops of diluted Victawet in the center of a microscope slide and then spin it rapidly with a high-speed motor about an axis perpendicular to the plane of the slide. This throws the Victawet off and rapidly dries it to form a smooth thin film. This seems to work better when the Victawet is applied to the slide before spinning it rather than while it is spinning. It is often a good idea to wet the entire slide before spinning it. Victawet works better than other water-soluble materials such as NaCl, which when applied to a microscope slide as a thin film by evaporation, allows the sample to be floated off, but tends to break it up into small pieces.

b. Thin Organic Film Substrate

Organic films, consisting mainly of carbon and hydrogen, have fairly small elastic and inelastic scattering cross sections; so if a thin film is made, it can often be used as a substrate and left in place without interfering much with the experiment. Useful materials are Formvar dissolved in ethylene dichloride, or Collodion dissolved in alcohol. A film is formed by allowing a drop of a dilute solution of the organic to fall gently on the surface of a bowl of water. The liquid quickly spreads out on the water surface and solidifies. It can then be picked up on a sample holder and used as a substrate for evaporation. By successively

[29] Stauffer Chemical Co. Industrial Div., 380 Madison Ave., New York, New York 10017.

diluting the solution one can make the thinnest possible film that can still be picked up. A little solvent added to the water slows the solidification process and also results in a thinner film. Films of Formvar have been made that can be picked up on a sample holder with a ¼ in. aperture and that transmit 95% of a 300 keV beam.

c. Thin Carbon Film Substrate

Carbon substrates, which can be used in a baked UHV system, can be made by sublimation using electron beam or carbon arc heating onto a Victawet coated slide as described above. Films as thin as 100 Å have been made that can be picked up on a sample holder with a ¼ in. diameter aperture. Films as thin as 20 Å have been made that can be picked up on an electron microscope grid.[30]

By far the easiest of the two methods of sublimation is the use of a carbon arc in vacuum. Two carbon rods are pressed together in a bell jar and current passed through them. The points of contact are extremely small, so when enough current flows (about 65 A) the regions of contact are vaporized and replaced by new contacts which vaporize, and so forth. An ordinary oil-pumped bell jar with the usual step-down transformer for evaporations is adequate.

One problem with carbon sublimation is that quite a lot of heat is radiated from the rods, so the quartz crystal thickness monitor, which is ordinarily used to determine film thickness as it is put down, may be heated enough to cause spurious frequency shifts. One method of avoiding this problem is to shield the substrate and monitor so they do not directly view the source, and arrange for a surface to be placed so carbon atoms may land on the substrate and monitor after bouncing once from the surface. M. Isaacson has successfully used a cylindrical surface with source and substrate close to the axis.[30]

d. Epitaxial Thin Film Single Crystals

By evaporation in good vacuum at a controlled rate onto a single crystal substrate whose temperature is controlled, it is possible to grow thin films that are single crystals. Alkali halides are often used as substrates because they are inexpensive, readily cleaved, and water soluble, making it easy to float the sample off. It is characteristically difficult to grow large-area single crystals. Crystallite size is often a few hundred microns. Practice and black magic are essential ingredients.

[30] M. Isaacson, private communication.

14. SPUTTERING

rf or dc sputtering is often used to produce films that for one reason or another cannot be evaporated. Refractory materials or compounds that decompose upon heating are examples. One of the unique features of sputtering is that films of alloys or mixtures can be made with the same composition as the cathode even though the different components may have different sputtering rates. This is because after the cathode has been "run in," the surface composition is altered so as to compensate for variations in sputtering rates. Another useful feature is that nonequilibrium mixtures can be produced by using two or more cathodes side by side.

15. THINNING MACROSCOPIC CRYSTALS

A completely different approach involves starting with a macroscopic single crystal and thinning it. The major difficulty is achieving a reasonably uniform thickness over a large enough area. Fortunately pinholes and thickness nonuniformities are not important as far as the shape of a spectrum is concerned; they do make absolute cross section measurements difficult or impossible however.

The first step involves cleaving or sawing and then mechanical grinding. Films as thin as 2 μm have been produced by mechanical polishing.[31] Final thinning must be done with more delicate means. Two methods have been used:

a. Electrochemical Thinning

For metallic samples this is a good choice. Usually an automatic shutoff is needed which turns off the current as soon as a very small pinhole is detected by a beam of light and a photocell. This method is quick and, combined with a means of measuring the thickness profile of the sample, should be a valuable technique of sample preparation. Commercial units are available.[32]

b. Ion Milling

By focusing a beam of argon, or ions of another inert element onto a surface, the atoms of the material can be removed one by one. Angle of incidence and beam kinetic energy are both important parameters. Milling rates are on the order of 1 μm per hour for a beam current density of 100

---

[31] R. B. Wilson, *J. Sci. Instrum.* **44**, 395 (1967).
[32] For example, South Bay Technology Model 550A, 4900 Santa Anita Ave., El Monte, California 91731.

$\mu A/cm^2$. This means of sample erosion works for almost all sample materials and so appears to have a bright future. A method of monitoring thickness during milling is not presently available and would be extremely desirable. A means of monitoring thickness profile after milling is desirable. Commercial units are available.[33]

VI. Experimental Examples

16. Plasmons in Simple Metals

The earliest use, and for many years the only use of inelastic electron scattering for studying solids, was plasmon excitation. The reason for this is that in a small q experiment, the easiest to perform, plasmons dominate the cross section in simple metals. In a thick sample, then, what is seen is a series of peaks at multiples of the plasma frequency. Before the nature of these peaks was understood, the technique of inelastic electron scattering came to be called "characteristic energy loss spectroscopy" since each metal had its own characteristic energy at which losses could be observed. This curiously alchemical phrase described the state of affairs up until about 1960 when David Pines introduced the ideas of collective modes, elementary excitations, and plasmons. Modern treatments of plasma properties of solids can now be readily found ranging from elementary to advanced. Here we shall give only the briefest description of the basic theoretical ideas.

The many-electron problem in the metallic density range is a particularly knotty one because the Coulomb force which acts between the electrons is of long range. This means that essentially an infinite number of electrons are simultaneously interacting together, and one is at first overwhelmed at the prospect of keeping track of where all these electrons are going. The problem can be considerably simplified using the idea of screening. Imagine one of the electrons in the system; other electrons in its vicinity move away in response to its Coulomb field. This electron carries with it a screening hole, so that from a distance it appears neutral. Therefore the interaction between "dressed" electrons is of short range and can be thought of in terms of Fermi liquid theory. This picture is easy enough to grasp as long as the charge being screened is static. If it is moving, then the time dependence as well as the spatial dependence of the screening cloud comes into question. There is one aspect of the time dependence that is easy enough to appreciate. As the neighboring electrons move to screen out the applied field, they will overshoot and

---

[33] For example, Commonwealth Scientific Corp., 500 Pendleton Street, Alexandria, Virginia 22314.

oscillate at a frequency that is easily shown to be the classical plasma frequency.[34] Thus plasma oscillations and screening are complementary aspects of the same problem; both arise from the long-range nature of the Coulomb force.

To understand the problem in detail, we need to know the space and time dependence of the screening cloud that surrounds a test particle or particular electron. This means knowing the space and time dependence of the polarizability of the interacting electron system, $\alpha(r, t)$ or its Fourier transform $\alpha(q, \omega)$. Earlier we showed that the inelastic electron scattering cross section is proportional to $\text{Im}[-1/\epsilon(q, \omega)]$. This is why inelastic electron scattering is so useful in studying interacting electron systems. The quantity measured by the experimentalist, $d^2\sigma/d\Omega\, d\omega$, is directly related to one of the most fundamental quantities needed in constructing theories of many electron systems, $\epsilon(q, \omega) \simeq 1 + 4\pi\alpha(q, \omega)$.

Simple metals are those metals for which it is a good approximation to smear out the positive ions into a uniform background charge, ignoring local field effects and forming the classic electron gas of the theorists dreams. These metals include the alkalis, Al, Ga, In, Be, and Mg. Some of these metals will be discussed in this section. They are the best testing ground for electron gas ideas, since to a greater extent than in any other materials the effects of band structure and atomic structure can be ignored.

The simplest model dielectric response function to be used in comparing with measurements is the Lindhard function appropriate for an interacting electron gas in the random phase approximation (RPA)[35]:

$$\epsilon(q, \omega) = 1 + \frac{3}{128\gamma^2 Z^2}\left\{4Z + [1 - (U - Z)^2]\ln\left(\frac{U - Z - 1}{U - Z + 1}\right)\right.$$
$$\left. - [1 - (U + Z)^2]\ln\left(\frac{U + Z - 1}{U + Z + 1}\right)\right\}, \quad (16.1)$$

where

$$\gamma = \frac{E_F}{\hbar\omega_p}, \quad Z = \frac{q}{2k_F}, \quad U = \frac{\hbar\omega}{4 Z E_F}, \quad \omega_p = \left(\frac{4\pi n e^2}{m}\right)^{1/2},$$

and k_F is the Fermi wave vector. This function reduces to the well-known Drude function in the long-wavelength limit ($u \gg z + 1$):

$$\epsilon(q, \omega) = 1 - (\omega_p^2/\omega^2). \quad (16.2)$$

[34] C. Kittel, "Introduction to Solid State Physics," 5th ed., Wiley, New York, 1975.
[35] J. Linhard, K. Dan. Vidensk. Selsk., Mat.-Fys. Medd. 28, No. 8, (1954).

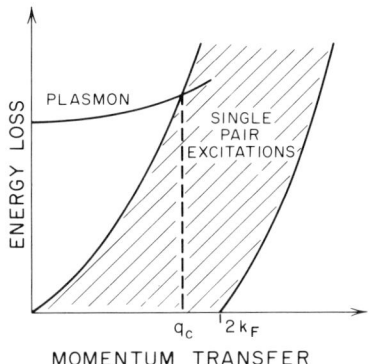

FIG. 10. Dispersion of the plasmon according to the random phase approximation.

In this limit, all the strength in the inelastic electron scattering appears in a single infinitely sharp peak at $\omega = \omega_p$. This is qualitatively consistent with the early observation of "characteristic energy losses" in metals. In fact the earliest evidence in support of the idea of plasma oscillations was the fact that the positions of these peaks agreed reasonably well with the calculated plasma frequency in many cases.

a. Plasmon Width for $q < q_c$

According to the Lindhard function, as q is increased, the peak in $\text{Im}[1/\epsilon(q, \omega)]$ moves to higher energy and loses strength but remains perfectly sharp until a cutoff wave vector, q_c, is reached, and single-particle excitations appear at low frequency. (See Fig. 10.) Experimental results[36] (see Figs. 11–14) differ from this primarily in that the plasmon peak has finite width at all wave vectors. At $q = 0$ this width is due to disorder in the sample (either thermal or static) and interband transitions. Both effects result in a nonzero value of $\epsilon_2(\omega_p)$ thereby broadening the plasmon peak. Figure 15 shows a plot of the $q = 0$ plasmon width versus the square of the lowest wave vector pseudopotential for the alkali metals.[37] This is the only pseudopotential that can cause interband transitions degenerate with ω_p in the alkali metals. The results are consistent with interband transitions plus something else, presumably disorder, causing the $q = 0$ plasmon width.

Sturm[38] describes a quantitative evaluation of the $q = 0$ width of the

[36] P. C. Gibbons, S.E. Schnatterly, J. J. Ritsko, and J. R. Fields, *Phys. Rev. B* **13**, 2451 (1976).
[37] P. C. Gibbons and S. E. Schnatterly, *Phys. Rev. B* **15**, 2420 (1977).
[38] K. Sturm, *Z. Phys. B* **25**, 247 (1976).

plasmon appropriate for simple metals. Good agreement with measurements is obtained using pseudopotentials obtained from Fermi surface measurements.

The increase in width with q (for small q values) is, however, quite a different problem. The Lindhard function (i.e., RPA for the electron gas) predicts zero width all the way to $q = q_c$, the cutoff wave vector at which point plasmons become degenerate with single-particle excitations and can decay directly into a single electron–hole pair without the need for disorder or a periodic field to provide the necessary momentum (see Fig. 10). What is observed instead is an increase in width proportional to q^2 in this region (see Figs. 16–18). In fact a direct analysis of data results in the width of the peak being proportional to q^2 at low q and q^4 at

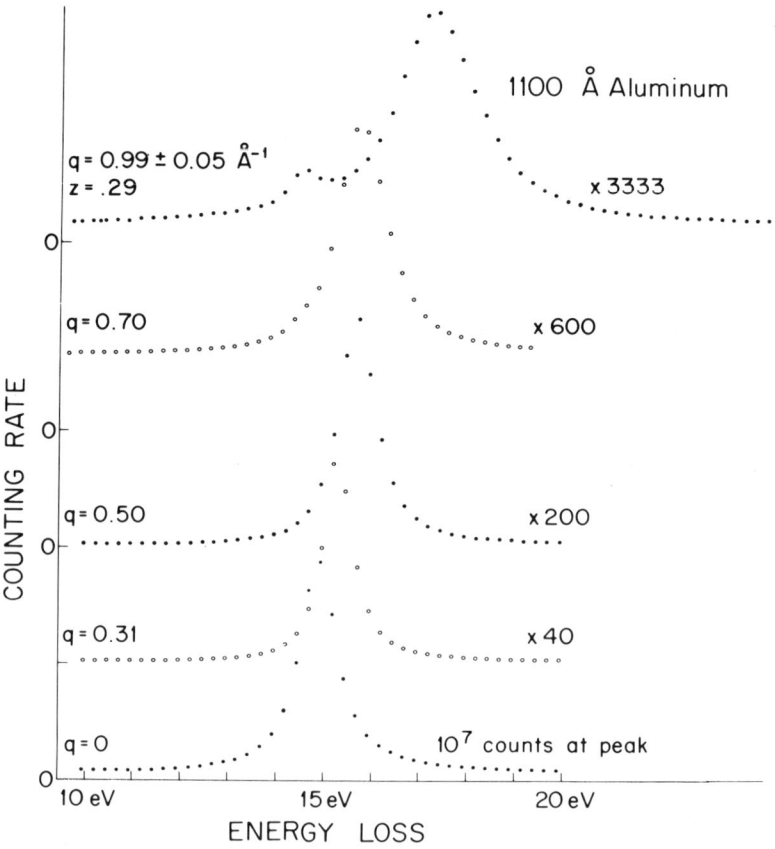

FIG. 11. Measured plasmon spectra of Al from $q = 0$ to 1 Å^{-1} [P. C. Gibbons, S. E. Schnatterly, J. J. Ritsko, and J. R. Fields, *Phys. Rev. B* **13**, 2451 (1976)].

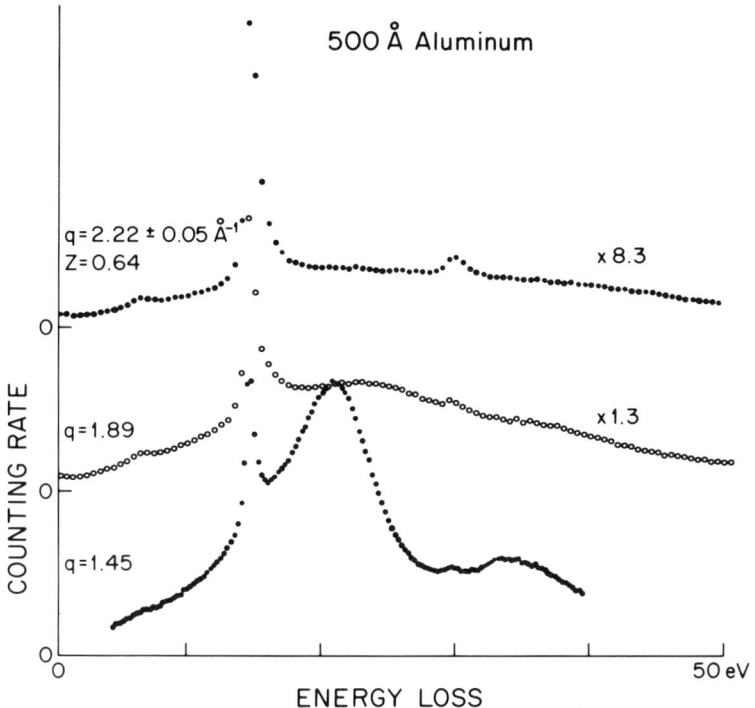

FIG. 12. Measured plasmon spectra of Al from q = 1.45 to 2.22 Å^{-1} [P. C. Gibbons, S. E. Schnatterly, J. J. Ritsko, and J. R. Fields, *Phys. Rev. B* **13**, 2451 (1976)].

intermediate values (Fig. 19) or to q^2 over the whole range, but with different slopes (Figs. 16–18). The changeover occurs at approximately $q = q_c$. The mystery occurs in the small q region. Three papers have appeared in the last 15 years attempting to calculate this increase in width with q.[39] At $q = 0$ the plasmon width must be rigorously zero in an electron gas due to momentum conservation. At small but finite $q < q_c$ two decay channels have been taken into account: decay of a plasmon into two electron–hole pairs, and decay into a smaller q plasmon plus one electron–hole pair. The first of the three papers obtained a result an order of magnitude too large, the second paper was about right, and the third, an order of magnitude too small. Each paper purported to be an improvement over previous work. A basic assumption is that the increase in width due to electron gas effects is to be simply added to the $q = 0$

[39] D. F. DuBois and M. G. Kivelson, *Phys. Rev.* **186**, 409 (1969), and references given therein.

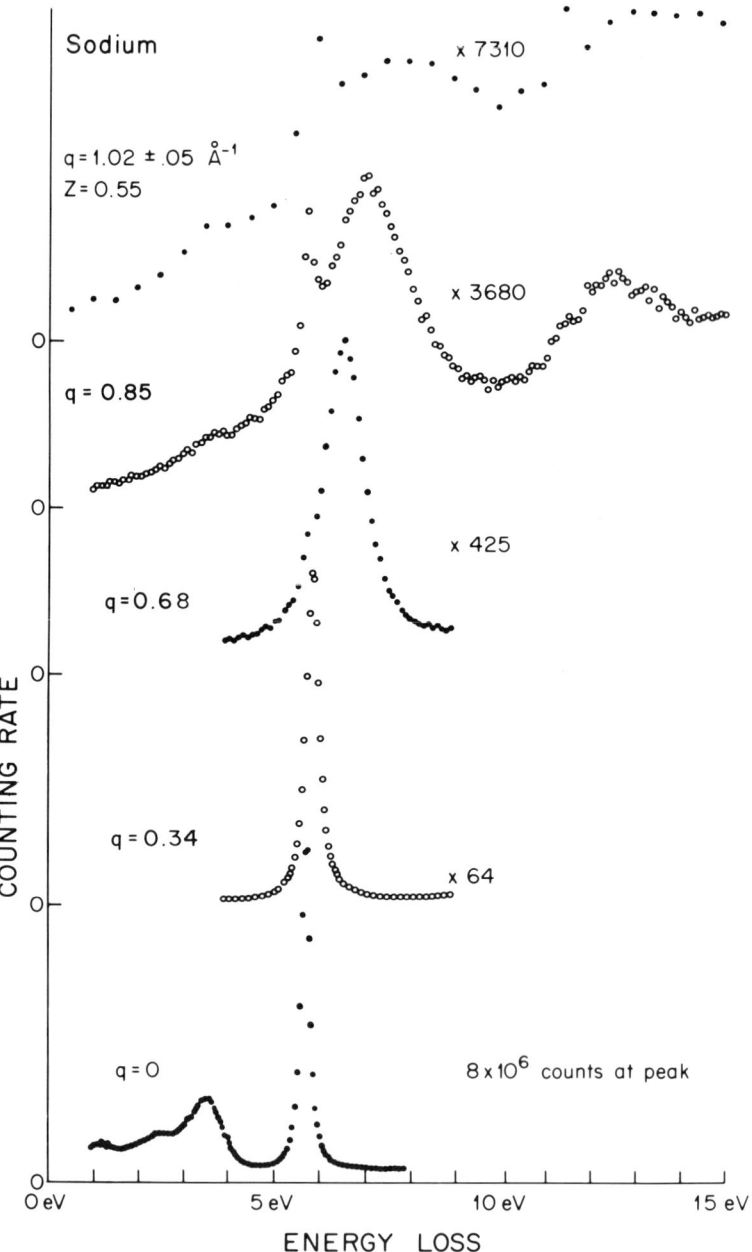

FIG. 13. Measured plasmon spectra of Na from $q = 0$ to 1 Å^{-1} [P. C. Gibbons, S. E. Schnatterly, J. J. Ritsko, and J. R. Fields, *Phys. Rev. B* **13**, 2451 (1976)].

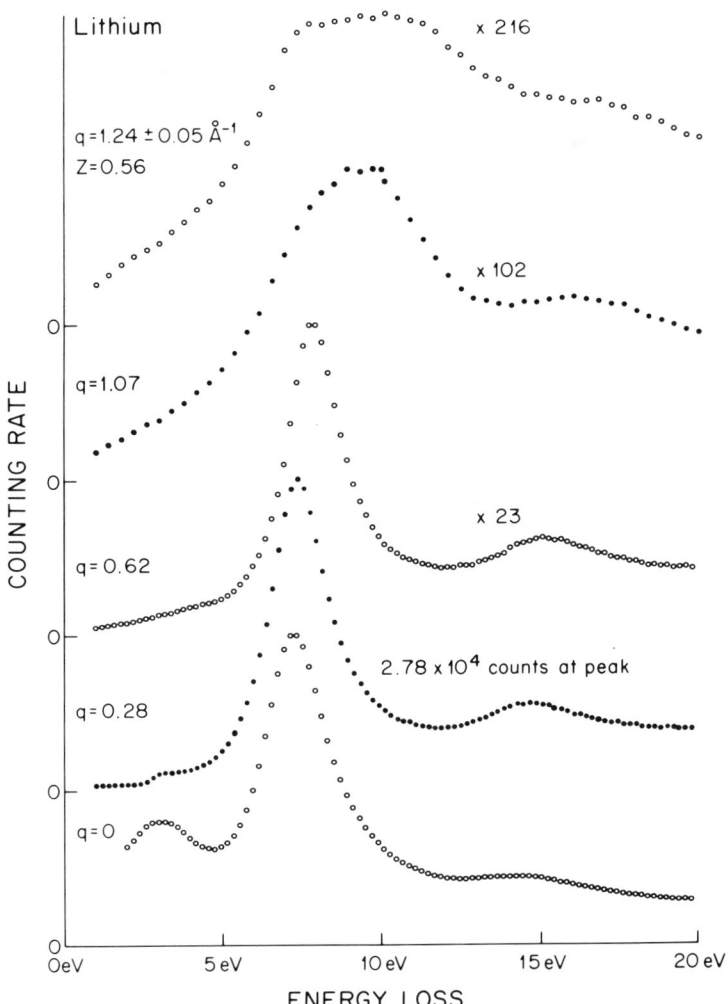

FIG. 14. Measured plasmon spectra of Li from $q = 0$ to 1.24 Å^{-1} [P. C. Gibbons, S. E. Schnatterly, J. J. Ritsko, and J. R. Fields, *Phys. Rev. B* **13**, 2451 (1976)].

width. In addition none of the calculations (mentioned above) include the effect of band structure.

Recent work[40] has shown that band structure parameters appropriate for some simple metals quantitatively describe the increase of plasmon width with q for $q < q_c$. The situation is complex in that the width may

[40] K. Sturm, *Z. Phys. B* **27**, No. 1 (1977) and P. C. Gibbons, to be published.

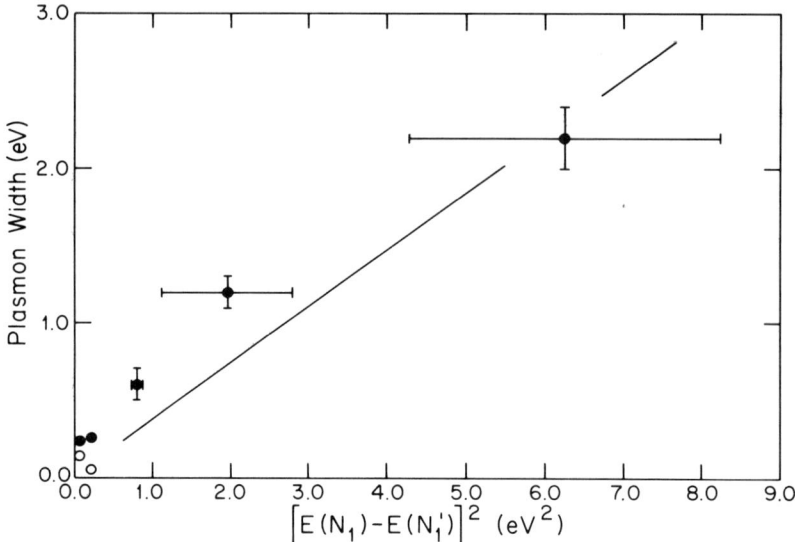

FIG. 15. Plasmon width at $q = 0$ vs square of nearest zone boundary pseudopotential for the alkali metals [P. C. Gibbons and S. E. Schnatterly, *Phys. Rev. B* **15**, 2420 (1977)].

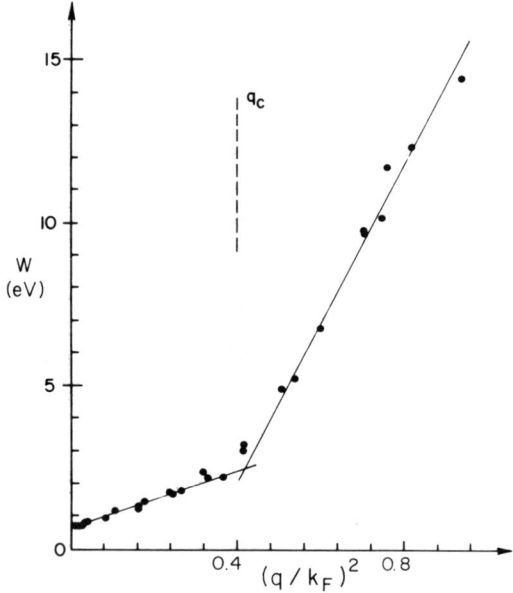

FIG. 16. Plasmon width in Al vs q^2 [P. C. Gibbons, S. E. Schnatterly, J. J. Ritsko, and J. R. Fields, *Phys. Rev. B* **13**, 2451 (1976)].

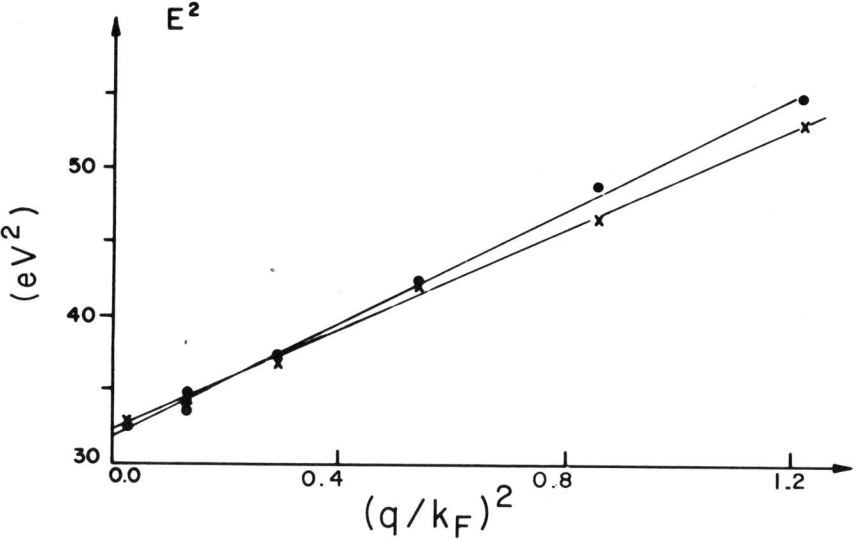

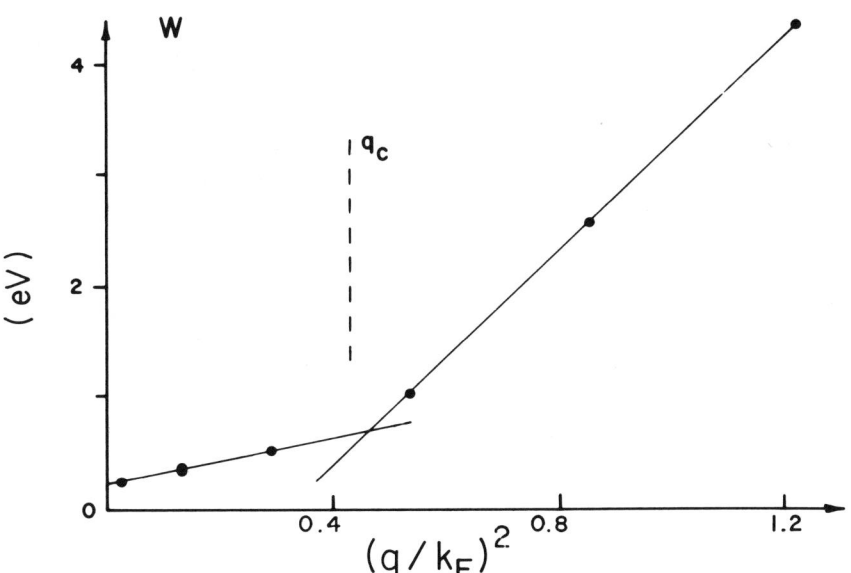

FIG. 17. Dispersion and width of the Na plasmon vs q^2 [P. C. Gibbons, S. E. Schnatterly, J. J. Ritsko, and J. R. Fields, *Phys. Rev. B* **13**, 2451 (1976)].

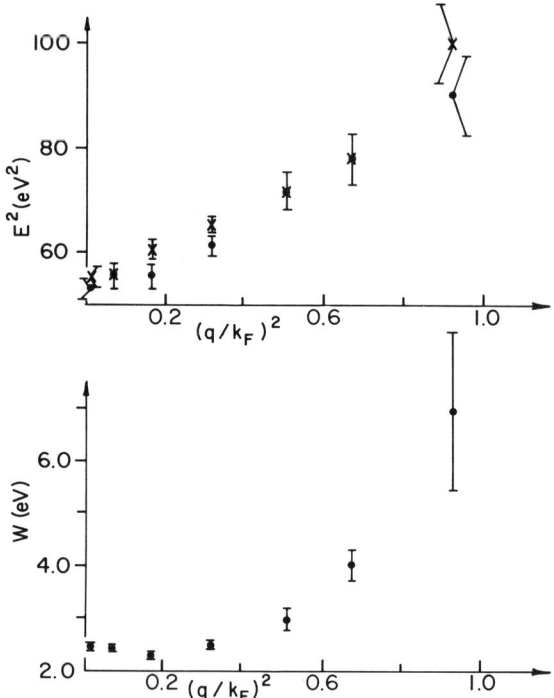

FIG. 18. Dispersion and width of the Li plasmon vs q^2 [P. C. Gibbons, S. E. Schnatterly, J. J. Ritsko, and J. R. Fields, *Phys. Rev. B* **13**, 2451 (1976)].

either increase or decrease as q is increased depending on the parameters. This may be the explanation of the initial decrease in width of the Li plasmon (see Fig. 18).

One of the interesting recent observations is that in the quasi one-dimensional metal TTF–TCNQ, the plasmon peak width increases *linearly* with q for momentum transfers parallel to the conducting axis[41] (see Fig. 20). Whether this is an electron gas effect related to dimensionally or whether it is due to the stronger influence of interband transitions (which cause a negative dispersion of the plasmon as shown in Fig. 20) than in the simple metals is not known at this time.

b. Modified Lindhard Function

It is sometimes useful, in approaching complex problems, to attempt a purely phenomenological description in order to condense the data and

[41] J. J. Ritsko, D. J. Sandman, A. J. Epstein, P. C. Gibbons, J. R. Fields, and S. E. Schnatterly, *Phys. Rev. Lett.* **34**, 1330 (1975).

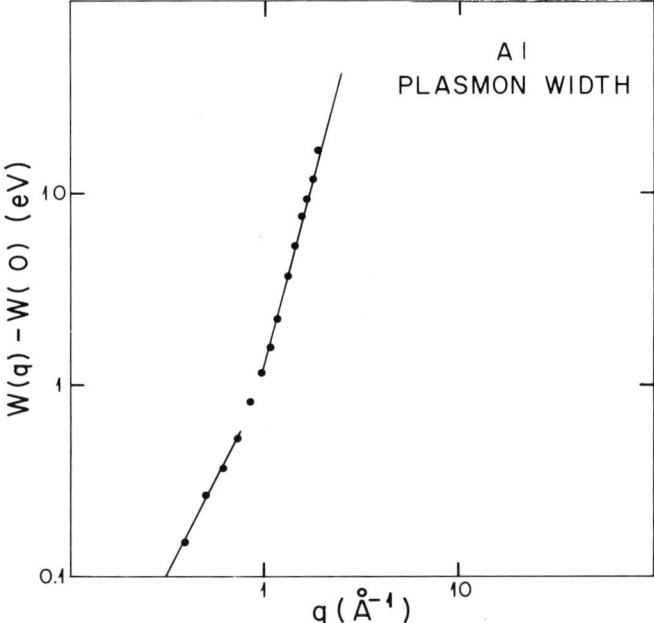

Fig. 19. Width of the Al plasmon vs q. The two straight lines have slopes of 2 and 4, respectively.

allow it to be viewed from a new perspective. In the case of plasmon excitation, i.e., the shape of $S(q, \omega)$ in simple metals, an obvious strategy is to start from the Lindhard expression, which we expect should work perfectly at infinite density and zero periodic field and hope to modify it in a straightforward way so as to describe real simple metals. The first change suggested by the data is the introduction of a finite lifetime; Mermin has described how this can be done.[42] Second, the effects of the electron–electron interaction beyond the RPA can be estimated using a Clausius–Mossotti-type local field correction:

$$\epsilon(q, \omega) = 1 + \frac{\epsilon_L(q, \omega) - 1}{1 - G(q, \omega)(\epsilon_L(q, \omega) - 1)}, \qquad (16.3)$$

where ϵ_L designates the Lindhard expression for $\epsilon(q, \omega)$. Here $G(q, \omega)$ represents that part of the interaction between electrons which is not included in the RPA result. For arbitrary $G(q, \omega)$ this form should be exact; the approximation lies in the form assumed for $G(q, \omega)$. Considerable work has been carried out assuming G to be real and a function

[42] N. D. Mermin, *Phys. Rev. B* **1**, 2362 (1970).

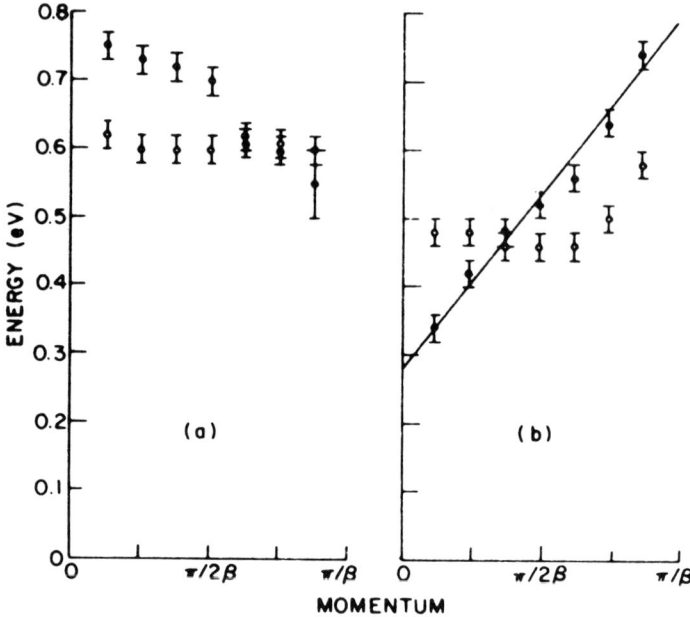

FIG. 20. (a) Dispersion, (b) Width of the plasmon in TTF–TCNQ. Filled circles: q parallel to the conducting axis. Open circles: q at 45° to the conducting axis [J. J. Ritsko, D. J. Sandman, A. J. Epstein, P. C. Gibbons, J. R. Fields, and S. E. Schnatterly, *Phys. Rev. Lett.* **34**, 1330 (1975)].

of q only.[43] Recently, experimental data were fitted with a function of this form allowing G to be complex but assuming it to depend only on q.[36] This is questionable since one might expect the "unscreening" of the electron–electron interaction at $\omega_p(q)$ to influence the value of G. The imaginary part of G produces damping but influences the shape of $S(q, \omega)$ in a different manner than the imaginary energy of Mermin. Qualitatively one might expect Mermin's damping to represent the electron–phonon interaction and interband transitions, and Im G to represent electron–electron interactions.

An example of a reasonably high q measurement of Al is shown in Fig. 21 for 500 Å and 1100 Å sample thicknesses. The effects of multiple scattering are evident and must be corrected before comparison with the modified Lindhard expression. First there is the narrow "thermal diffuse" peak at 15 eV. This is caused by a double scattering event in which the fast electron scatters from a phonon with momentum transfer $q \simeq$

[43] P. Vashishta and K. S. Singwi, *Phys. Rev. B* **6**, 875 (1972).

1.5 Å^{-1} and also creates a $q \simeq 0$ plasmon. This feature is apparent also in Figs. 11 and 12 and increases in importance with q relative to the plasmon, since elastic scattering is roughly q independent (see Fig. 3). Since the thermal diffuse plasmon is so narrow, it was simply left out of the data when comparing with Eq. (16.3). Double plasmon scattering is evident near the upper part of Fig. 21. This was corrected for by extrapolating it to zero energy following an E^3 form.

An example of the data with thermal diffuse region absent and the extrapolated double scattering are shown in Fig. 22a. In Fig. 22b are shown the resulting corrected data and some fits using Eq. (16.4).

The values of the three parameters that were varied to obtain the fits are shown in Figs. 23 and 24. Im E and Im G are piecewise linear in q^2, while Re G is simply linear. In this way it is possible to arrive at an analytic expression for a dielectric response function appropriate for real metals throughout the first Brillouin zone and including all energies of

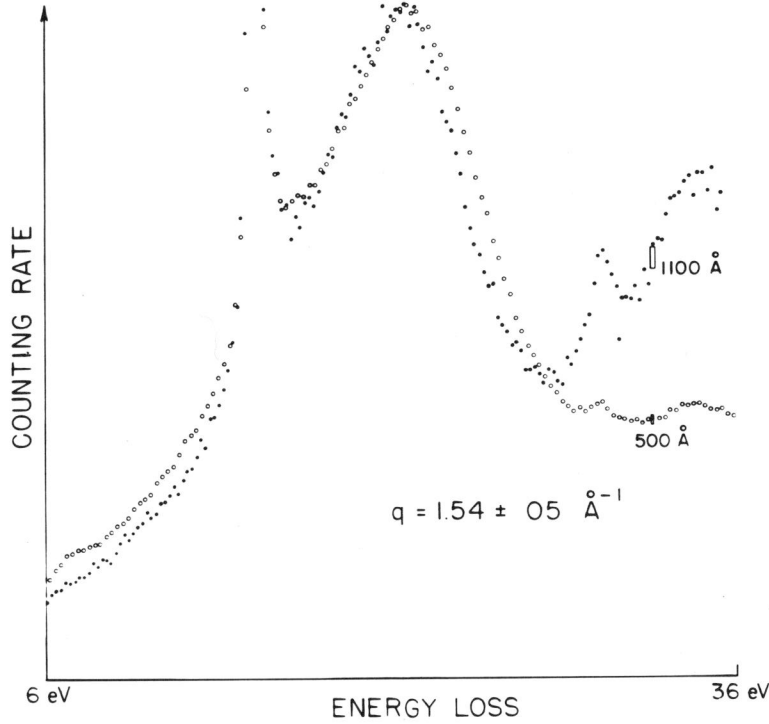

FIG. 21. Plasmon spectrum of Al at $q = 1.54$ Å^{-1} for two different sample thicknesses [P. C. Gibbons, S. E. Schnatterly, J. J. Ritsko, and J. R. Fields, *Phys. Rev. B* **13**, 2451 (1976)].

interest. Table I shows the functional forms that correspond to the solid lines in Figs. 23 and 24.

Note that for $q < q_c$, Im E and Im G increase with q as expected since the Lindhard expression by itself produces no damping in this region. For $q > q_c$, however, no additional increase is needed. Therefore the often observed striking increase in plasmon width for $q > q_c$ is

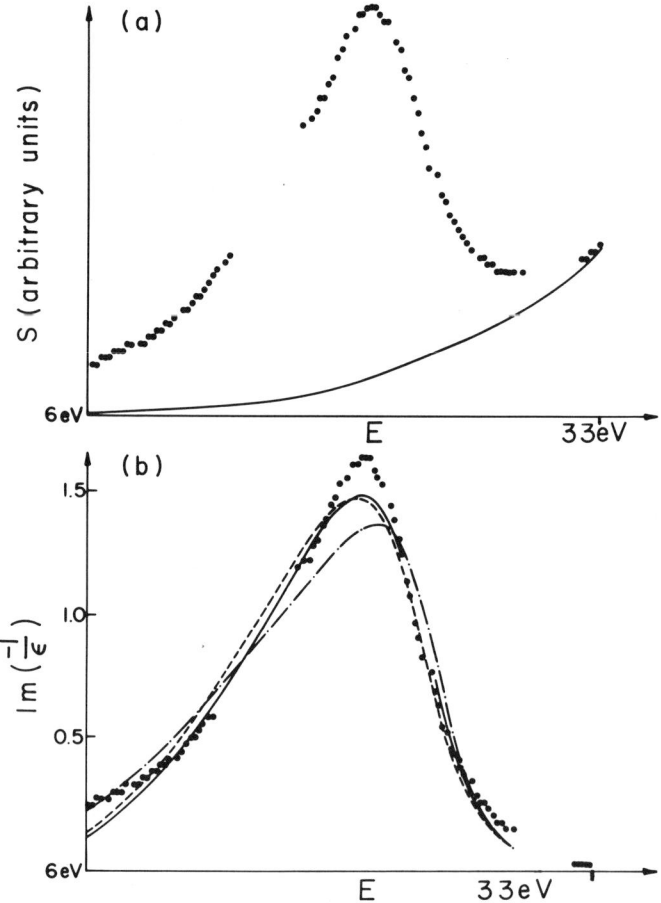

FIG. 22. Measured and calculated spectra for $q = 1.47$ Å^{-1} in Al. (a) Data with thermal diffuse region excised, and the extrapolated double scattering estimate. (b) Data after double scattering subtraction and calculated spectra. Solid line: best fit with complex exchange parameter. Dashed line: using parameters from Table II. Dot–dashed line: best fit with real exchange parameter. [P. C. Gibbons, S. E. Schnatterly, J. J. Ritsko, and J. R. Fields, *Phys. Rev. B* **13**, 2451 (1976)].

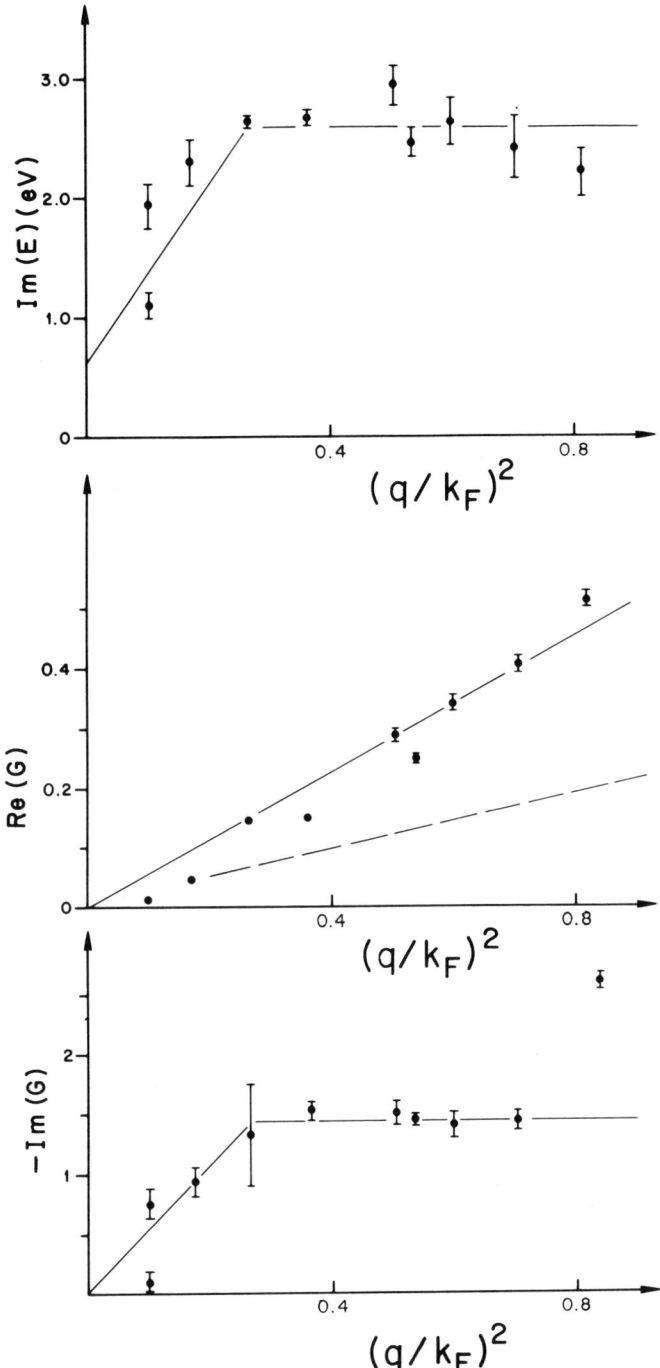

FIG. 23. Values of Im E, Re G, and Im G which provide the best fits of Eq. (16.4) to the measured spectra in Al. Error bars indicate standard deviations. The dashed line for Re G is from P Vashishta and K. S. Singwi, *Phys. Rev.* B **6**, 875 (1972).

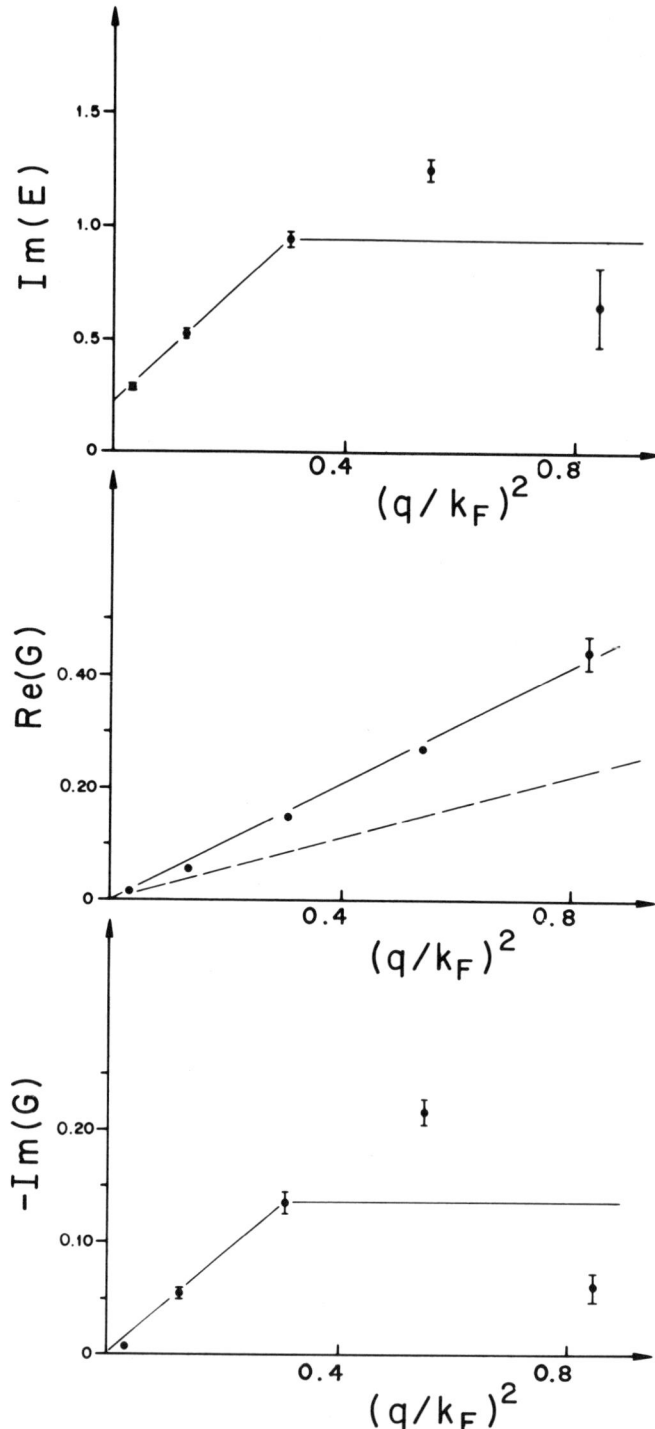

TABLE I. PARAMETER VALUES CORRESPONDING TO THE SOLID LINES IN FIGS. 23 AND 24[a]

Aluminum

$$\mathrm{Im}\ E = \begin{cases} 0.53 + 30.9Z^2 \pm 6.2\ \mathrm{eV} & Z^2 < 0.067 \\ 2.60 \pm 0.2\ \mathrm{eV} & Z^2 \geq 0.067 \end{cases}$$

$$\mathrm{Re}\ G = 2.5Z^2 \pm 0.05$$

$$\mathrm{Im}\ G = \begin{cases} -2.12Z^2 \pm 0.2 & Z^2 \leq 0.067 \\ -0.142 \pm 0.2 & Z^2 \geq 0.067 \end{cases}$$

Sodium

$$\mathrm{Im}\ E = \begin{cases} 0.22 + 9.6Z^2\ \mathrm{eV} & Z^2 < 0.076 \\ 0.95\ \mathrm{eV} & Z^2 \geq 0.076 \end{cases}$$

$$\mathrm{Re}\ G = 2.02Z^2$$

$$\mathrm{Im}\ G = \begin{cases} -1.79Z^2 & Z^2 < 0.076 \\ -0.136 & Z^2 \geq 0.076 \end{cases}$$

[a] P. C. Gibbons, S. E. Schnatterly, J. J. Ritsko, and J. R. Fields, *Phys. Rev. B* **13**, 2451 (1976).

correctly described by the Lindhard function: it is only the small q width that is missing.

A major question is: Is this formula simply a means of conveniently describing data, or do these parameters have physical meaning? The most direct way to pursue this question is to carry out similar analyses for data obtained from metals of widely differing electron density. If these parameters have meaning, they should vary in a simple, systematic manner with density.

c. Double Plasmon Excitation

As a fast electron passes through a sample, it can interact more than once, creating several plasma excitations. If these creation events take place far enough apart in the sample, they will be independent of one another, and therefore the probability of multiple plasmon excitation follows a Poisson distribution:

$$P_n = \frac{1}{n!}\left(\frac{t}{\lambda}\right)^n e^{-t/\lambda}, \tag{16.4}$$

where t is the sample thickness and λ the mean free path for plasmon excitation.

FIG. 24. Values of Im E, Re G, and Im G which provide the best fits of Eq. 16.4 to the measured spectra in Na. Error bars are standard deviations. The dashed line from Re G is from P. Vashishta and K. S. Singwi, *Phys. Rev. B* **6**, 875 (1972).

In addition to this incoherent multiple plasmon excitation, it is possible to create two (or more) plasmons in a single interaction. The probability of double plasmon creation has been calculated,[44] and experiments have been carried out purporting to observe such events.[45] Here we examine this question carefully in the light of recent experiments which indicate that double plasmon coherent scattering is a weaker process than the above-mentioned work indicates.[46]

The total probability for creating two plasmons is the sum of two terms describing the incoherent Poisson distributed probability and the coherent double scattering event:

$$P_2 = \frac{1}{2}\left(\frac{t}{\lambda_1}\right)^2 e^{-t/\lambda_1} + \frac{t}{\lambda_2} e^{-t/\lambda_2}, \qquad (16.5)$$

where λ_1 is the mean free path for single plasmon creation and λ_2 the mean free path for coherent double plasmon excitation. Ashley and Ritchie have evaluated λ_2 for a free electron gas in the random phase approximation and find

$$\frac{1}{\lambda_2} = 0.0103 r_s^2 \frac{1}{\lambda_1}. \qquad (16.6)$$

So they predict that the double creation probability ranges between 4% and 37% of the single plasmon scattering on going from Al to Cs. These events can best be observed by looking for a deviation from the Poisson distribution of the two-plasmon creation probability as the sample thickness is varied.

There are some complications. If as in the above-cited theoretical and experimental work, we ignore plasmon dispersion, then the double plasmon coherent and incoherent spectra will be identical, each being a peak at $2\omega_p$. Including plasmon dispersion smears out the coherent double plasmon spectrum significantly. For example in Al, $\omega_p(q_c) - \omega_p(0) \simeq$ 3 eV, as can be seen in Figs. 11 and 12. Ignoring final state interactions between the plasmons, this converts the double plasmon coherent spectrum into a square-root shape beginning at $2\omega_p(0)$ and extending up approximately 6 eV. Interactions between the plasmons could alter this shape, even resulting in a bound state of two plasmons producing a sharp peak near threshold if the interaction is strong enough.[47]

To make the coherent double scattering event as visible as possible on

---

[44] J. C. Ashley and R. N. Ritchie, *Phys. Status Solidi* **38**, 425 (1970).

[45] J. C. H. Spence and A. E. C. Spargo, *Phys. Rev. Lett.* **26**, 895 (1971); *Electron Microsc., Proc. Int. Congr., 8th, 1974* Vol. 1, p. 390 (1975).

[46] W. Crutchfield, "Graduate Generals Project." Princeton University, Princeton, New Jersey, 1975 (unpublished).

[47] J. Ruvalds, A. K. Rajagopaland, J. Carballo, and G. S. Grest, *Phys. Rev. Lett.* **36**, 274 (1976).

top of the incoherent peak, one needs a very thin sample and high beam energy to make the ratio (t/λ_1) as small as possible. Recently Crutchfield[46] has analyzed data on 300 Å Al films and 100 Å Mg films using a beam energy of 300 keV. The analysis was carried out making two alternative assumptions about the shape of the coherent double plasmon spectrum: (1) It is a sharp peak near $2\omega_p(0)$, and (2) it is a broad spectrum spread over several electron volts beginning at $2\omega_p$.

The experiment was carried out by measuring the probabilities for creation of 0, 1, and 2 plasmons at a definite wave vector for a thick sample (for which incoherent events should dominate) and a very thin sample. If no double coherent events are present, then the ratio of thin to thick intensities will be proportional to

$$R(n) = \frac{I_1(n)}{I_2(n)} = \exp\left[-\frac{(t_1 - t_2)}{\lambda_1}\right] \left(\frac{t_1}{t_2}\right)^n \qquad (16.7)$$

Therefore a graph of the logarithm of this ratio versus n should yield a straight line. Any deviation of the $n = 2$ ratio from the straight line could be due to double plasmon coherent scattering events. No such deviations were observed, so only upper limits could be placed on their magnitudes.

Assuming the double coherent plasmon spectrum to be a peak near $2\omega_p$, the result for Mg is $1/\lambda_2 \leq 2 \times 10^{-3}(1/\lambda_1)$. For Al, $1/\lambda_2 \leq 3 \times 10^{-2}(1/\lambda_1)$. Assuming the double coherent spectrum to be spread over several electron volts, upper limits were obtained by examining $q = 0$ spectra. The result for Mg is $1/\lambda_2 \leq 0.01(1/\lambda_1)$ and for Al, $1/\lambda_2 \leq 0.04(1/\lambda_1)$.

The Ashley–Ritchie calculation[44] predicts $1/\lambda_2 = 0.07(1/\lambda_1)$ for Mg and $1/\lambda_2 = 0.04(1/\lambda_1)$ for Al. The predicted probability is clearly too high for Mg; for Al the results of these measurements are inconclusive. Since the predicted double coherent intensity varies as the plasmon cutoff wave vector raised to the fifth power, the result could be reduced by a factor of 7 by using a cutoff wave vector smaller by a factor of 1.5 than that used by Ashley and Ritchie ($q_c = \omega_p/v_f$).

There is a direct disagreement between this work and that of Spence and Spargo.[45] They used the first assumption listed above, assuming that all the strength in the coherent double plasmon event, should appear in a peak at $2\omega_p$. Their result for Mg is $1/\lambda_2 \simeq 0.07(1/\lambda_1)$ which should be compared with the Crutchfield result $1/\lambda_2 \leq 2 \times 10^{-3}(1/\lambda_1)$. Judging from the data presented, Spence and Spargo used significantly larger values of t/λ_1, and relied on a Fourier transform deconvolution technique to remove the incoherent multiple plasmon events from the data. The Crutchfield experiments were carried out with very thin films and high beam energy, so the double plasmon incoherent scattering was either barely visible, or in the case of the thin Mg film, too small to be detected in the measured spectrum. No complex analysis was needed to obtain the above upper limits for $1/\lambda_2$.

At present it appears that experimental observations of multiple coherent scattering events are not firmly established. The most recent experimental results indicate that the probability of these events is significantly less than predicted.

d. Influence of Sample Crystallinity

Simple metals are almost universally studied using inelastic electron scattering by evaporating a thin film of the metal onto an amorphous substrate, usually carbon. This means that the sample is by no means a single crystal, and may even be amorphous if deposited on a cooled substrate. Even after annealing, the mean crystallite size may be quite small.

Kunz[48] and later von Festenberg[49] measured both the width of the 15 eV plasma resonance in Al and the mean crystallite size in a series of samples. They found that the linewidth decreased as the samples were annealed. Later Krishan and Ritchie[50] carried out a calculation of the effect of density inhomogenieties on the plasmon width. They found that elastic scattering of plasmons from density variations can increase the plasmon width, and that a measurement of the plasmon width as a function of momentum transfer provides a direct measure of the autocorrelation function of the density variations. The unique feature of this effect is that the plasmon width *decreases* as q is increased from zero; all other known effects tend to *increase* the width with q.

Krishan and Ritchie assumed a Gaussian autocorrelation function and adjusted its range to fit Von Festenberg's data. They found a range of 210 A in a sample with a mean crystallite size of 70 A. This seems to indicate a coherence length for crystallite size of about three mean crystallite sizes. In other words a crystallite and its adjacent neighbors tend to be the same size.

Later Nagel and Schnatterly[51] pointed out that variations in mean free time with position in the sample can also cause an increase in plasmon width. The mean free time in a grain should be the bulk value, while within a grain boundary it could be much shorter. A value for the grain boundary mean free time which was obtained from optical data on other samples was found to yield approximately the correct plasmon widths as

---

[48] C. Kunz, *Z. Phys.* **167**, 53 (1962).
[49] C. von Festenberg, *Phys. Lett.* **23**, 293 (1966); *Z. Phys.* **207**, 47 (1967).
[50] V. Krishan and R. H. Ritchie, *Phys. Rev. Lett.* **24**, 1117 (1970).
[51] S. R. Nagel and S. E. Schnatterly, *Phys. Rev. B* **12**, 6002 (1975).

measured by von Festenberg. Therefore it is not known at this time whether the dominant mechanism for scattering of plasmons in a polycrystalline sample is density variations or mean free time variations.

This is one of the few examples in which plasmons themselves are used as a probe to scatter from an object of interest.

17. CORE SPECTRA OF SIMPLE METALS

a. Threshold Region: MND

One of the most elegant and intriguing developments among the theories of many-body problems in recent years is the soft X-ray threshold theory of Mahan,[52] and Nozières and De Dominicis. Imagine creating a core excitation in a simple metal. The excited state created includes an extra positive charge located on one of the ion cores in the crystal and a screening charge density which surrounds the excited ion, canceling the long-range part of the coulomb field of the ion. So although one might at first think of this transition as a single particle excitation, in fact, many electrons other than the one excited from the core have changed their states—namely all those which are involved in building up the screening charge. A method for describing the effect of these electrons on soft X-ray absorption and emission shapes was worked out by the above-mentioned authors, with many contributions from others.[52]

Here we shall describe the basic elements of this theory in order to see how it is that inelastic electron scattering can provide a direct test of the validity of the theory. First, consider the simplest view one can take of the core excitation process in which a core electron changes its state, becoming a plane wave in the conduction band, and all other electrons remained unchanged. We shall write the wave functions as simple products, ignoring exchange between core and conduction states. Then the transition matrix element for inelastic electron scattering [Eq. (3.9)] factors in a simple way. Let

$$|\lambda_0\rangle = |\phi_1(r_1)\phi_2(r_2)\cdots\phi_N(r_N)\rangle \quad \text{and} \quad |\lambda\rangle \qquad (17.1)$$
$$= |\phi_1'(r_1)\phi_2'(r_2)\cdots\phi_N'(r_N)\rangle.$$

Then since all the ϕ's remain unchanged except ϕ_1, which we take to be the core state,

$$\sum_i \langle\lambda|e^{iq\cdot r_i}|\lambda_0\rangle = \langle\phi_1'(r_1)|e^{iq\cdot r_1}|\phi_1(r_1)\rangle, \qquad (17.2)$$

where $\phi_1'(r_1)$ is a plane wave in the conduction band. Figure 25 shows

[52] For a recent review of this subject, see G. D. Mahan, *Solid State Phys.* **29**, 75 (1974).

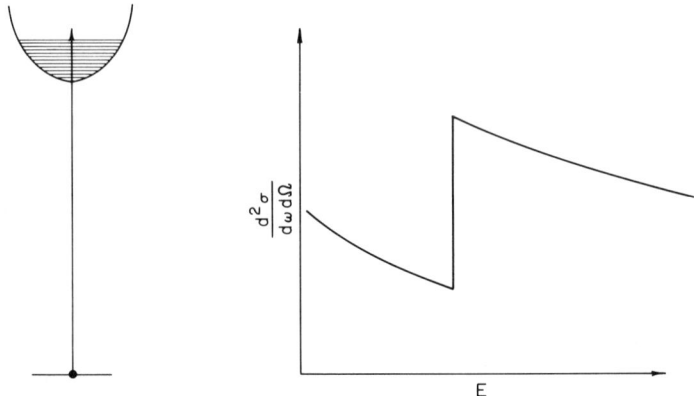

FIG. 25. Schematic representation of a core threshold in a metal. On the left is shown a one-electron energy level diagram with the core to conduction band transition indicated. On the right is the expected threshold shape in the one-electron approximation.

schematically the expected shape of the inelastic electron scattering cross section in this approximation near the threshold energy at which transitions can begin to occur from the core state to empty states above the Fermi energy. Below threshold there is a background due to conduction electron excitations. At threshold there is a sharp increase in scattering due to core excitations. The Mahan–Nozières–De Dominicis theory describes the shape of the spectrum just at threshold and for a certain energy range above. In the one electron approximation this shape is a simple step function, or more precisely a Fermi function at finite temperature. The inclusion of lifetime broadening will round off the threshold, and band structure effects will further modify the shape.

Now let us allow all the conduction electrons to readjust their states slightly so as to pile up extra charge around the excited core to screen its Coulomb field. We must then multiply Eq. (17.2) by $N - 1$ terms, each of the form

$$\langle \phi_i'(r_i) | \phi_i(r_i) \rangle \tag{17.3}$$

where ϕ_i is the ith conduction state before the transition and ϕ_i' the ith state after readjustment to the core potential. Each term in the product is unity in the one-electron approximation when $\phi_i = \phi_i'$, and each term is slightly less than unity when ϕ_i and ϕ_i' are slightly different. The product of 10^{23} terms, each slightly less than unity, is of course zero. Since primarily states near the Fermi energy readjust, we expect the threshold strength to be annihilated just at threshold by this effect. This is the orthogonality catastrophy, first described by Anderson.[53]

[53] P. W. Anderson, *Phys. Rev. Lett.* **18**, 1049 (1967).

Only a finite number of electrons exist within a short distance from the core. In order for the number of electrons that readjust to be truly enormous, we must include electrons that are far from the core. But since the screened potential is of short range, we must wait some time for the distant electrons to move to the excited atom and scatter from it. This corresponds to a modification of the spectrum very close to threshold. The further we go from the core, the longer we must wait, but there are more and more electrons to wait for. As was pointed out by Hopfield,[54] this is a classic example of an infrared divergence. The striking strength of the effect at threshold is not a feature of a particular assumed model Hamiltonian, but is a general feature of the impurity problem in metals, coming directly from the high degeneracy of the ground state.

There is a second effect of the readjustment which has the opposite result; it tends to enhance the transition strength at threshold. Since the electrons involved in the problem are identical particles, it is not possible to say which of them ends up where. The core electron could make a direct transition to an empty state above the Fermi energy, or one of the conduction electrons could go there, with the core electron replacing it. All possible replacements involving conduction electrons must be summed over. Each enhances the transition amplitude slightly. This attempt of the metal to form an exciton, enhancing the transition strength at threshold was first described by Mahan.[55] This enhancement also has the form of an infrared divergence.

So we expect two competing effects near threshold, one tending to annihilate the transition strength, the other producing a divergence in strength. The calculations leading to both effects are expected to be asymptotically exact as threshold is approached. Too much lifetime broadening or lack of resolution in a spectrometer could prevent these effects from being observed. It is thought, however, that the energy range away from threshold over which the theory should be valid greatly exceeds typical lifetime broadenings on the order of 0.1 eV.

Nozieres and DeDominicis found an exactly soluble one-electron model of the problem which includes both of these effects.[56] Their result for $\epsilon_2(q, \omega)$ is

$$\epsilon_2(q, \omega) \propto \sum_l A_l \left(\frac{\xi}{\omega - \omega_T}\right)^{\alpha_l} \theta(\omega - \omega_T), \qquad (17.4)$$

where

$$A_l(q) = |\langle \phi(r) | e^{i q \cdot r} | \phi_l'(r) \rangle|^2 \qquad (17.5)$$

[54] J. J. Hopfield, *Comments Solid State Phys.* **2**, 40 (1969).
[55] G. D. Mahan, *Phys. Rev.* **163**, 612 (1967).
[56] P. Nozières and C. T. De Dominicis, *Phys. Rev.* **178**, 1097 (1969).

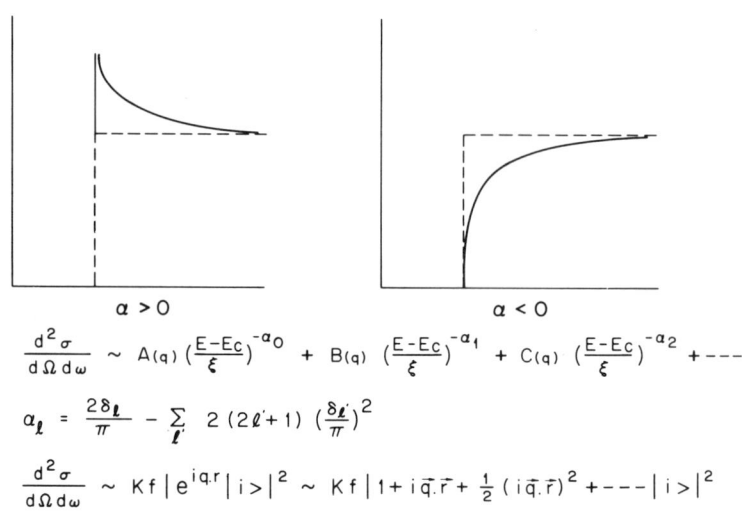

FIG. 26. Expected threshold shapes in simple metals according to the MND theory.

and ϕ is the core wave function, ϕ_l' the lth partial wave of a screened conduction band state, ξ a range parameter, $\hbar\omega_T$ the threshold energy. Figure 26 shows schematically the expected threshold shapes.

In soft X-ray absorption experiments the operator $e^{iq\cdot r}$ is replaced by $\hat{\epsilon}\cdot r$, where $\hat{\epsilon}$ is the polarization vector of the incident photon. In this case, using dipole selection rules, the important partial waves at the Fermi energy are determined by the orbital angular momentum of the core state. Most calculations of the exponents result in $\alpha_0 > \alpha_1 > \alpha_2 \cdots$ with α_0 positive, α_1 close to zero, and all higher exponents negative. Therefore we expect peaked thresholds only for p cores, e.g., $L_{II,III}$, $M_{II,III}$ edges, etc., with the spectra of all other core states rounded or suppressed. This is generally what is observed. Therefore the theory is in qualitative agreement with existing soft X-ray absorption measurements.

As explained earlier, inelastic electron scattering measurements offer greater flexibility than optical absorption as far as selection rules are concerned. Since the core states are well localized, the results described earlier apply directly to this case and "forbidden" transitions can be caused by measuring spectra at sufficiently large values of q. Thus different partial waves may be selected in transitions from a single core state. According to Eq. (17.4) this means that the threshold shape should change with q in a predictable manner if the exponents and matrix elements are known. This means of testing the Mahan–Nozieres–De-Dominicis theory was first proposed by Doniach, Platzman, and Yue.[57]

[57] S. Doniach, P. M. Platzman, and J. T. Yue, *Phys. Rev. B* **10**, 3345 (1971).

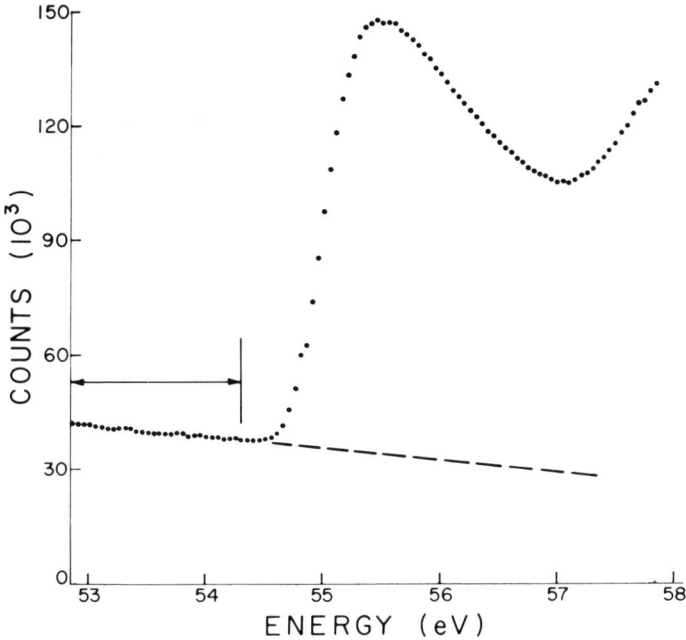

FIG. 27. Measurement of the K-edge region in Li metal. The arrow indicates the data used in determining the background to be subtracted [P. C. Gibbons, J. J. Ritsko, and S. E. Schnatterly, *Phys. Rev. B* **10**, 5017 (1974)].

The specific experiment they described was X-ray raman scattering rather than electron scattering, but the theory is the same in both cases since each scattering process can be treated in the Born approximation. Sufficient energy resolution to carry out this test using X-ray scattering has not yet been achieved.

The K edge of Li was the first threshold to be studied using inelastic scattering and the result was completely unexpected.[58] For years, Li had been taken to be a classic example of the success of the threshold theory. The threshold shape is strongly rounded in striking contrast to the sharply peak $L_{II,III}$ edges of the other alkali metals. Lithium was the most discussed since it stood alone as being different. Moreover the exponent values needed to fit the edge shape were close to values calculated using a Thomas–Fermi screened point charge.

Figure 27 shows measurements of the K threshold of Li for $q = 0$. It is important in testing a theory of spectral shape to have high-quality data which is carefully analyzed. Figure 27 shows the threshold on top

[58] P. C. Gibbons, J. J. Ritsko, and S. E. Schnatterly, *Phys. Rev. B* **10**, 5017 (1974).

of a background due to conduction electron excitations and the substrate. In this case the background is essentially linear in the edge region. Sometimes a polynomial with higher power terms must be used in fitting the background. The arrow indicates the region used in fitting to the background.

Figure 28 shows the subtracted edge region. The shape is not at all like that shown in Fig. 26 for a negative exponent. In fact, the threshold can be described very well using a Gaussian convoluted with a step. This is the shape that would be expected if the broadening is dominated by phonons rather than lifetime effects, or the MND theory. The solid line in Fig. 28 represents a least squares fit to the data in the region shown by the arrow, of a Gaussian convoluted with the step function shown. This function has been extrapolated to higher energies and continues to describe the data well through the edge region. The open circles show the result of a least squares fit of a Lorentzian convoluted with a step. This is the shape that would be expected if lifetime effects dominate the broadening. As is apparent from the figure, the Lorentzian shape does not describe the data nearly as well as the Gaussian. This strongly suggests that phonons, rather than Auger or other lifetime effects, or the

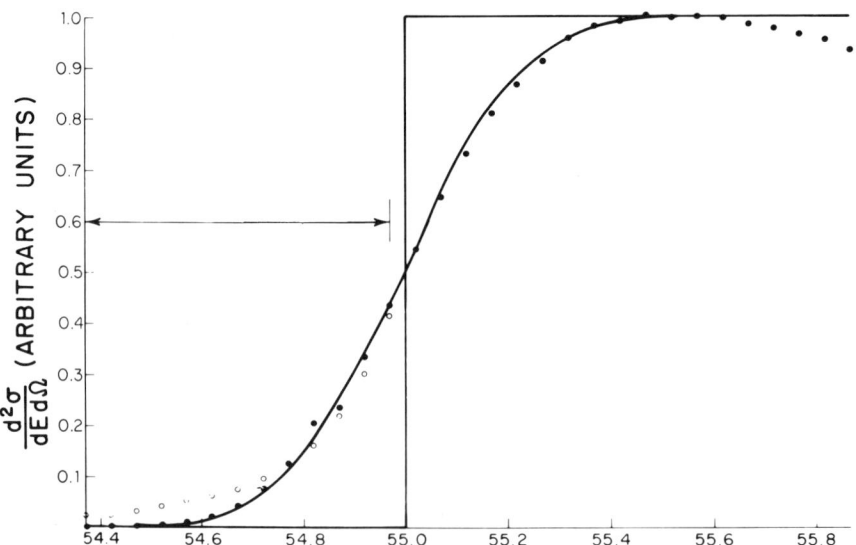

FIG. 28. K-edge data in Li after background subtraction. Filled circles are the data. The curved line going through the data is a least squares fit using a Gaussian convoluted with the step function shown. The parameters were adjusted to fit the data in the region shown by the arrows. Open circles are a best fit using a Lorentzian convoluted with a step [P. C. Gibbons, J. J. Ritsko, and S. E. Schnatterly, *Phys. Rev. B* **10**, 5017 (1974)].

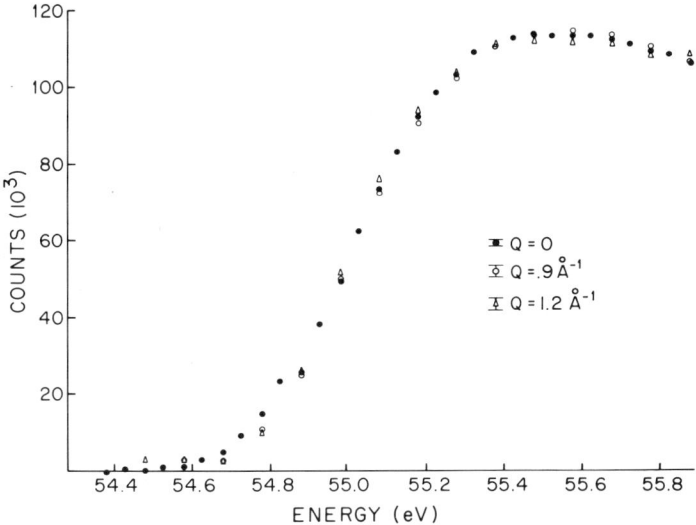

FIG. 29. Momentum transfer dependence of the K threshold of Li [P. C. Gibbons, J. J. Ritsko, and S. E. Schnatterly, *Phys. Rev. B* **10**, 5017 (1974)].

MND theory, predominantly produce the broadening. This conclusion confirms McAlister's earlier proposal[59] and has been reconfirmed by more recent work.[60]

Recently Petersen used the photoyield method to study bulk samples at DESY.[61] At the lowest temperatures studied, near liquid nitrogen, which are difficult to achieve with thin film samples, he found the measured threshold shape to be more consistent with McAlister's calculated density of states than the previously used step function. The corresponding observation in X-ray emission spectroscopy has recently been carried out.[62] The results are consistent with a one-electron density of states broadened by phonons which do not have time to fully relax around the core hole before emission takes place.[63]

The ultimate test, of course, is to observe how the shape of the threshold changes with momentum transfer. Figure 29 shows the measured result. Surprisingly, no change in shape occurs at all within the experimental errors. For small q, transitions to p waves are occurring. At larger

[59] A. J. McAlister, *Phys. Rev.* **186**, 595 (1969).
[60] Y. Baer, P. H. Citrin, and G. K. Wertheim, *Phys. Rev. Lett.* **37**, 49 (1976).
[61] H. Petersen, *Phys. Rev. Lett.* **35**, 1363 (1975).
[62] T. A. Calcott and E. T. Arakawa, *Phys. Rev. Lett.* **38**, 442 (1977).
[63] C. O. Almbladh, *Int. Conf. Vacuum Ultraviolet Radiat. Phys. 5th, 1977* Vol. II, p. 6 (1977).

q, s waves are also involved, with a small contribution from d waves. In order for the shape to remain invariant as is observed, it is necessary that $|\alpha_0 - \alpha_1| \lesssim 0.05$. This differs significantly from the Thomas–Fermi result, for which $\alpha_0 - \alpha_1 \simeq 0.5$.[64] Recently Girvin and Hopfield have pointed out that the exchange interaction between conduction electrons and the core hole can significantly modify the phase shifts.[65] The primary effect is to reduce α_0 and increase α_1. Their numerical results are still not in agreement with the above inequality, but the correction is in the right direction.

Magnesium was the second metal to have been measured and analyzed carefully.[66] Figure 30 shows the measured spectrum at $q = 0$ along with a least squares fit using the MND theory. The exponent needed to fit the $q = 0$ shape is $\alpha_0 = 0.18$. Using this value of α_0 and Dow's compatability relation to determine α_1,[67] the expected change in threshold shape can be computed. The result is shown in Fig. 30 with the observed change between $q = 0$ and $q = 1.8$ Å^{-1} superimposed on it. As is apparent from the figure, too much change in shape was observed. To produce the observed change in shape, $\alpha_0 = 0.4$ is required. Thus the $q = 0$ shape and the $q = 1.8$ Å^{-1} shape are inconsistent.

At the time of this writing this problem is being vigorously pursued, both experimentally and theoretically. It is not possible now to pass final judgment as to the ability of the MND theory to describe soft X-ray thresholds in real metals. In the spectra measured so far, it appears that other effects are of at least comparable importance in determining threshold shapes.

b. Spectra above Threshold

Usually the most prominent feature present in soft X-ray core spectra of metals is not the edge itself but the broad hump which is almost always present 10–50 eV above the edge. If atomic potential effects are not important, this can be understood in very simple terms. Transitions are occurring from the core state to plane wave–like final states in the conduction band. The matrix element for the transition involves an integral of the radial wave function for the core multiplied by a Bessel function whose argument is kr, where k is the excited electron's momentum and

[64] G. A. Ausman and A. J. Glick, *Phys. Rev.* **183**, 687 (1969).
[65] S. M. Girvin and J. J. Hopfield, *Phys. Rev. Lett.* **37**, 1091 (1976).
[66] S. G. Slusky, P. C. Gibbons, S. E. Schnatterly, and J. R. Fields, *Phys. Rev. Lett.* **36**, 326 (1976).
[67] J. D. Dow, J. E. Robinson, J. H. Slowick, and B. F. Sonntag, *Phys. Rev. B* **10**, 432 (1974).

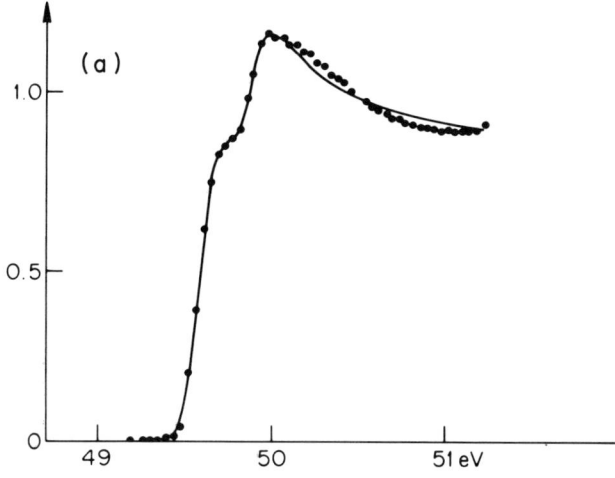

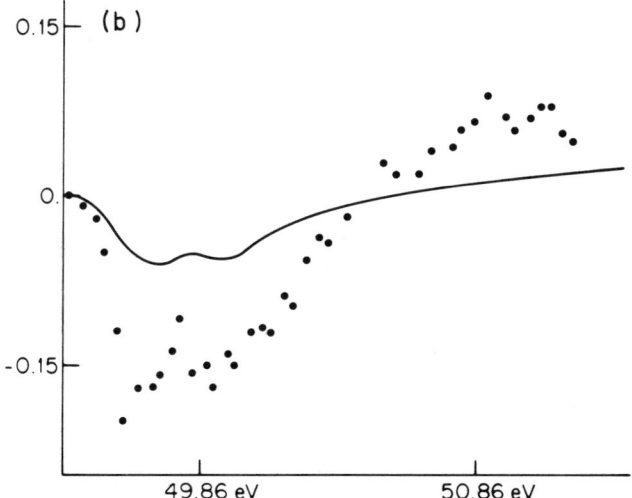

FIG. 30. (a) $q = 0$ shape of the Mg $L_{II,III}$ edge. Dots are the data; line is a least squares fit using the MND theory. (b) Dots are the observed differences in shape between $q = 0$ and $q = 1.8$ Å^{-1}. Line is the calculated change in shape using the MND theory and $\alpha^\circ = 0.18$. See text [S. G. Slusky, P. C. Gibbons, S. E. Schnatterly, and J. R. Fields, *Phys. Rev. Lett.* **36**, 326 (1976)].

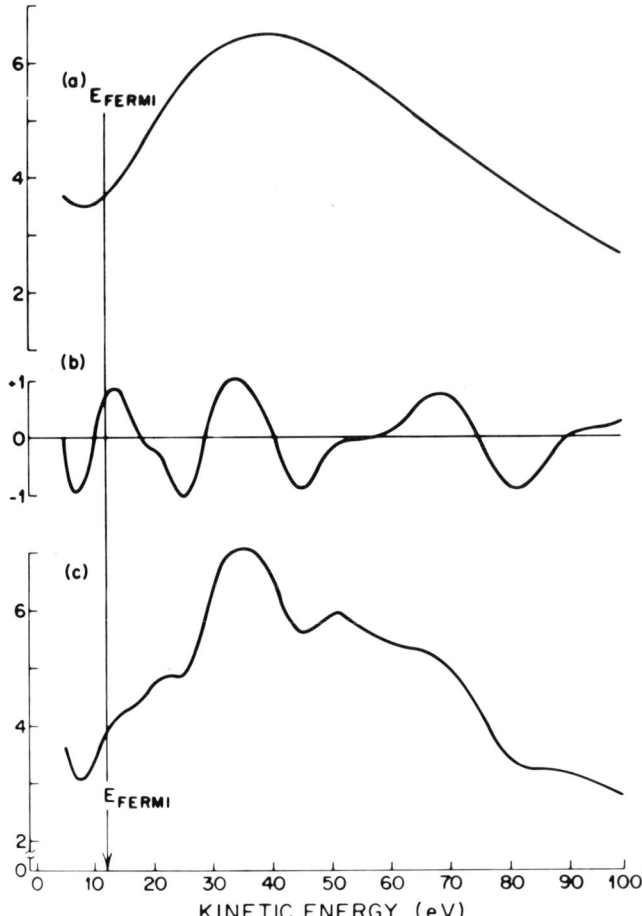

FIG. 31. (a) Transition density for excitations from the Al 2p core to plane wave continuum states. (b) Expected modification of the spectrum due to backscattering from neighboring atoms. (c) The resulting modified spectral shape [J. J. Ritsko, P. C. Gibbons, and S. E. Schnatterly, *Phys. Rev. Lett.* **32**, 671 (1974)].

r the distance from the nucleus. As k varies the magnitude of the matrix element changes, producing broad structure in the spectrum.

Figure 31 shows an example, calculated with parameters appropriate for the $L_{II,III}$ edge of Al.[68] The upper curve shows the spectrum calculated as described above, i.e., assuming a plane wave final state. The middle curve shows the expected modification of the spectrum due to the excited

[68] J. J. Ritsko, P. C. Gibbons, and S. E. Schnatterly, *Phys. Rev. Lett.* **32**, 671 (1974).

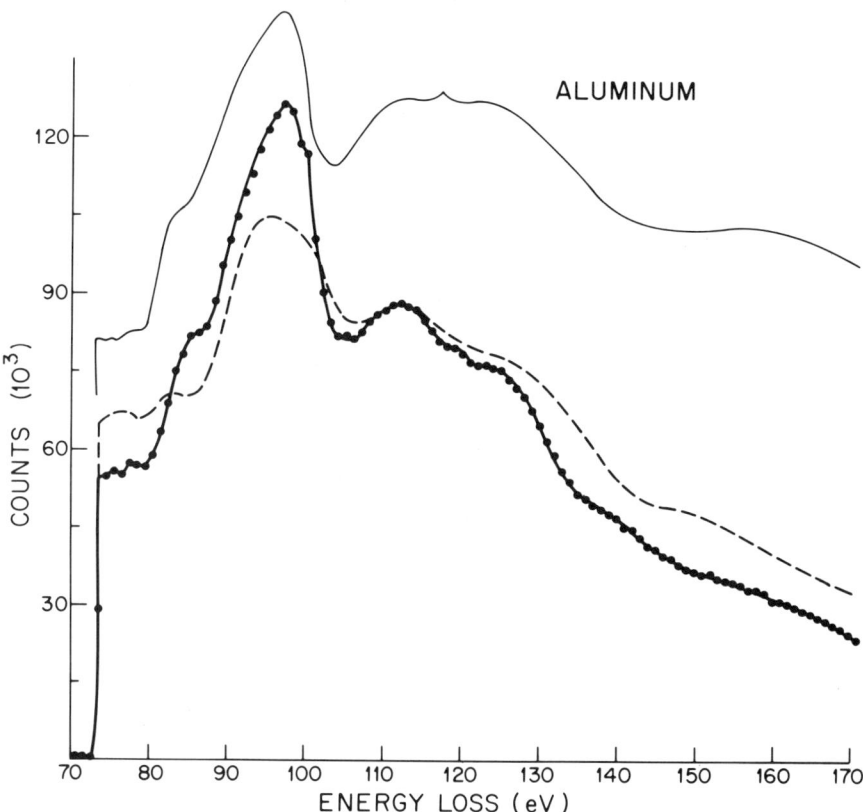

FIG. 32. Dots with solid line: measured Al $L_{II,III}$ spectrum. Dashed line: calculation shown in Fig. 31 and described in text. Upper solid line: optical absorption spectrum measured with synchrotron radiation [J. J. Ritsko, P. C. Gibbons, and S. E. Schnatterly, *Phys. Rev. Lett.* **32**, 671 (1974)].

electron backscattering from neighboring atoms. The bottom curve is the resulting modified spectrum. Figure 32 shows a comparison of this calculated shape with measurements. The lower solid line with dots is the shape measured with inelastic electron scattering, and the upper solid line is the optical absorption measured with synchrotron radiation.[68]

Recently Hayes and Sen[69] have carried out a spherical solid model calculation of EXAFS (extended X-ray absorption fine structure), which is the modulation of the absorption coefficient well above threshold due to backscattering from neighboring atoms. According to this model the excited electron is treated as an outgoing spherical wave centered on the

[69] T. M. Hayes and P. N. Sen, *Phys. Rev. Lett.* **34**, 956 (1975).

excited atom, and backscatters from spherical shells representing the neighboring atoms. Not only does the phase of the outgoing wave as it arrives at a shell depend on the energy of the electron, but it also depends on its angular momentum. This is because the coefficient of each partial wave is $j_l(ka)$, where l is the angular momentum, k the electron's momentum, and a the distance to the shell. Therefore, by varying q and changing emphasis from one partial wave to another as described above, the shape of the EXAFS structure should change in a predictable way. In particular, well above threshold where the Bessel functions can be replaced with their asymptotic values, adjacent partial waves should differ in phase by $\pi/2$. Experiments of this type will require a low-resolution, high-current spectrometer since the counting rates are likely to be low, and very high momentum transfers will be needed to see the changes.

18. SEMICONDUCTORS AND INSULATORS: *Valence Excitations*

Semiconductors and insulators differ from metals in that they lack the highly degenerate ground state which dominates so many metallic properties. The shape of the elementary excitation spectrum is strikingly altered: in semiconductors and insulators a minimum energy gap exists, due to the stronger periodic field, so that no electronic excitations with energy much less than the gap can be created. The kinds of electronic excitations that can be created include excitons, with energy somewhat less than the gap, interband transitions above the gap, and plasmons, generally well above the gap.

a. Universal Semiconductor Model

Recently a simple model was developed which includes both excitons and plasmons,[70] and which allows one to explore the relative importance of these two kinds of elementary excitations in the measured spectrum. A coupling between the exciton and plasmon is used which is simply the product of the fluctuating dipole moment associated with the exciton and the electric field due to the plasmon. The Lorentz–Lorentz local field treatment is used, and the energies and strengths of longitudinal and transverse modes are derived. The input parameters needed are $\omega_p^2 = 4\pi n e^2/m$, Δ is the gap energy, and f_0 the oscillator strength of the transverse exciton.

To obtain a broad-brush view of these excitations for a range of materials, we shall use the Phillips–VanVechten[71] universal semiconductor

[70] P. V. Giaquinta, E. Tosatti, and M. P. Tosi, *J. Phys. C* **9**, 2031 (1976).
[71] J. C. Phillips, *Rev. Mod. Phys.* **42**, 317 (1970).

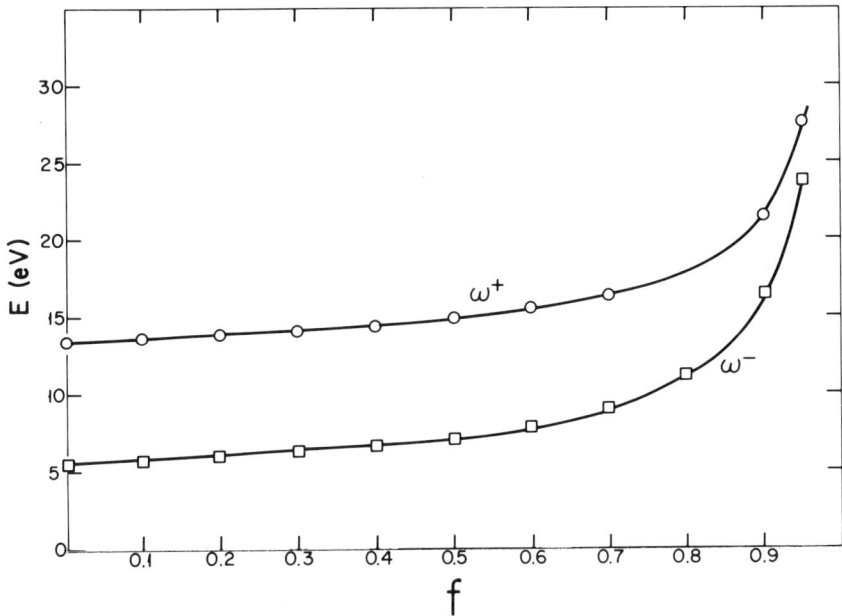

FIG. 33. Exciton and plasmon energies as a function of ionicity f according to the model described in the text.

model to express Δ and f_0 in terms of the Phillips ionicity

$$f = \frac{c^2}{E_h^2 + c^2}, \tag{18.1}$$

where c is the ionic contribution to the energy gap, and E_h the homopolar contribution. The energy gap is then

$$\Delta^2 = E^2 + c^2 = E_h^2/(1 - f). \tag{18.2}$$

The exciton oscillator strength should vary as f^2. Fitting this form to the recent accurate measurements[72] of the oscillator strength of the LiF fundamental transverse exciton ($f_0 = 0.08$) we have

$$f_0 = 0.113 f^2. \tag{18.3}$$

If we assume the typical valves $\omega_p = 15$ eV and $E_h = 5.5$ eV, we can evaluate the energies and strengths of the exciton and plasmon according to this model. Figure 33 shows the predicted longitudinal exciton and plasmon energies as a function of the ionicity f. Both energies increase

[72] J. R. Fields, Ph.D. Thesis, Princeton University, Princeton, New Jersey (1975); J. R. Fields, P. C. Gibbons, and S. E. Schnatterly, *Phys. Rev. Lett.* **38**, 430 (1977).

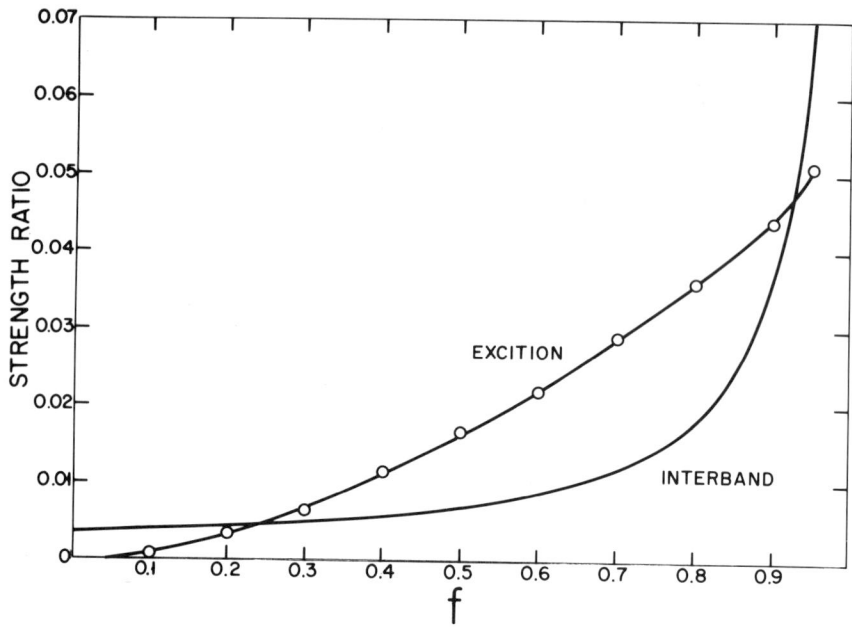

FIG. 34. Ratio of the strength of the exciton and interband transitions to the plasmon according to the model described in the text.

with f as expected. Figure 34 shows the exciton to plasmon strength ratio as a function of f. Also shown is an estimate of the interband transition strengths as a function of f. This is obtained as follows: Since the interband transition rate is proportional to the square of an interband matrix element, we assume

$$\Delta^2 \propto R_b = B/(1 - f), \tag{18.4}$$

where R_b is the interband transition rate, and B is a constant. Normalizing the total transition rate to 1 and ignoring the exciton contribution,

$$R_p + R_b = 1, \tag{18.5}$$

where R_p is the plasmon transition rate. So

$$R_p = 1 - \frac{B}{1 - f} = \frac{1 - f - B}{1 - f}. \tag{18.6}$$

Solving for R_b/R_p we find

$$\frac{R_b}{R_p} = \frac{B}{1 - f - B}. \tag{18.7}$$

The constant B can be determined by fitting this result to measurements. In doing so we have chosen to regard easily visible features in measured spectra as a measure of R_b. Thus the result is not a measure of total interband scattering, but indicates the visibility of interband effects in the measured spectra. The value of B determined in this way from LiF measurements (see Figs. 37 and 38) is $B = 0.0035$. This value gives rise to the curve shown along with the exciton strength in Fig. 34.

The first conclusion to be drawn from Fig. 34 is that in all solids, plasma excitations dominate the spectrum in the region of valence excitations. (In the core excitation region, the plasma or collective contribution is generally much less.) For binary compounds, crystal structure is correlated strongly with ionicity,[71] so we can summarize these results by saying that the NaCl structures ($f \geq 0.79$), exciton and interband strengths are comparable, each about 4% of the plasmon; for wurtzite structures ($0.5 \leq f \leq 0.79$), the exciton strength is about twice that of interband features and is about 2% of the plasmon; for zincblende structures ($f \leq 0.50$) interband features are about 0.5% and excitons 0–1.5% of the plasmon.

b. Single Crystal Spectra: LiF

It is clear from these approximate results that exciton and interband effects will be considerably more difficult to study than plasmons, because they usually appear as small features on a large and sloping background. In principle one can observe spectral features varying continuously in energy as the momentum transfer is varied throughout the Brillouin zone, and map out an entire band structure diagram, or exciton dispersion. In practice, however, the weakness of these features, combined with the requirements of good energy and momentum resolution means that the experiments are difficult, and have only been carried out in a few cases.

We present one example here of such a study, namely LiF, for which exciton and band effects are relatively strong and readily seen in the spectra.[72] Figure 35 shows a series of spectra obtained from a self-supporting single-crystal thin film which was prepared by mechanical polishing followed by ion milling. The portion of the sample measured was very likely wedge-shaped, with an average thickness of about 3500 Å. No corrections have been made for multiple scattering. The data show considerable dependence on both magnitude and direction of momentum transfer.

The dispersion of the fundamental exciton can be read directly from a series of spectra such as shown in Fig. 35. Figure 36 shows the results

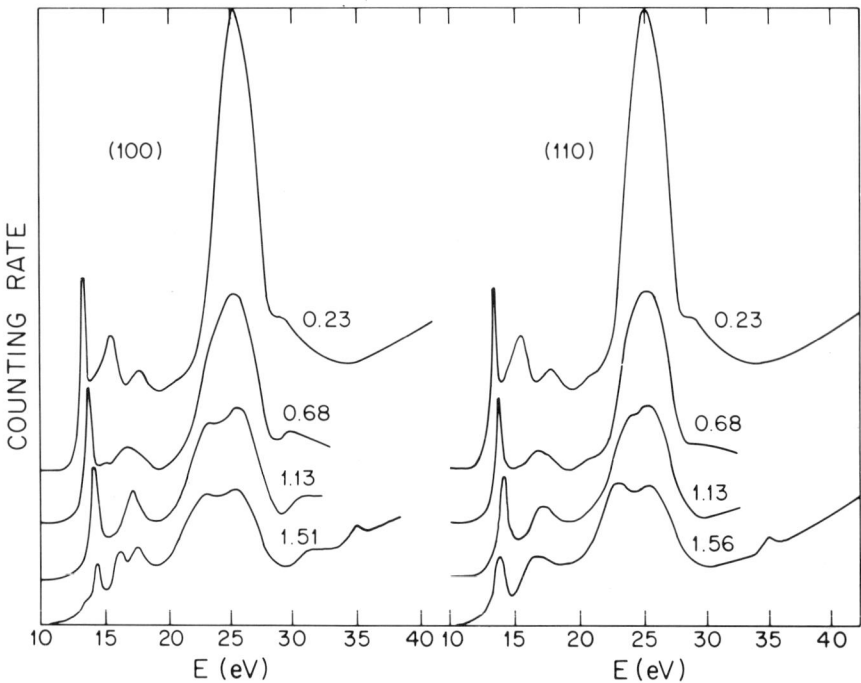

FIG. 35. Inelastic electron scattering spectra for single crystal LiF samples with momentum transfers in (100) and (110) directions [J. R. Fields, Ph.D. Thesis, Princeton University, Princeton, New Jersey (1975); J. R. Fields, P. C. Gibbons, and S. E. Schnatterly, *Phys. Rev. Lett.* **38**, 430 (1977)].

of a series of careful measurements. There is a considerable dispersion (~1 eV) and it is quite anisotropic. A good Frenkel exciton, being highly localized, would have little or no dispersion at all. A good Wannier exciton, on the other hand, should disperse approximately following the electron and hole energy bands out of which it is made. The fact that the dispersion is isotropic for small q suggests that for $q = 0$ the exciton is made up of electron and hole states near the zone center, in contrast to some proposals.[73]

If the exciton is following the electron and hole bands in its dispersion, then it is easy to show that in the effective mass approximation, the dispersion is given by

$$E(q) - E(0) = \frac{\hbar^2}{2} \frac{1}{(m_e + m_h)} q^2. \qquad (18.8)$$

The measured small q dispersion gives $m_e + m_h = 3.8m$, where m is the

[73] D. J. Mickish, A. B. Kunz, and T. C. Collins, *Phys. Rev. B* **9**, 4461 (1974).

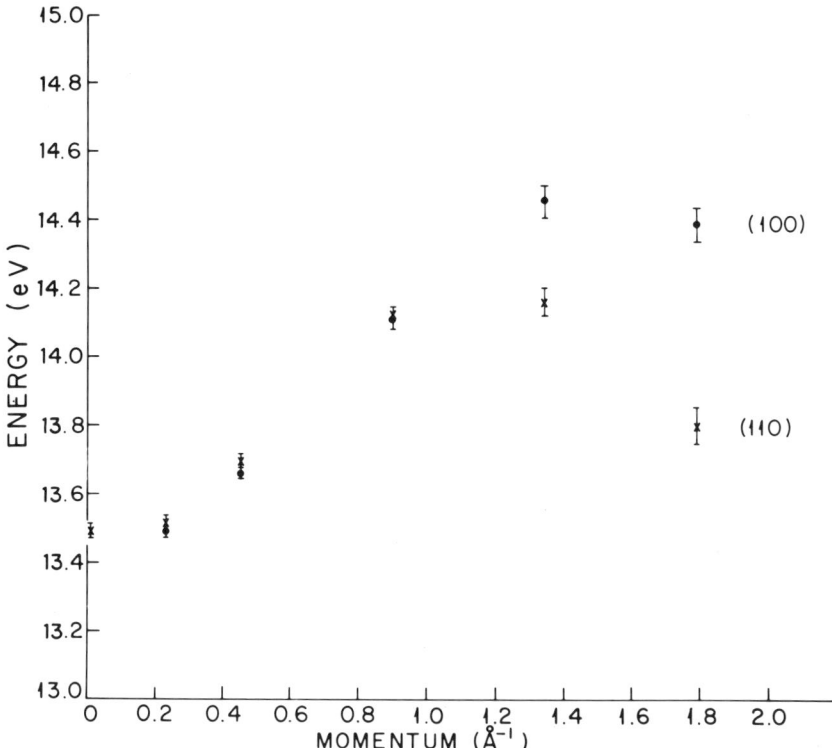

FIG. 36. Dispersion of the fundamental exciton of LiF in the (100) and (110) directions. The zone boundary in the (100) direction is at $q = 1.57$ Å^{-1}, and in the (110) direction it is at $\bar{q} = 1.66$ Å^{-1} [J. R. Fields, Ph.D. Thesis, Princeton University, Princeton, New Jersey (1975); J. R. Fields, P. C. Gibbons, and S. E. Schnatterly, *Phys. Rev. Lett.* **38**, 430 (1977)].

electron mass. Most band structure calculations for the alkali halides predict $0.5 \leq m_e/m \leq 1.0$. This results in $1/\mu = 1.83 \pm 0.5$, where μ is the reduced mass. We shall use $R = h^2/2\mu a^2$, where R is the exciton Rydberg, recently determined to be 1.6 ± 0.3 eV.[74] These numbers result in $a = 2.1 \pm 0.3$ Å. The corresponding rms value for the momentum width in the $n = 1$ exciton state is $\bar{q} = 1/a = 0.48 \pm 0.06$ Å^{-1}. To test the consistency of these numbers we can use the relation for the dielectric constant ϵ:

$$\epsilon = a(h^2/\mu e^2)^{-1} = 2.2 \pm 0.5. \quad (18.9)$$

The value of ϵ at $q = 0$ is 1.93. Thus within the uncertainties this Wannier model of the exciton is self-consistent: The exciton radius is approxi-

---

[74] M. Piacentini, *Solid State Commun.* **17**, 697 (1975).

mately equal to the nearest-neighbor separation so it is not completely localized, the momentum width is about one third of the Brillouin zone radius, and the effective dielectric constant is approximately equal to the macroscopic value. These results do not prove that the Wannier picture is the only appropriate one to use for LiF, but they do show that it is a reasonable choice, in support of earlier work.[74]

Let us now move to the more highly localized core spectra of LiF. In Fig. 35 notice the small peak near 35 eV which is present in the high q data but absent at low q. This feature appears at approximately the right energy to be a transition from the F 2s state to the Li conduction band. This transition should be optically forbidden at threshold and so, as explained earlier, should be absent in low q data and should increase in strength relative to dipole transitions in proportion to q^2. To further verify identification, the oscillator strength for this transition has been evaluated[72] using Hermann–Skillman wave functions. The calculated and measured results are, respectively, 0.064 q^2a^2 and 0.050 q^2a^2, where a is the Bohr radius.

From XPS data[75] the position of the F 2s level in LiF relative to the top of the valence band is 23.7 eV. Adding to this the optical band gap of 14.2 eV,[74] results in a predicted threshold energy of 37.9 eV. The major correction to this should be the exciton binding energy which is, then 2.9 eV, which can be compared with the binding energy of the F 2p exciton of 1.6 ± 0.3 eV. This is the only case known to the author in which binding energies of excitons associated with adjacent core levels has been determined. The striking increase in exciton binding energy with core binding energy is qualitatively consistent with a decrease in central cell corrections, but a quantitative understanding of these results is lacking at present.

Figure 37 shows spectra measured at $q = 0.34$ Å^{-1} and $q = 2.25$ Å^{-1} in the 52–67 eV range. As explained in reference 72, the growth with q of the 61 eV shoulder preceding the main 62 eV peak allowed a definite identification of this feature as being an atomic 1s–2s transition. At $q = 0$ it would be completely forbidden except for phonon mixing with the 2p state which gives rise to the 62 eV peak, and its increase in strength with q relative to dipole transitions follows from reasons given earlier.

Recently Zunger and Freeman[76] have carried out a calculation of the excitation energies of the core electrons in LiF using a cluster model. A 16-atom unit cell is used with an excited ion at the center of each unit cell. Only energies (not spectral shapes) are calculated. The results are in excellent agreement with experiments, suggesting that this type of approach may be generally useful for calculating core spectra in solids.

[75] W. Gudat, C. Kunz, and H. Peterson, *Phys. Rev. Lett.* **32**, 1370 (1974).

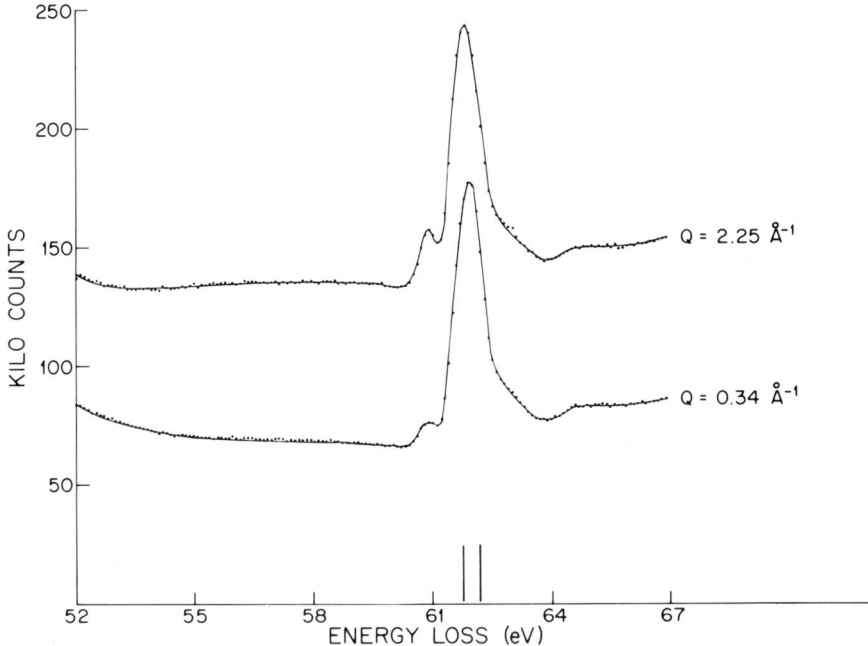

FIG. 37. Spectra measured at high and low q values in the K-threshold region of Li in LiF [J. R. Fields, Ph.D. Thesis, Princeton University, Princeton, New Jersey (1975); J. R. Fields, P. C. Gibbons, and S. E. Schnatterly, *Phys. Rev. Lett.* **38**, 430 (1977)].

Hopefully a version of this method can be extended to include spectral shapes which still provide the ultimate test of its validity.

It is clear that Fig. 35 contains a wealth of additional information which requires a serious calculation to be dealt with adequately. A few comments can be made, however. First, we see no hint of the splitting of the low q plasma peak at 25 eV which has been previously reported.[77] At high q, especially near the zone boundaries, we do see a partially resolved splitting of 2–3 eV. Whether this is a genuine splitting of the plasma collective mode due to the periodic field, as has been suggested,[78] or whether it is an interband transition, is not known.

It is tempting to use the spectra in Fig. 35 to associate spectral features with particular interband transitions. Based on past experience with optical data, this is a game one plays at one's own peril. However, since several distinct spectra are available, the number of constraints is much

---

[76] A. Zunger and A. Freeman, *Phys. Rev. B* **16**, 2901 (1977).
[77] C. Gout and F. Pradal, *J. Phys. Chem. Solids* **29**, 581 (1968).
[78] R. Girlanda, M. Parrinello, and E. Tosatti, *Phys. Rev. Lett.* **36**, 1386 (1976).

larger than if there were only one measurement. We make two postulates: (a) Except for the plasmon, peaks in the spectra are association with M_0 or M_1 critical points. (b) Features in the spectrum taken with q near the zone boundary in the (100) direction are dominated by regions near Γ and X. Postulate (a) is supported by the fact that excitonic effects enhance M_0 and M_1 thresholds and suppress M_2 and M_3 features. Postulate (b) is reasonable since the regions near Γ and X will be nearly congruent for this momentum transfer, while other regions are not likely to be.

Using these postulates and adding 1 eV to all peak positions to represent longitudinal exciton binding energies, we find we can consistently describe the qualitative features of these spectra with the following energy differences: $\Gamma_{1c} - \Gamma_{15v} = 14.3$ eV, $X_{1c} - X_{5'v} = 15.3$ eV, $X_{5'v} - X_{4'v} = 1.8$ eV, $X_{3c} - X_{1'c} = 5.5$ eV, $X_{1c} - \Gamma_{1c} = 4.2$ eV, $\Gamma_{15v} - X_{5'v} = 1.0$ eV, and $K'_c - \Gamma_{1c} \simeq 3$ eV, where c and v refer to conduction and valence bands, respectively. These results differ from the available calculations [73,79] and previous assignments [80] primarily in the conduction band being more isotropic than previously proposed.

c. Kramers–Kronig Analysis and Comparison with Optical Measurements: LiF

Measurements of LiF spectra out to 200 eV have recently been Kramers–Kronig analyzed using the multiple scattering corrections described earlier.[72] Figure 38 shows the resulting real and imaginary parts of the dielectric response function from 0–90 eV. These can be compared with earlier measurements. In the vacuum UV region, taking the magnitude of Im$[-1/\epsilon(0,\omega)]$ at the peaks of the exciton and plasmon as a measure, we find differences with previous work of factors of two as shown in Table II. Using our values of the optical constants, however, we have computed the optical reflectivity $R(\omega)$ and find good agreement with the reported measured values.[72] A possible explanation for this discrepancy is that in analyzing their reflectivity measurements to obtain a dielectric function, earlier workers attributed too little strength in $\epsilon_2(\omega)$ to the energy region above their upper cutoff (~ 30 eV). We find the measured spectrum does not fall as $1/\omega^3$ until above 150 eV.

Also shown in Table II is a comparison of measured optical absorption coefficients in the soft X-ray region with our computed values. There is good agreement between our determinations and those obtained using synchrotron radiation except in the regions of high absorption, near peaks in the spectrum. In these strongly absorbing regions our values are a

---

[79] W. P. Menzel, C. C. Lin, D. F. Fouquet, E. A. Lafon, and R. C. Chaney, *Phys. Rev. Lett.* **30**, 1313 (1973).

[80] M. Piacentini, D. W. Lynch, and C. G. Olson, *Phys. Rev. B* **13**, 5530 (1976).

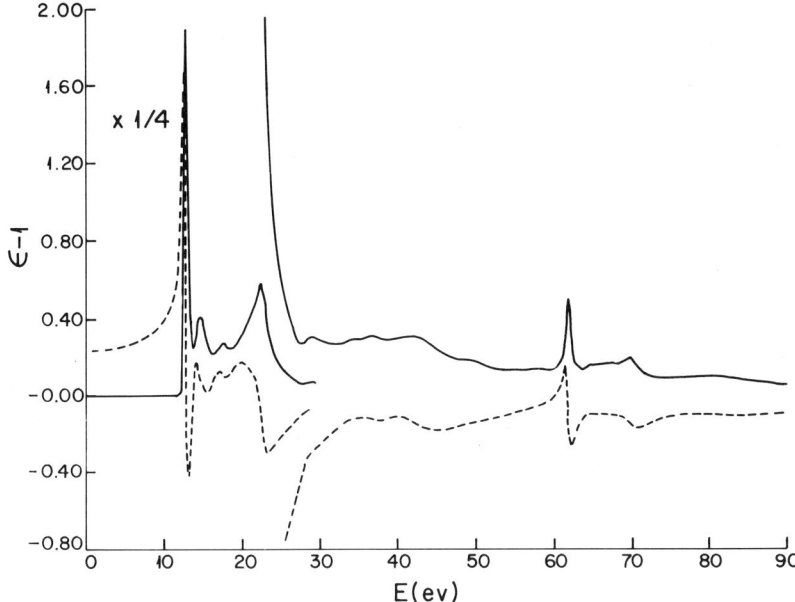

FIG. 38. Real and imaginary parts of the dielectric response function for LiF determined from electron scattering measurements [J. R. Fields, Ph.D. Thesis, Princeton University, Princeton, New Jersey (1975); J. R. Fields, P. C. Gibbons, and S. E. Schnatterly, *Phys. Rev. Lett.* **38**, 430 (1977)]. Real part of ϵ is the dashed line, imaginary part the solid line.

factor of two to three higher than the optical results. Since our background values are in good agreement, the discrepancy must be due to a systematic error in one of the experimental methods which is important in strongly absorbing regions.

This comparison points out one of the great strengths of inelastic electron scattering: Irregularities in sample thickness, even pinholes, do not affect the measured spectral shape. Electrons that pass through a pinhole do not scatter at all and so do not contribute at all to the inelastic spectrum. By contrast, pinholes or other sources of leakage such as higher order diffraction from the grating, do affect the shape of a measured optical absorption spectrum. The general effect is to decrease the apparent strength of strong features in the spectrum by allowing too many photons to be transmitted through the sample.

d. Core Excitations: Elliott Model

Many examples of core spectra of semiconductors and insulators have been reported in the literature, but in no case has the kind of precision

TABLE II. COMPARISON OF OPTICAL AND INELASTIC ELECTRON SCATTERING DETERMINATIONS OF OPTICAL PARAMETERS IN LiF

Energy (eV)	Loss function peak heights			
	R & W[a]	Rao[b]	Creuzberg[c]	Fields[d]
13.5	0.87	1.30	0.3	0.76
25	3.25	2.85	0.85	1.60

Energy (eV)	Optical absorption coefficients (10^5 cm^{-1})		
	Lukirskii et al.[e]	Brown et al.[f]	Fields[d]
62	6.2	5.0	15.2
64	4.3	2.6	4.6
70	5.7	2.9	7.3
75	2.9	2.0	3.5
100	2.4	2.0	2.3
160	—	0.7	1.0

[a] D. M. Roessler and W. C. Walker, *J. Phys. Chem. Solids* **28**, 1507 (1967).
[b] K. K. Rao, T. J. Moranec, J. C. Rife, and R. M. Dexter, *Phys. Rev. B* **12**, 5937 (1975).
[c] M. Creuzberg, *Z. Phys.* **196**, 433 (1966).
[d] J. R. Fields, Ph.D. Thesis, Princeton University, Princeton, New Jersey (1975); and to be published.
[e] A. P. Lukirski, O. A. Ershov, T. M. Zimkina, and E. P. Savinov, *Sov. Phys.—Solid State, (Engl. Transl.)* **8**, 1422 (1966).
[f] F. C. Brown, C. Gähwiller, A. B. Kunz, and N. O. Lipari, *Phys. Rev. Lett.* **25**, 927 (1970).

comparison between theoretical and experimental shapes which is now being carried out for the simple metals been accomplished. As in the case of the metals, there is no theory of core spectral shapes that includes the effect of the ion core interaction (excitonic effects) and the periodic field (band structure) at the same time. Far above threshold, where the Coulomb interaction can be included in its asymptotic form and the excited electron lifetime is short, and EXAFS-type model should be adequate. Near threshold, excitonic and band effects may be comparable and the problem more complex.

To make a start, we shall compare measured spectra near threshold with the Elliott model which treats the Coulomb interaction in the effective mass approximation.[81] The general results of this model are well known and are summarized in Fig. 39. If we are dealing with an allowed transition, then in the absence of the Coulomb interaction the absorption

[81] R. J. Elliott, *Phys. Rev.* **108**, 1384 (1957).

has a square-root shape as shown in the upper figure. If we now turn on the Coulomb interaction between the electron being excited and the core hole, the amplitude of absorption near threshold is greatly enhanced, and a Wannier exciton is formed with a hydrogenic Rydberg series converging at threshold. The observed spectral shape depends strongly on the total broadening of the spectrum due to phonons, lifetime, etc. In fact, if this broadening is energy independent, the only thing that the spectral shape depends on is the ratio of the exciton Rydberg to the broadening. Figure 39 shows two examples; in the upper figure this ratio is two, in the lower figure it is four. It is clear that the possible shapes range between a rounded step function with no visible exciton peak at all, to a fully resolved Rydberg series as this ratio varies from 0 to ∞. An important

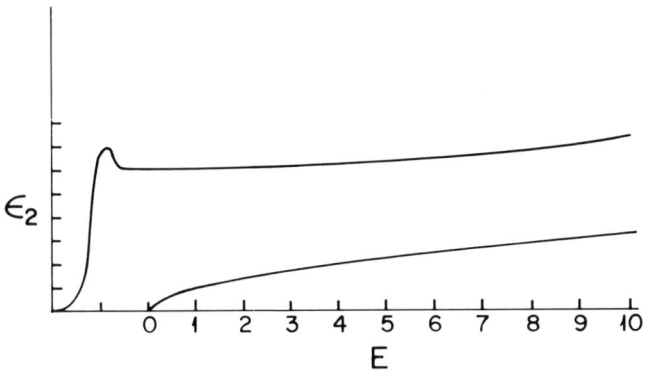

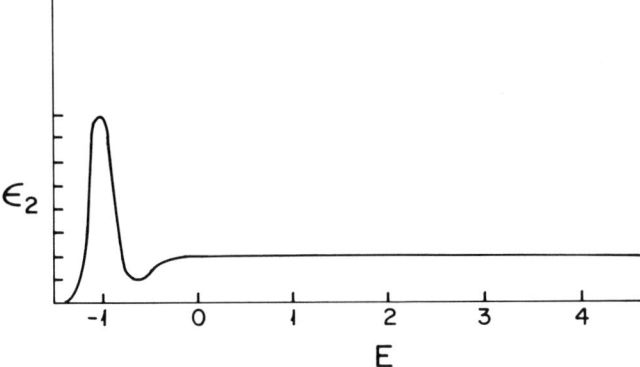

FIG. 39. Threshold shapes according to the Elliott model [R. J. Elliott, *Phys. Rev.* **108**, 1384 (1957); and text].

point to keep in mind is that there is absolutely no sign of the threshold itself in the spectrum for any finite value of the ratio. The intensity in this region is flat, and the value below threshold, obtained by smearing together the members of the Rydberg series, is equal to the value above threshold.

Figure 40 shows, for comparison, the measured spectrum in LiF obtained from the data presented above. Except for the extra modulations near 14.5 eV and 17.5 eV, this shape looks very Elliott-like; indeed, quantitative fits to spectra obtained from optical measurements have been obtained.[74] Figure 41 shows the spectrum of LiF in the region of the Li K threshold. It was argued above that the forbidden Li 1s–2s exciton appears at 61 eV, with the interband threshold at 64 eV. The dominant peak at 62 eV is due to 1s–2p transitions, and p states do not contribute to the conductions band until approximately 8 eV above threshold. It is reasonable to assume then that the only effect of the Li 2p states is to contribute the large peak at 62 eV.

It is possible to describe the observed spectrum with a peak 0.5 eV

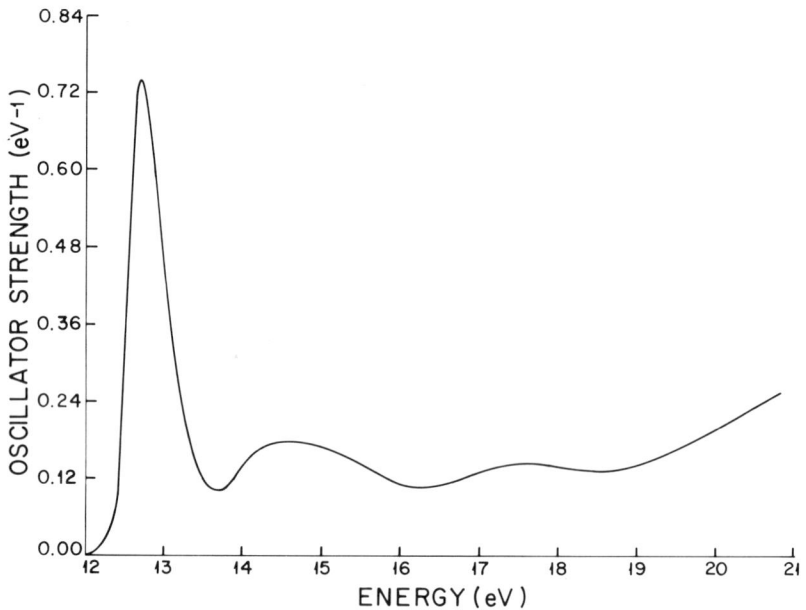

FIG. 40. Oscillator strength density as a function of energy in the region of the fundamental exciton of LiF [J. R. Fields, Ph.D. Thesis, Princeton University, Princeton, New Jersey (1975); J. R. Fields, P. C. Gibbons, and S. E. Schnatterly, *Phys. Rev. Lett.* **38**, 430 (1977)].

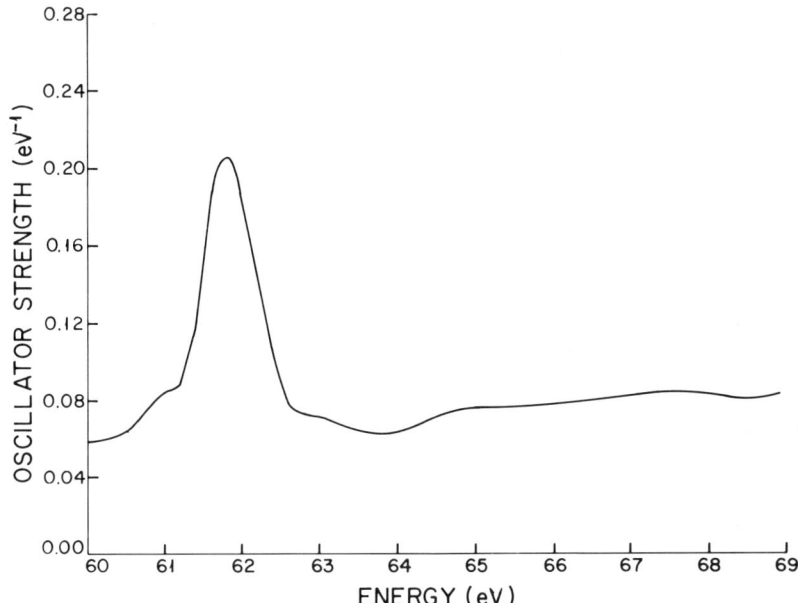

FIG. 41. Oscillator strength density in the region of the K threshold of Li in LiF [J. R. Fields, Ph.D. Thesis, Princeton University, Princeton, New Jersey (1975); J. R. Fields, P. C. Gibbons, and S. E. Schnatterly, *Phys. Rev. Lett.* **38**, 430 (1977)].

wide at 61 eV, the 2p exciton at 62 eV, and a broadened Rydberg series converging to 64 eV, including a partially resolved $n = 2$ peak at 63 eV. It may be straining the Wannier model to use it for such a highly localized single-ion exciton, but given the absence of simple alternatives, it is encouraging to see the model work so well.

Figure 42 shows the measured spectral shape in the region of the F K threshold in LiF. This transition should be forbidden near threshold and so would be expected to resemble the F 2s spectrum. Instead there is a dramatic peak at 692 eV rising above background with numerous features at higher energies. The peak near 700 eV might be due to the strong p-wave density of states in the conduction band. The peak near 718 eV resembles the shape of $\text{Im}[-1/\epsilon(q,\omega)]$ shown for comparison in Fig. 42, indicating that it might be a plasmon sideband on the strong peak at threshold. If it is, the strength of the sideband relative to the main peak (0.3 ± 0.1) implies a relaxation energy of 7.5 ± 2.5 eV, which is consisent with Fowler's calculation of approximately 5 eV for alkali halides generally.[82] These assignments are highly speculative and are really beside

[82] W. B. Fowler, *Phys. Rev.* **151**, 657 (1966).

the point. The real question is: Why does the F K spectrum look the way it does? It resembles neither the F 2s shape nor any possible version of the Elliott model. Relaxation and sideband effects should not depend strongly on core binding energy, and certainly the selection rules are energy independent. One thing that does change with energy is the strength of the background. It is possible that the F 2s spectrum has all the features seen in the 1s shape, but they are superimposed on a much stronger background due to the nearby F 2p excitations.

e. Crystal Structure Dependence of Core Spectra: Si $L_{II,III}$

One of the most disturbing of recent experimental results is shown in Fig. 43. If any of the simple models of spectral shapes are to be accurate, then band structure effects must not be very important. Figure 43 shows a comparison of the $SiL_{II,III}$ threshold region in a single crystal and an amorphous sample of Si. There are two major differences in overall shape. First, the EXAFS-like modulations which are present in the single-crystal results are absent in the amorphous sample. Second, the peak of the broad hump above threshold is about 7 eV higher in the single crystal than in the amorphous case.

The first observation can be compared with the results of Sayers and Stern[83] who found that for GeO_2 the variation in nearest-neighbor separation between amorphous and crystalline phases is too small to observe. In order to reduce the size of the apparent modulation in the single-crystal data near 20 eV above threshold by a factor of 4 to make it essentially unobservable in the amorphous case, an rms spread in nearest-neighbor separations of 0.35 Å is required. At 50 eV above threshold 0.15 Å spread is needed. Both of these numbers are too large. The only conclusion possible is that an EXAFS-type model is inadequate and that the potential energy must be altered also. This conclusion is supported by the second observation. If the maximum in the broad hump above threshold is due to free-electron-like matrix element effects, then the wave vector of the excited electron is the same at the maximum in each case. This means that the average potential energy of an excited electron in an amorphous sample differs by about 7 eV from that of a crystalline sample 20–40 eV above threshold. A highly excited conduction electron sees a very different mean lattice potential depending on the details of the crystal structure. If the states giving rise to the broad peak are thought of as antibonding orbitals. then these results suggest that the bonding–antibonding splitting is greater in a single crystal than an amorphous sample.

[83] D. E. Sayers and E. A. Stern, *Phys. Rev. Lett.* **35**, 384 (1975).

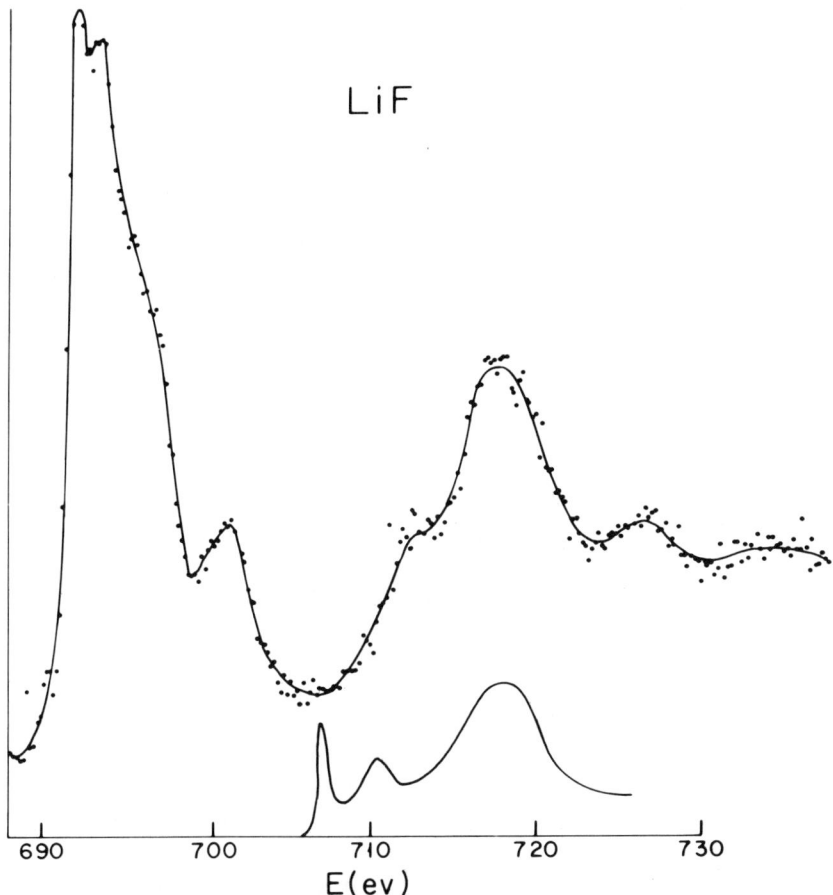

Fig. 42. Inelastic electron scattering spectrum of LiF in the region of the K threshold of F.

Figure 44 shows that Si $L_{II,III}$ spectrum on a finer scale near threshold for the same two samples. These results are similar to those obtained using synchrotron radiation.[84] Again there is a dramatic difference in shape, and furthermore, no version of the Elliott model corresponds to either shape. In addition, Van Dyke has shown that the transition density including band structure effects but no excitonic effects rises much too slowly to agree with the measurements.[85] Altarelli and Dexter have used an Elliott approach to fit the leading edge of the spectrum, but they did

[84] F. C. Brown and O. Rustgi, *Phys. Rev. Lett.* **28**, 497 (1972).
[85] J. P. Van Dyke, *Phys. Rev. B* **5**, 4206 (1972).

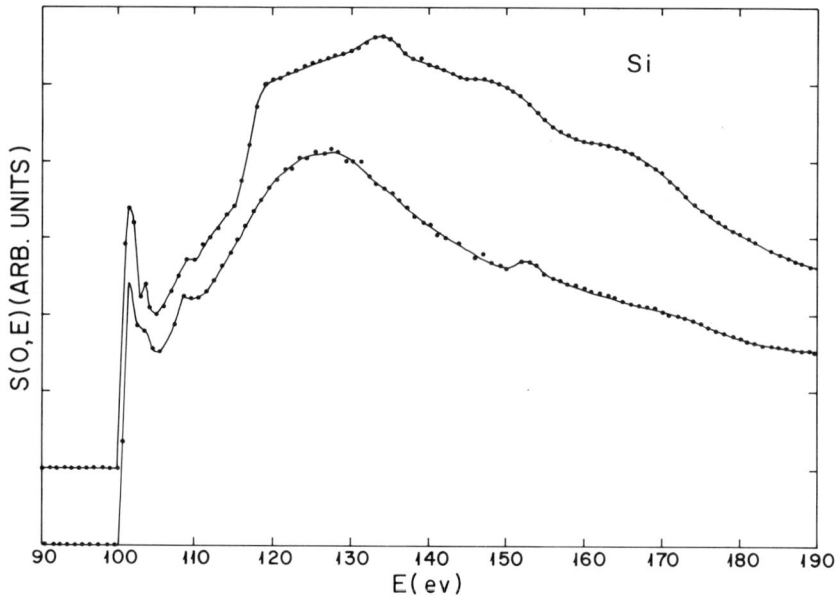

FIG. 43. Inelastic electron scattering spectra of single crystal (upper curve) and amorphous (lower curve) Si in the region of the $L_{II,III}$ threshold and above.

not extend their comparison to higher energies where the major anomalies appear.[86]

The single-crystal data appear at first to have an L_{III} to L_{II} strength ratio approximately equal to one, instead of the value one-half, which is expected and which is observed in the amorphous case. Notice however that the single-crystal threshold can be described as a series of three thresholds each split by the 0.7 eV spin–orbit splitting of the 2p core. The first pair, which appear as edges, are at 99.8 eV and 100.5 eV; the second, which appear as peaks, are at 100.8 eV and 101.4 eV; and the third, which appear as peaks, are at 102.6 eV and 103.3eV. The superposition of the first and second pairs produces the apparent anomalous L_{III} to L_{II} strength ratio.

If this is an accurate description of the threshold shape, the question remains, What are these separate thresholds due to? One possibility is that the electron–hole interaction strongly modifies the conduction band states which are made up of 3s and 3p atomic states. The 3s states will feel the core potential most strongly, and transitions to these states are allowed. If these states are reduced in size by their interaction with the core, then their antibonding repulsive interaction with the Si neighbors

[86] M. Altarelli and D. L. Dexter, *Phys. Rev. Lett.* **29**, 1100 (1972).

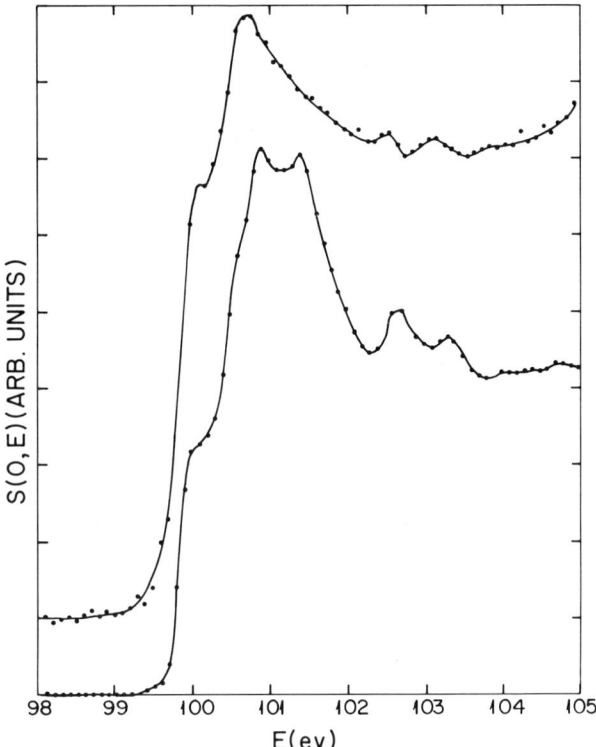

FIG. 44. Inelastic electron scattering spectra of single crystal (lower curve) and amorphous (upper curve) Si in the region of the $L_{II,III}$ threshold.

will be reduced, and energy shifts larger than those expected on a purely coulombic basis could occur. This is very similar to the case of the Li 1s–2p exciton in LiF described earlier.

The spectral shape of the amorphous sample might then be described by saying that the energies of these quasi-atomic states are smeared out by variations in the local potential. Most dramatic is the complete loss of the second pair, 1 eV above the first.

If these ideas are in any way close to reality, then the major conclusion to be drawn from both the LiF and Si results is that atomic effects, band structure, and excitonic effects all play an important role in determining core spectral shapes in insulators and semiconductors. Perhaps a cluster approach similar to that mentioned above in connection with LiF, or like the recent work on core spectra of molecules[87] should be applied also to semiconductors, and possibly even to metals.

[87] See, for example, J. L. Dehmer and D. Dill, *J. Chem. Phys.* **65**, 5327 (1977).

ACKNOWLEDGMENTS

I have the good fortune to be working with a spirited group of highly able and energetic colleagues who have taught me most of the ideas that appear in this article. Those who have contributed the most and with whom my work on electron scattering was begun are Patrick Gibbons, John Ritsko, and John Fields. Many others have been extremely helpful including Susan Slusky, Carl Franck, Steve Girvin, Larry Fleischman, and John Hopfield.

Author Index

Numbers in parentheses are reference numbers and indicate that an author's work is referred to although his name is not cited in the text.

A

Abrikosov, N. Kh., 14, 19(170), 20
Adler, S. L., 295
Aggarwal, S. K., 164
Agrawal, B. K., 9(59), 10
Airapetyan, V. M., 133, 138(116)
Akhiezer, A., 3
Akhundov, T. S., 48(316), 49, 52(316)
Albany, H. J., 15, 33(105)
Alefeld, G., 147
Alckseeva, G. T., 63
Alfrey, G. F., 9(54), 10
Ali, M., 25
Aliev, S. A., 15
Alieva, M. A., 15
Allain, J., 217
Allen, S. M., 115, 161, 185(71, 72), 189, 191(71, 72), 194(71,72), 199
Alm, O., 35
Almbladh, C. O., 335
Als-Nielson, J., 239
Altarelli, M., 356
Alton, W. J., 28
Amelinckx, S., 116, 163, 194(180), 250(75)
Amirkhanov, Kh. I., 35
Amit, D., 217
Anderson, A. C., 68, 69
Anderson, H. J., 64
Anderson, M. S., 7, 36(27)
Anderson, O. L., 12, 19, 27, 28(218)
Anderson, P. W., 330
Andersson, P., 34, 35, 36(25), 50, 51(341, 342)
Animalu, A. E. O., 91
Anthony, A. M., 64, 66(218)
Arakawa, E. T., 335
Arasly, D. G., 15
Ardell, A. J., 130, 136
Ascher, E., 87, 100, 237
Ashley, J. C., 326, 327(44)

Auerback, B. L., 140
Ausman, G. A., 336
Austerman, S. B., 27, 28(217), 32(217)
Axe, J. D., 163

B

Bachmann, R., 52(363), 54
Backstrom, G., 35, 36(262)
Badalyan, D. A., 154
Baer, Y., 335
Bailey, A. C., 25
Bailey, B. J., 47, 48(301)
Balzer, R., 58
Banerjee, R., 9(57, 58), 10
Bardhan, P., 115, 142, 223, 242, 243
Barker, J. A., 165
Barker, R. E., 34
Barlow, A. J., 28
Bar-on, M., 253
Barrett, H. H., 17
Barron, T. H. K., 12
Barsch, G. R., 27, 28(220), 30(225), 35, 67(225)
Bartram, S. F., 28, 43
Basu, A. N., 22
Batchelder, D. N., 5, 7(24), 8(24), 9(24), 35(24), 37(24), 38(24, 25), 48(24)
Bates, J. L., 64
Batterman, B. W., 242
Bauer, G. S., 138, 143(122), 143
Baumann, F. C., 20
Bausch, W., 20
Beeler, J. R., 116, 143, 145
Behari, J., 8
Belayev, L. M., 27
Belson, H. S., 60
Benedek, F., 90
Benin, D. B., 3(16), 4, 70
Bennett, B. I., 3(19), 4

Berg, H., Jr., 88, 242
Berman, R., 52
Berné, A., 8
Bernstein, H., 150, 194(155), 205(155)
Berry, B. S., 120
Bertman, B., 52
Berwaldt, O. E., 49
Bewilogua, L., 48
Bhandari, C. M., 33
Billard, D., 10, 12(63), 13(63), 14(63), 15(63)
Binder, K., 212, 253
Bingham, J. L., 18
Birch, J. A., 14
Blackman, M., 16
Blander, M., 56
Bluthardt, W., 19
Boato, G., 8
Boccara, N., 179
Boersch, H., 300, 301
Bogatov, G. F., 48(314), 49, 52(314)
Bohlin, L., 34, 36(258)
Bondi, A., 32
Borcherds, P. H., 9(54), 10
Borie, B., 141
Bortz, A. B., 253
Botaki, A. A., 60
Bouchaert, L. P., 106, 107
Bouchard, M., 228, 229, 265(255)
Bourret, A., 133
Boutard, M., 15, 33(107)
Bowen, D. K., 145
Boyer, L. L., 118
Boyle, W. F., 14
Bracker, J. M., 45
Brafman, O., 12
Bratby, P., 45, 47(290)
Bridgman, P. W., 34, 46(251), 50(251), 56
Briggs, A. G., 63
Briscoe, C. V., 18, 35
Brockhouse, B. N., 9, 19, 24
Broerman, J. G., 15, 16(113)
Brophy, T. T., 32
Brout, R., 154, 176
Brower, G. D., 24, 27(198)
Brown, F. C., 350, 355
Brunel, M., 242
Brunner, L. J., 60
Buck, O., 245
Bührer, W., 19, 60
Bullough, R., 118, 123, 130, 143, 144
Burch, J. L., 51

Burenkov, Yu. A., 60
Burford, R. J., 239
Burlew, J. S., 51
Burley, D. M., 164, 166(190), 173(190), 174
Burton, J. T. A., 47, 48(299)
Bury, P., 48(311), 49
Butler, E. P., 265
Buttrey, T. W., 32
Buyers, W. J. L., 19

C

Cabannes, F., 10, 12(63), 13(63), 14(63), 15(63), 64, 66(218)
Cadoret, R., 78, 149
Cahn, J. W., 113, 115, 120, 157, 161, 179, 185(71, 72), 188, 189, 191(71, 72), 194(71, 72), 197, 199, 210(65, 207), 211(207), 216(207), 217(170, 217), 217, 225, 234(217), 235(217), 252(169), 259(170)
Calcott, T. A., 335
Callen, H., 154
Capps, W., 63
Carballo, J., 326
Cardona, M., 12
Carlson, R. O., 15
Carter, G. C., 203
Castelli, V. J., 48(311), 49
Castles, J. R., 103
Chadderton, L. T., 144
Chakravarti, B., 220, 242(250)
Challis, L. J., 8, 63
Chaney, R. C., 348
Chang, S. S., 19
Chang, Z. P., 27, 28(220), 35
Chasmar, R. P., 63
Chau, C. K., 20
Chechelnitskii, A. Z., 20, 68
Chelikowsky, J. R., 296
Chellman, D., 136
Chen, H., 242, 243(280), 243, 262
Cheng, V. M., 49
Chizhevskaya, S. N., 43
Chopra, K. L., 63
Christen, D. K., 8
Chudnovskii, A. F., 52(362), 52, 54, 57
Churg, P. L., 24, 25
Citrin, P. H., 335
Clapp, P. C., 81, 87, 108(7), 109(7), 109, 115, 116(60), 161, 237, 238(66), 239, 240, 246(9, 10, 11), 249(9, 289)

Clayton, F., 5, 7(24), 8(24), 9(24), 35(24), 37(24), 38(24), 48(24)
Cleavelin, C. R., 18
Clemans, J. E., 8, 35(30)
Cline, C. F., 28
Cochran, W., 19
Cocking, S. J., 45
Cohen, J. B., 88, 115, 142, 243(80), 243, 249(12), 262
Cohen, M. L., 91, 296
Collins, J. G., 19
Collins, T. C., 344
Connolly, T. J., 68
Cook, G. A., 49
Cook, H. E., 95, 96(36), 100, 117, 118, 119(83), 119(36), 121, 140(36), 140, 162, 163(178), 179, 197(178), 198, 216, 217, 242, 246(84, 85), 246, 250(84), 252(38), 252, 256(283, 298), 261(38, 294), 262, 263(294), 264, 266
Cooley, J. W., 121
Corbett, J. W., 202
Cornwell, K., 56, 70(376)
Cottam, R. I., 14
Coufal, H. J., 38
Cowley, J. M., 103
Cowley, R. A., 3(15), 4, 9, 11(47), 12(47), 19, 23(188), 24, 62(188)
Cracknell, A. P., 158
Crandall, P. B., 28
Crawford, R. K., 47, 49(304), 49
Creuzberg, M., 350
Crewe, A. V., 302
Croft, W. J., 28
Crooks, M. J., 52
Croxton, C. A., 45
Crutchfield, W., 326

D

Damon, D. H., 32
Daney, D. E., 8
Daniels, J., 277
Daniels, W. B., 8, 12, 47, 49(304), 49
Danielson, G. C., 24, 25
Das, S. K., 220
Davis, H. T., 56
Davis, L. C., 25
Dawihl, W., 64
Day, C. R., 52

de Bergevin, F., 242
de Boer, J., 54, 165
Debye, P., 2, 45(2)
Dederichs, P. H., 127, 138, 142(122), 142, 143
De Dominicis, C. T., 331
de Fontaine, D., 77(1), 78, 87, 95, 96(36), 98(1), 100, 104, 106, 107, 107(49), 109(49), 110(49), 118, 119(36), 121, 122, 127, 132(93), 132, 133, 134(101), 134, 135(101), 137(101), 138, 140(36), 153, 156(149), 156, 157(49), 160(49), 163(49), 164(15), 165(15), 169, 170(15), 173, 174, 194(49), 195, 197, 198(219), 199(199), 203, 204(202), 205, 206, 207(202, 203), 207, 209, 218(49), 218, 219, 221, 222(49, 199), 224, 225(167), 227, 237, 238(49), 244, 245, 247, 250(1, 3), 251, 252(38), 252, 253, 254(3), 255(3), 256(3), 259(304), 261(38, 294), 262, 263(38, 294), 264(49), 264, 265, 266, 267
Dehmer, J. L., 357
Delafond, J., 217
Delavignette, P., 163, 194(180)
Demarest, H. H., 36
DePaz, M., 8
De Ridder, R., 116, 163, 249, 250(75)
Devyatkova, E. D., 16, 17(118), 20, 60
Dexter, D. L., 356
Dexter, R. M., 350
Dienes, G. J., 252
Dietrich, O. W., 239
Dill, D., 357
Diller, D. E., 48, 49(307)
Din, F., 49
DiSalvo, F. J., 102
Dolling, G., 9, 11(47), 12(47), 19, 23(188), 24, 62(188)
Domb, C., 5
Donecker, J., 18
Doniach, S., 332
Dorner, B., 60
Dorsey, N. E., 56
Dow, J. D., 336
Drabble, J. R., 4
DuBois, D. F., 313
Dubroff, W., 88
Ducastelle, F., 89, 91(28)
Dugdale, J. S., 3
Dunegan, H. L., 28
Dupuy, E., 28
Durham, P. J., 145

Dwyer, D. F., 49
Dzhabbarov, R. M., 15

E

Ecsedy, D. J., 34, 40(259), 40
Edeskuty, F. J., 49
Efimova, B. A., 63
Egelstaff, P. A., 45
Egloff, G., 49, 51(336), 52(336)
Ehrenreich, H., 89
Ehrhart, P., 138, 143(123), 143
Ekstrom, L., 63
Elcombe, M. E., 18, 19(139)
Elcombe, M. M., 19, 23, 24(185, 186), 25(185, 186)
Elliot, R. J., 350
Emirov, S. N., 35
Endoh, Y., 8
Epstein, A. J., 318
Ershov, O. A., 350
Eshelby, J. D., 124, 129(97), 130(97), 130, 145
Espinosa, G. P., 28
Ettenberg, M., 14
Eucken, A., 20, 31(174), 52, 62(174)
Eurin, P., 133
Evgen'ev, S. B., 43
Ewing, C. T., 70

F

Fairbank, H. A., 52
Farag, B. S., 17
Farr, M. K., 9(53), 10
Farrell, G., 63
Fauche, M., 64, 66(218)
Feder, J., 163
Fedorov, V. I., 56
Fields, J. R., 293, 312, 314, 317, 318, 321, 322, 325, 337, 341, 345, 345(72), 350, 352
Fisher, P. M. J., 220, 239
Flocken, J. W., 135, 142
Flubacher, P., 14
Fodor, J., 205, 209, 265, 267
Fouquet, D. F., 348
Fowler, W. B., 353
Frazer, B. C., 8
Freeman, A., 347
Friedel, J., 89, 124, 129(98)
Frisillo, A. L., 24
Froitzhein, H., 302

Fugate, R. Q., 7, 8, 36(27)
Fujii, Y., 8, 24(41)

G

Gähwiller, C., 350
Galetskaya, A. D., 20
Gaskell, T., 45
Gautier, F., 89, 91(28)
Gavrilko, V. G., 8, 38(46)
Gayton, W. R., 14
Gehlen, P. C., 116, 143, 145, 249(12)
Geller, S., 28
Geller, V. Z., 48(313), 49, 52(313)
Gerlich, D. A., 14, 19, 25, 28
Gerstein, B. C., 25
Getting, I. C., 50
Gewurtz, S., 58
Giaquinta, P. V., 340
Gibbons, P. C., 298, 304(316), 311, 314, 316, 317, 318, 321, 322, 325, 333, 334, 336, 337, 338, 339, 341, 348(72), 349, 352
Gibbs, J. W., 210
Gildebrandt, E. M., 69
Giner, J., 89
Girlanda, R., 347
Girvin, S. M., 336
Gladun, C., 51
Glansdorff, P., 256
Glassbrenner, C. J., 15, 43(102), 60(102)
Glauber, R. J., 251
Glazov, V. M., 43
Glick, A. J., 336
Gluck, P., 3(14), 4
Gmelin, E., 18
Gobrecht, K. H., 18
Goldsmid, H. J., 4
Golosov, N. S., 210
Goodman, J. W., 132
Goodwin, R. D., 49
Gornall, W. S., 58
Gorodilov, B. Ya., 52
Gosh, A., 22
Goulder, D. P., 52
Gout, C., 347
Gragg, J. E., Jr., 140, 142
Graves, R. S., 16, 20(125)
Grest, G. S., 326
Grigor'ev, B. A., 48(314), 49, 52(314)
Greenough, R. D., 43, 60
Grest, G. S., 326

Grilhé, J., 217
Grilly, E. R., 49
Grimsditch, M. H., 14
Grivet, P., 300
Groves, W. O., 14
Guckenbiehl, F., 20
Gudat, W., 346
Gürmen, E., 19
Guinier, A., 82
Guptill, E. W., 8
Gurevich, V. L., 16, 17(116)
Guthrie, G. L., 33
Guttman, L., 160

H

Hackspill, L., 28
Hamilton, R. A., 3(17), 4, 10(17), 12(17)
Hansen, M., 199, 200
Harada, J., 163
Hardy, J. R., 118, 123, 130, 135, 142, 150, 231
Harrison, W. A., 88, 89, 91(19), 101
Hart, K. R., 51
Harth, W., 301
Haubold, H. G., 138, 143(123), 143
Hayakawa, M., 142
Hayes, T. M., 339
Heine, V., 88, 91, 103
Henderson, G. W., 28
Henry, N. F. M., 105, 106, 107, 159(55)
Henshaw, D. G., 45
Hickernall, F. S., 14
Hidshaw, W., 35
Hijmans, J., 165
Hill, R. W., 52
Hilliard, J. E., 113, 114(65), 127, 179, 198, 210(65, 207), 211(207), 216, 217, 252, 258(104, 314), 259, 261(294), 262, 263(294), 266
Ho, C. Y., 25
Ho, P. S., 90
Hoffman, D. W., 81, 145, 237
Holder, G. A., 49, 51(339)
Holland, M. G., 17, 33(10), 33
Holste, J. C., 8, 14, 49
Hopfield, J. J., 331, 336
Hopkins, R. H., 32
Horn, F. H., 24, 28(197)
Horner, H., 143, 145
Horowitz, N. C., 43
Horrocks, J., 46

Horwitz, G., 154
Houska, C. R., 152, 242
Houston, B., 60
Houston, W. V., 142
Huang, K., 141
Hughes, D. S., 35, 36(261)
Hunklinger, S., 69
Hunter, O., Jr., 27, 28
Hurrell, J. P., 23, 24(187)

I

Ianniello, L. C., 202
Ibach, H., 302
Ikenberry, L. D., 47, 48(300)
Inden, G., 199
Inglesfield, J. A., 89
Isaaacson, M., 302, 307
Ishida, I., 12, 13(71), 14(71)
Iyengar, P. K., 9

J

Jackson, H. E., 58
Jaffee, R. I., 87, 100, 116, 143, 145, 237
Jelinek, G. E., 8, 24(40), 25
Jex, H., 12
Johnson, E., 144, 302
Jones, R. M., 14
Joshi, Y. P., 33
Julian, C. L., 3, 4(7)
Junqua, A., 217
Just, W., 138

K

Kaburagi, M., 189, 192
Kajitani, T., 162, 163, 179, 197(178)
Kakehashi, Y., 111, 191(63), 192(63), 193
Kalos, M. H., 179, 217(208), 253
Kamal, I., 46
Kanamori, J., 111, 115(63), 188, 189, 191(62, 63), 192(62, 63), 193
Kanzaki, H., 117, 135(81), 142(81), 143(81)
Kappus, W., 27, 30(226), 66(226)
Katsura, S., 115, 190, 191(73), 191
Kaufman, L., 150, 194(71), 204(155)
Kawasaki, K., 251
Kearney, R. J., 24
Keating, D. T., 237, 239
Keeson, P. H., 49

Kell, G. S., 49, 51(335)
Keller, W. E., 48
Kellner, K., 47, 48(301)
Kelly, A., 132
Kennedy, G. C., 50
Kerrisk, J. F., 48
Keyes, R. W., 3, 32, 57
Khachaturyan, A. G., 80, 82(5), 104(5), 105, 131, 133(108), 133, 135(108), 135, 137(108), 138(116), 139, 145, 153(5, 54), 153, 154, 156, 159(5), 161, 181(5, 54), 181, 182(5), 182, 183, 184(212), 191(5), 244
Khantadze, V., 28
Kieffer, S. W., 50
Kiefte, H., 58
Kikuchi, R., 87, 132, 147, 153, 164(14), 165(14), 165, 166(14), 166, 167, 170(188, 189), 172(189), 173(14), 173, 174, 175(197), 176, 195, 199(199), 199, 202, 203, 204(201, 202), 207(202), 207, 213, 217(239), 218, 219, 221(180), 222(199), 268, 269(322), 272
Kimber, R. M., 8
King, J. S., 9(55), 10
Kingery, W. D., 66, 69
Kirillova, I. V., 16, 17(120)
Kittel, C., 69, 310
Kivelson, M. G., 313
Klein, M. V., 20
Klein, P. H., 28
Klein, R., 17
Klemens, P. G., 2, 3(18), 4, 10, 16(66), 25, 34, 31(66), 40(259), 40, 57(64), 62(61)
Knobler, C. M., 260
Koch, F., 242
Kohn, W., 102
Kohnstamm, P., 217
Koloskova, L. A., 52
Kontorova, T. A., 3
Korobova, I. L., 16, 17(120)
Korpiun, P., 8, 38
Kotelnikova, G. A., 69
Kovalev, N. N., 20
Kramers, H. A., 168
Krishan, V., 328
Krivoglaz, M. A., 117, 158, 237
Krupskii, I. N., 8, 38(34), 52(352–354)
Kudman, I., 10, 12(60), 14(60), 15, 16(60, 106), 17(60), 63
Kudo, T., 115, 191
Kuhn, G., 20, 31(174), 62(174)

Kunc, K., 9, 24(52)
Kunz, A. B., 350
Kunz, C., 328, 344, 346
Kupperman, D. S., 58
Kurata, M., 164
Kurdjumov, G. V., 145
Kuyatt, C. E., 299

L

Lafon, E. A., 348
Lam, D. G., 66
Landau, L. D., 152, 154(157), 158(157), 158, 159(157), 233, 278, 283
Landheer, D., 58
Langer, J. S., 253, 256(299), 259(314)
Larsson, K. E., 45
Lašek, J., 197, 198
Lasjaunias, J. C., 69
Laughlin, D. E., 220
Lawrence, D. J., 8
Lawson, A. W., 3, 34, 60
Leadbetter, A. J., 14
Leake, J. A., 8
Lebowitz, J. L., 179, 217(208), 253
Lee, B. H., 14
Legler, W., 301
LeGuillou, G., 15, 33(105)
Lehoczky, A., 14, 18
Lehwald, S., 302
Lejček, L., 147
LeNeindte, B., 48(311), 49
Leonhard, F., 301
Leroux-Hugon, P., 16, 40(119)
Lewandoski, D. T., 259
Lewis, J. T., 18, 35
Lewis, M. F., 24
Liebermann, R. C., 27, 28(218)
Liebfried, G., 3, 4(7)
Lifschitz, E. M., 105, 137, 152, 154(157), 158(157), 158, 159(157), 160, 233, 278, 283
Liley, P. E., 46
Lin, C. C., 348
Lin, W., 242
Linhard, J., 310
Lipari, N. O., 350
Lochtermann, E., 48
Logachev, A. Yu., 16, 17(121–124), 20(124), 23(121), 34(121, 122), 39(124), 40(123), 56(124)
Loje, K. F., 19

Author Index

Lonsdale, K., 105, 107, 159
Louie, S. G., 296
Low, L. F., 62
Likina, V. I., 14
Lukirski, A. P., 350
Lurie, N. A., 8, 24(41)
Luscher, E., 8
Lutz, H. D., 24
Lynch, D. W., 348
Lyubarskii, G. Ya., 158

M

McAlister, A. J., 335
MacDonald, D. K. C., 3
McDonald, J., 56
Macedo, P. M., 63
MacEwan, J. R., 64
Machlin, E. S., 115
Machlin, G. E., 88
Machuev, V. I., 56
McLaren, R. A., 58
McLaughlin, E., 45, 46(294), 46, 48(312), 49, 52(312)
McMillan, W. L., 102, 164(43)
McNeely, J. B., 14
McNelly, T. F., 20
McTaggart, J. H., 67, 68(426)
Magomeedov, B. Ya., 35
Mahajan, S., 102
Mahan, G. D., 329, 331
Makishima, P., 269
Manenc, J., 220
Manning, W. R., 27, 28
Manzhelii, V. G., 8, 38(34, 46), 49, 52(352–354)
March, N. H., 45
Marro, J., 179, 217(208), 253
Marshall, B. J., 18
Martin, G., 256
Martin, J. J., 25
Martin, R. M., 296
Massalski, B., 103
Mastoor, M. A., 49
Matsubara, T. J., 117, 179
Matsuda, H., 179
Mazdiyashi, K. S., 63
Meijering, J. L., 202, 204(221), 205(228), 206, 231
Men, A. A., 20, 68
Mendelssohn, K., 2
Merchant, H. D., 19, 38(159)

Mermin, N. D., 319
Merz, J. W., 17
Meyer, B., 52
Michelson, A., 160
Mickish, D. J., 344, 348(73)
Migoni, R., 19
Millar, D. J., 132
Miller, H. D., 253
Miller, R. R., 70(440)
Mills, R. E., 87, 100
Mimault, J., 217
Minkiewicz, V. J., 23
Missenard, A., 31
Mitra, S. S., 12
Mizushima, S., 3
Möllenstedt, G., 301
Mogilevskii, B. M., 25, 52(362, 364), 53, 54, 57
Moizhes, B. Ya., 16, 17(120, 121), 23(121), 34(121), 37
Mooney, D. L., 34
Moore, J. P., 16, 20(125)
Moranec, T. J., 350
Morinaga, M., 109, 110(61)
Morral, J. E., 225, 230, 231
Morrison, J. A., 14
Moser, J., 8
Moss, S. C., 81, 102, 103, 108(7), 109(7), 115, 161, 238(66), 239(67), 240, 241, 249(270)
Mozer, B., 237
Mozhaev, V. M., 56
Mozhaeva, V. N., 69
Müller-Krumbhaar, H., 253
Muhlestein, L. D., 19
Mura, T., 125
Murakami, Y., 153, 205(203), 205, 206(202), 207, 209, 265, 267
Muto, T., 152
Muzhdaba, V. M., 20

N

Nabarro, F. R. N., 118, 129
Nagel, S. R., 295, 296, 297, 328
Nagels, P., 25
Nagornykh, L. G., 3
Nakamura, T., 173, 205(202), 206, 207(202), 207
Nakanishi, N., 153
Namjoshi, K. V., 12, 14(72)
Narita, A., 191
Nath, P., 63

Nathan, B. D., 62
Naudon, A., 217
Needham, D. P., 48(310), 49
Nelin, G., 9
Nelson, D. A., 14, 15, 16(113, 114)
Nelson, R. S., 144
Neuberger, J., 22
Nevins, H. E., 49
Nicholson, R. B., 130, 132
Nicklow, R. M., 9(56), 10, 19
Niemyski, T., 27
Nikanorov, S. P., 60
Nikitin, E. N., 25
Nilsson, G., 9
Norton, F. H., 64
Novikova, S. I., 14, 20, 25, 43(98)
Nowick, A. S., 120, 252
Nozieres, P., 292, 331
Nusimovici, M. A., 27, 30(224)

O

Ogilvie, R. E., 205
Ogita, N., 179
O'Horo, M. P. O., 24
Okamoto, P. R., 220
Oliver, D. W., 24, 27(198), 27, 28(197, 222), 31
Olson, C. G., 348
Onissimova, T. A., 244
Onsager, L., 168, 175(194)
Ostrouskaya, L. M., 63
O'Toole, J. T., 253
Overhauser, A. W., 102

P

Packard, J. R., 49
Paff, R. J., 14
Palmer, S. B., 43, 60
Panter, C. H., 19
Papapetru, A., 3
Parfenjeva, L. S., 25
Parker, D. L., 14
Parrinello, M., 347
Parrott, J. E., 3(17), 4, 10(17), 12(17)
Parthé, E., 116, 250(74)
Paton, N. E., 244
Paulson, W. M., 262, 264
Paxhia, E. C., 15, 16(113)
Payne, R. T., 12

Pearson, W. B., 187
Peckham, G. E., 19
Pederson, D. O., 18
Penisson, J. M., 133
Penn, D. R., 292
Peter, M., 19
Peterson, H., 335, 346
Peterson, O. G., 7, 38(25)
Petrov, A. V., 16, 17(124), 20(124), 20, 39(124), 56(124)
Phillipi, C. M., 63
Phillips, J. C., 14, 105, 340
Piacentini, M., 345, 346(74), 348, 352(74)
Pieringer, F. J., 8
Piermarini, G. J., 12
Pierron, E. D., 14
Pinard, P., 15, 33(107)
Pines, D., 292
Piotrowski, P., 32
Pirious, B., 64, 66(418)
Pisarevski, V. V., 27
Pittman, J. F. T., 48(312), 49, 52(312)
Platzman, P. M., 285, 332
Plesset, F., 121
Pohl, R. O., 20, 68
Pollack, G. L., 8, 36
Pollman, J., 127
Polyakov, P. V., 56, 69, 70(373)
Pomeranchuk, I., 3
Popchapsky, T. E., 56
Poplavnoi, A. S., 32
Potter, R. F., 60
Powell, B. R., Jr., 27
Powell, R. L., 58
Powell, R. W., 25, 49
Pradal, F., 347
Price, D. L., 9(56), 10
Prigogine, I., 203
Prince, A., 202
Prydz, R., 49
Pryor, A. W., 23, 24(185)
Pugach, V. V., 48(315), 49, 52(315)
Pynn, R., 8, 24(41)
Pytte, E., 163

R

Radelaar, S., 242, 260(282)
Raether, H., 277
Rajagopaland, A. K., 326
Ralph, B., 88

Ramdas, A. K., 14
Rampton, V. W., 8, 34
Ranninger, J., 37
Rao, K. K., 350
Rao, R. R., 19, 36, 43(275), 57
Rapp, J. E., 19, 38(159)
Rastorguev, Yu. L., 48(314, 315), 49, 52(314, 315)
Ratcliffe, E. H., 52
Raunio, G., 19, 23(142, 148), 24(142)
Ravev, A., 69
Reeber, R. R., 14
Regel, A. R., 45
Reid, J. S., 19
Reiss, H., 213
Renker, B., 19
Renucci, J. B., 12
Rice, S. A., 47, 48(300)
Richards, M. J., 151115, 161, 194(71)
Richter, W., 12
Rieder, K. H., 19
Rife, J. C., 350
Ringermacher, H. I., 51
Riste, T., 163
Ritchie, R. N., 326, 327(44), 328
Ritsko, J. J., 298, 304(16), 312, 314, 316, 317, 318, 321, 322, 325, 333, 334, 338, 339, 345
Rivaud, L., 179
Roberts, B. W., 140, 242
Roberts, R. W., 36, 38(173)
Robinson, J. E., 336
Robinson, M. T., 143, 144
Roder, H. M., 48, 49(307), 49
Rodionov, K. P., 34
Roessler, D. M., 350
Rogers, S. J., 8, 20
Rogers, W. M., 58
Rolandson, S., 19, 23(142, 148), 24(142)
Romanova, M. V., 32
Romashin, A. G., 68
Rosander, T., 35, 36(262)
Rosenbaum, R. L., 20
Rosenberg, H. M., 2
Roufosse, M., 10, 16(66), 31(66), 57(64), 62
Rowe, J. M., 9(56), 10
Rundman, K. B., 258(314), 259
Ruppin, 19(168), 20, 36, 38(273)
Rusakov, A. P., 14
Russell, K. C., 212
Rustgi, O., 355
Ruvalds, J., 326

S

Saakyan, V. A., 63
Sagan, L. S., 49, 51(330)
Saint-Paul, M. E., 18
Salter, L., 5
Samuelson, E. J., 163
Sanchez, J. M., 87, 164(15), 165(15), 169, 170(15), 170, 173, 174, 205, 209, 244(196), 245(197), 247, 265, 267
Sandman, D. J., 318
Sangster, M. J. L., 19
Sar-El, H. Z., 302
Sato, H., 103, 173, 173(197), 177, 272
Sauerwald, F., 56
Saunders, G. A., 14
Saunderson, D. H., 9(54), 10, 19
Sauvage, M., 116, 250(74)
Savinov, E. P., 350
Sawin, F., 35, 36(261)
Savena, S. C., 46
Sayers, D. E., 354
Schatz, J. F., 66, 67
Scheule, D. E., 25
Schilling, W., 138, 143(123), 143
Schinke, H., 56
Schittenhelm, C., 19
Schlömann, E., 3, 4(7)
Schmatz, W., 143, 236
Schmunk, R. E., 24
Schnatterly, S. E., 298, 304(16), 311, 312, 314, 316, 317, 318, 321, 322, 325, 328, 333, 334, 336, 337, 338, 339, 341, 345, 348(72), 349, 352
Schneider, W., 19, 179
Schneidmesser, B., 52
Schreiber, E., 27, 28(218)
Schröder, E., 52
Schuele, D. E., 19
Schwahn, D., 236
Schwartz, L. M., 89
Seidel, T., 63
Seitz, E., 121, 132, 12334, 138, 143
Semenouskaya, S. V., 152, 199
Sen, P. N., 39
Sengupta, S., 22
Serebryannikova, O. S., 63
Shadrichev, E. V., 45
Shalyt, S. S., 20
Sham, L. J., 296
Shapiro, S. M., 163

Sharko, A. V., 60
Shatalov, G. A., 131, 135(108), 135, 137(108), 183, 184(212)
Sheard, F. W., 63
Shendalman, L. H., 253
Sherman, R. H., 49
Schickell, W. D., 25
Shirane, G., 8, 24(41), 163
Shirley, G. G., 94, 95(35), 115
Shively, J. E., 88
Schockley, W., 199, 201
Siems, R., 120
Silverman, S. J., 15
Silvestrova, I. M., 27
Simmons, G., 66, 67
Simmons, R. O., 7, 38(25), 58
Simonato, J., 64, 66(418)
Simpson, A., 70
Simpson, J. A., 299
Simum, E. N., 20
Sinclair, R., 273
Sines, G., 132
Singh, R. K., 22
Singwi, K. S., 320, 323, 324
Sinha, S. K., 9(53), 10
Skal, A. S., 16, 17(121), 23(121), 34(121)
Skalyo, J., Jr., 8
Sköld, K., 45
Slack, G. A., 10, 12(62), 15, 16(112), 17(62), 20, 24, 25, 27(198), 27, 28(179, 194, 197, 217, 222), 28, 31(67), 31, 32(217, 222), 32, 34(67), 39(177), 43(102, 112, 282), 56(171), 60(102, 112), 67, 68(426), 70(441)
Sladek, R. J., 14
Slowick, J. H., 336
Slusky, S. G., 336, 337
Slyozov, V. V., 137
Smirnov, I. A., 16, 17(118), 17, 20, 25, 32, 37, 45, 60
Smith, C. S., 19
Smith, H. G., 19
Smith, T. F., 14, 19
Smoluchowski, R., 106, 107
Soga, N., 27, 28(218)
Sokolov, V. A., 20
Somenkov, V. A., 143
Sonnenberg, K., 143
Sonntag, B. F., 336
Sorokin, O. V., 20
Southworth, H. N., 88

Spann, J. R., 70
Spargo, A. E. C., 103, 326, 327(46)
Sparks, C. J., 141, 220, 242(250)
Sparks, P. W., 13
Spence, J. C. H., 326, 327(45)
Spitzer, D. P., 32
Squires, G. L., 106
Srivastava, K. K., 19
Staes, K., 47
Stalinsky, B., 145
Stanley, E. M., 48(311), 49
Stanley, H. E., 175, 176, 213
Starke, E. A., 220, 242
Stauffer, D., 212
Staveley, L. A. K., 47, 49(305), 49, 51(305, 327, 330), 51
Steg, R. G., 34
Steigmeier, E. F., 10, 12(60, 61), 13(81), 14(60), 15(60), 15, 16(60, 106), 17(60), 17
Stepanov, A. V., 60
Stephenson, J., 51
Stern, E. A., 354
Stewart, A. T., 8
Stinchfield, R. P., 66
Stoicheff, B. P., 58
Stoll, E., 179
Stoneham, A. M., 144, 145
Stoute, R. L., 64
Straka, R. E., 60
Streett, W. B., 47, 49(305), 49, 51(305, 327, 330), 51
Striefler, M. E., 27, 30(225), 66(225)
Stryland, J. C., 47, 49(303), 49
Stuckes, A. D., 63, 70(442)
Sturham, H. H., 64
Sturm, K., 311, 315
Suezawa, M., 179
Sugawara, A., 52(365), 54
Sundqvist, B., 50
Sur, A., 179, 217(208)
Surin, V. G., 52(362, 364), 52, 54, 57
Susse, C., 60
Swenson, C. A., 7, 8, 13, 36(27), 36, 49, 52(320)

T

Tait, R. H., 62
Takagi, Y., 152
Talley, P., 111
Talwar, D. N., 9(59), 10

Tanaka, T., 164
Tanner, L. E., 133, 163
Tannhauser, D. S., 60
Tavadze, F. N., 28
Taylor, A., 14
Tefft, W. E., 66
Teubner, B. G., 2
Tewary, V. R., 118, 126(100), 127, 143, 144, 145
Thamerus, G., 64
Thom, R., 218
Thomas, G., 220, 228, 229, 265(255), 265
Thompson, C. J., 175
Thorson, I. M., 19
Tikhonov, V. V., 25, 32, 63
Tikhonova, E. A., 117
Tilford, C. R., 8
Tjuterev, V. G., 32
Tolkachev, A. M., 49
Tomlinson, R. D., 43
Tompson, C. W., 19
Tosatti, E., 340, 347
Tosi, M. P., 340
Toth, R. S., 103
Touloukian, Y. S., 25, 46
Traylor, J. G., 9(53), 10
Trinkhaus, H., 142
Tripathi, B. B., 8
Tsagareishvili, G. V., 28
Tsederberg, N. V., 46, 69(266)
Tsong, T. T., 88
Tsypin, M. I., 63
Tsypkina, N. S., 16, 17(124), 20(124), 39(124), 56(124)
Tuckey, J. W., 121
Tufeu, R., 48(311), 49
Tumpurova, V. F., 25
Turnbull, A. G., 52, 70

U

Ubbelholde, A. R., 49, 51(337), 52(337)
Ueda, A., 179
Upadhyaya, K. S., 22
Usikov, M. P., 181

V

Vagelatos, N., 9(55), 10
van Baal, C. M., 164, 170(188), 173, 199(198), 199, 202

van der Waals, J. D., 217
Van Dijk, D., 116
van Dijk, C., 242, 260(282)
Vandorpe, M., 69
Van Dyke, J. P., 355
Van Festenberg, C., 277, 328
Van Hove, L., 105, 283(8), 285
Van Itterbeek, A., 47, 49(302)
van Landuyt, J., 163, 194(180)
van Royen, E. W., 242, 260(282)
Van Tendeloo, G., 116, 163, 194(180), 250(75)
Van Vechten, J. A., 296
van Witzenberg, W., 47, 49(303)
Varshni, Y. P., 9(57,58), 10
Vashishta, P., 320, 323, 324
Vasil'ev, L. N., 16, 17(123), 40(123)
Vekilov, Yu. Kh., 14
Venart, J. E. S., 52
Verbeke, O., 47, 49(302)
Verble, J. L., 19
Verma, G. S., 33
Vetelino, J. F., 12, 14(72)
Veyssie, J. J., 16, 40(119)
Vijayaraghavan, P. R., 19
Villain, J., 108
Vineyard, G. H., 252
Vitek, V., 147
Vodar, B., 48(311), 49
Voitovich, E. I., 8, 38(46)
vonderOsten, W., 60
von Lenz, F., 280, 287(5)
Vos, J. E., 52
Vrijen, J., 242, 262(282)

W

Wachtmann, J. B., Jr., 63, 64, 66
Waeber, W. B., 89
Wagini, H., 10, 15(65), 17(65), 60
Wagner, C., 137
Wagner, M., 19
Waidelich, W., 20
Walker, C. B., 238
Walker, C. T., 20
Walker, M. S., 32
Walker, R. H., 102
Walker, W. C., 350
Walton, D., 69
Wannier, G. H., 168
Warren, B. E., 78, 140(2), 140

Warren, J. L., 9, 19
Watanabe, H., 24, 305
Watari, T., 164
Weaire, D., 88, 103
Weber, L. A., 49
Wehe, D., 9(55), 10
Weil, R., 14
Weinstein, B. A., 12
Weisenberg, L. R., 252
Wentzel, G., 279, 287(4)
Wertheim, G. K., 335
Whalley, E., 49(335, 339)
Wheat, M. L., 64
White, C. W., 52
White, G. K., 4, 8, 14, 19
White, W. B., 24
Whitsett, C. R., 14, 15, 16(113,114)
Whitten, W. B., 24, 25
Wigner, E., 106, 107
Wilkins, S. W., 115, 239(67), 240, 241
Wilkinson, K. R., 35
Wilkinson, M. K., 19
Wilks, J., 35
Williams, J. C., 244
Williams, M. G., 43
Williams, R. K., 16, 20(125)
Williams, R. O., 220, 242
Williams, W. S., 62
Wilmshurst, J. K., 45
Wilson, J. A., 102
Wilson, R. B., 308
Winslow, G. H., 25
Witten, T. A., 295, 296, 297
Wohlenberg, T., 144
Wolff, P. A., 285
Wong, C., 25
Wong, N.-C., 260
Woods, A. D. B., 9, 10, 19, 23(188), 24, 45, 62(188)
Woods, S. B., 4, 8

Wooster, W. A., 70
Worlton, T. G., 24
Wostenholm, G. H., 18
Wray, K. L., 68
Wurst, J. C., 43, 60(281)
Wyatt, A. F. G., 8, 34
Wyckoff, R. W. G., 106, 107

Y

Yamauchi, H., 122, 126, 134, 135(101), 137(101), 138, 251, 252(290, 291), 256(290), 260
Yarnell, J. L., 9, 19
Yates, B., 18, 25
Yonezawa, F., 179
Yoshimura, T., 48
Young, F. W., 143, 144
Young, J. D., 24, 27(198)
Yousef, Y. L., 17
Yue, J. T., 332
Yur'ev, M. S., 16, 17(122), 34(122)

Z

Zachariasen, W. H., 127
Zaitlin, M. P., 68, 69
Zaitsev, V. K., 25
Zanio, K., 9(56), 10
Zaporzhan, G. V., 48(313), 49, 52(313)
Zeller, R. C., 68
Zendejas, M., 253
Zeppenfeld, K., 277
Zhdanova, V. V., 14
Ziabland, H., 47, 48(299, 310), 49
Ziebold, T. O., 205
Zimam, J. M., 4, 16, 40(117), 89, 102
Zimkina, T. M., 350
Zunger, A., 347

Subject Index

A

Adamantine crystals, thermal conductivity, 9–17
 minimum, 59–63
 values, 13
Amplification rates, 255
 curve, 262, 264–266
 for spinodal ordering, 261
 vs. wave vector, 258
Angular momentum selection rule, 294
Antiphase boundaries, spinodal modulations, 229
Ashley–Ritchie calculation, 327
Asymptotic direction, 224
 for ternaries, 225
Atomic arrangement, 232–350
Atomic interchange rate, 252

B

Bain-type defect, see Transformation defect, Bain-type
Band structure energy, 91
Bethe pair approximation, 174–175
Binary clustering, 233–237
Binary ordering, 197–201, 218–224, 237–244
Binary phase separation, 210, 216–218
 grand vs. chemical potential, 218–219
 and ordering spinodals, 218–220
Binodal points, in miscibility gaps, 205
Born approximation, 278–280, 282–285
Bragg–Williams method
 vs. cluster variation method, 173–175, 197
Bragg–Williams model, 148–154
 concentration wave, 153–154
 nonequilibrium free energy, 149
 order–disorder, 151–153
 regular solution model, 150–151
 and Taylor's expansion, 214
Brillouin zone center, special point wave vector, 216
Burgers-type defect, see Transformation defect, Burgers-type
BW, see Bragg–Williams method, Bragg–Williams model

C

Cahn and Hilliard free energy expression, 114
Cahn–Hilliard nucleus, 210–211
Cahn's spinodal decomposition theory, 257–258, 260
Characteristic lines, 230–231
Charge density waves, 102
Clausius–Mossotti-type local field correction, 319
Cluster variables
 linear chains, 165–166
 square lattices, 167–168
 tetrahedron, 169
Cluster variation method, 147, 164–175, 177–178, 273
 vs. Bragg–Williams method, 173–175, 197
 clusters, 270–271
 configurational entropy, 165–170
 free energy, 170–173
 grand potential, 170–171
 ground state, 184–194
 internal energy, 171, 185
 for local order, 244–250
 with many-body interactions, 200
 pairwise energy parameter, 185
 phase equilibrium, 198
 vs. probability variation method, 247–248
 and spinodals, 213
 vs. wave method, 244
Coaxial cylindrical electrostatic energy selector, 302–303
Coherent double scattering, 326
Coherent potential approximation, 89
Collodion, 306

371

Subject Index

Compton scattering, *see* Inelastic X-ray scattering
Concentration intensity, 81–82
   spectrum, 86
Concentration modes, normal, 214
Concentration plane wave, effect on lattice, 221
Concentration wave, 80–82
   Bragg–Williams model, 153–154
   ground state, 181–184
   order structure, 182–183
   special points, 159
   static, 80–81
   thermodynamic potential, 153
   use in ordering reactions, 82–83
   wavelength, 104
Configurational energy
   defined, 113
   elastic, 124
   free, 146–147
Configurational partition function, 146
Continuum elastic energy, 196
Continuum elastic theory, 121–122, 124–130
   for cubic crystals, 127
   formulas, 125–127
Continuum gradient expansion, 112–114
   internal energy, 112–113
Core excitation, 349–354
Core spectra
   crystal structure dependence, 354–357
   above threshold, 336–340
Counting rate, 289–290
CPA, *see* Coherent potential approximation
Critical point, ternary, 203–204
Critical surface
   in Brillouin zone, 263
   in concentration wave, 261
Crum's theorem, 129–130
Crystallinity, influence of sample, 328–329
Crystallite size, 328
Crystals
   average intensity, 85
   nonmetallic, thermal conductivity, 66–69
   single, spectra, 343–348
CVM, *see* Cluster variation method
Cylindrical electrostatic energy selector, 302

D

Debye model, 57
Debye temperature, 2, 5, 12, 27
   transverse, 33

Defect
   site of, 75–77
   in solid solutions, 75–88
Detectors, 304–305, *see also* specific detectors
Dielectric function, 348–349
   in Fermi surface, 101–102
Diffuse intensity, 140
   contour, 240
Dipole tensor, 120, 142–143
Disordus lines, 223
Displacive scattering, 138–143
Displacive transformation, omega, 244–247
Distortion tensor, 122
Double-force tensor, *see* Dipole tensor
Double plasmon excitation, 325–328
Double scattering, errors, 292–293
Dyson equation, 143

E

Einzel lens, cutoff, 300–301
Elastic cross section, function of momentum transfer, 287
Elastic energy, 119, 128–130
   average value, 123–124
   coefficient, k-space, 121
   correction, 216
   of isotropic solution, 128
   k-space, 120–123
   long-wavelength limit, 234
   real space, 123
Elastic interaction, 116–145, *see also* specific elastic theory
Elastic potential, interparticle, 136
Elastic scattering, 278–281
Elastic strain tensor, symmetric, 125
Electron diffraction, 280–281
   vs. X-ray diffraction, 280–281
Electron multipliers, 305
Electron scattering spectrometer, basic description, 296–305
Electronic detection, 304–305
Elliott model, 349–354
Energy difference, bulk contribution, 196
Energy selector
   dispersion of, 298
   for spectrometer, 298–303
Energy-wavenumber characteristic, 92
Epitaxial thin film single crystals, 307
Excitation spectrum, 277–278
Exciton, dispersion, 344–345

Subject Index

Exciton energy, function of ionicity, 341
Exciton strength ratio, vs. plasmon, 342
Exponential factor, vs. temperature, 39–41, 43–44
Extended X-ray absorption fine structure, 339–340

F

Fermi–Dirac distribution function, 100–102, 153
Fermi surface, singularity at, 101–104
Fermi's golden rule, 283
Film, as detector, 305
Fluctuation, 231–272
  in disordered phase, 232–250
  heterophase, 243
  multicomponent, 250
  profile, 260–261
Fluctuation-dissipation theorem, 285
Fluorite structure crystal, thermal conductivity, 63–66
Force-constant matrix, 95
Formvar, 206–207
Fourier expansion, 154–157
Fourier operators, 81
Free energy
  coarse-grained, 253
  in concentration amplitudes, 232
  function, 211
  formulation, 197
Free energy models, 146–179, see also specific models
Friedel oscillation, 101
  formula, 241

G

Gaussian, convoluted with step function, 334
Gradient energy coefficient, 113–114
Green–Christoffel tensor, 126
Green's function, 90, 143–144
Green's tensor, for cubic crystals, 127
Ground state
  diagram, 189
  ordered, 180–194
  pair correlation function, 180
  pair frequency, 180
  structure, 191, 193
Grüneisen constant, 12–13
Grüneisen parameter, liquids, 51

Grüneisen's equation, 7
Guinier–Preston zone, 132

H

Hamiltonian, for relaxation energy, 97
Harrison treatment, 90
Helmholtz free energy, 146–147
Hemispherical electrostatic deflector, 299–300
  optimum energy, 300
  schematic, 299
Hermann–Skillman wave function, 346
Heusler alloy structure, 227
Huang peak, 141
Huang scattering, 142–143
Hydrostatic stress, 34–36

I

Incoherent Poisson distributed probability, 326
Inelastic electron scattering, 282–296
Inelastic cross section
  double, 293
  as function of momentum transfer, 287
  single, 293
  threshold region, 330
Inelastic electron scattering spectrometer, schematic, 298
Inelastic electron scattering spectroscopy, 275–357
  experimental examples, 309–357
Inelastic neutron scattering, 276
Inelastic X-ray scattering, 276
Instability direction, and polarization, 224
Instability temperature, 224
Insulator, 340–357
Intensity
  decay, 266–267
  diffuse, 239, 242
  equation, 256
Interband transition
  strength ratio vs. plasmon, 342
  postulates, 348
  rate, 342
Internal energy, 114–116
  with bulk entropy, 234
  total, 129
Interstitial phases, 182–184
  energy, 184

Ion milling, 308–309
Isointensity contour, 238
Ising ferromagnet, approximations, 174
Ising model, specific heat, 176

K

k-space potential
  for cubic crystals, 105–112
  elastic, 100, 123
  internal energy coefficient, 104
  maxima, 104–105
  minima, 104–105, 108–109
  perspective plots, 111
  properties, 99–104
  singularity, 99–104
  saddle points, 104–105
  special points, 104–107, 109–110
  symmetry, 104–112
K threshold, 333
  after background subtraction, 334
  momentum transfer dependence, 335
  oscillator strength density, 353
  shape, of F, 355
  spectra, 347
Kaburagi–Kanamori method, 189–190
Kanamori's inequalities, 189–192
Kanzaki force, 117, 142
Kanzaki force model, 121–122, 143
KCM, see Krivoglaz–Clapp–Moss
Kikuchi cluster variation method, 87
Kinetic equation, linear, 254
Kinetic path, 269
Kinetic principal directions, 255
Kinetic probability function, 268–269
Kinetic stability, analysis, 260
Kinetics, 231–272
  master equation, 251–253
  ternary systems, 264–268
Kirkwood superposition approximation, 87
Kohn anomaly, 102
Kramers–Kronig analysis, 348–349
Kramers–Kronig relations, 292–294
Kramers–Wannier approximation, 168
Krivoglaz–Clapp–Moss formula, 237
  in interstitial solutions, 244
  for multicomponents, 250
Krivoglaz correction, 138–142
Krivoglaz function, 142
Kuyatt–Simpson monochromator, 300

L

Landau formula, 234–235
Landau–Lifshitz criteria, and ordering, 222
Landau–Lifshitz rules, 158–161
Landau–Lifshitz theory, 157–164
Landau-type phenomenological expansion, 163
Langer's equation, intensity structure, 258–259
Lattice pair interactions, 93–95, see also Microscopic elastic theory
Lattice positions, 93
Lattice statics, see Microscopic elastic theory
Laue monotonic scattering, 85
Lifshitz criterion, 159–160
Lindhard function, 310–312
  modified, 318–325
Liquids, thermal conductivity, 69–70
LL, see Landau–Lifshitz
Local field effects, 294–296
  change in spectral shape, 297
Localized state rule, 294
Long-range order, 77–80
  Bragg–Williams model, 152
  parameter, 78–80, 175, 177–178, 223
Long-wave relations, 119
Longitudinal dielectric response function, 285
Lorentzian, convoluted with step function, 334
Lorenz–Lorenz formula, 295
Lorenzian shape, 234–235
LRO, see Long-range order

M

Macroscopic single crystal, thinning of, 308
Mahan–Nozières–DeDominicis theory, 329–337
Martensitic transformation, 250
Mass absorption coefficient
  defined, 288
  as function of atomic number, 289
Mean free path, 57–58
  approximate, 288–290
  random phase approximation, 326
Meijering categories, ternary, 226, 230
Meijering's classification, 204, 207–208, 225
Mermin's damping, 320

MG, *see* Miscibility gap
Microcouples, diffusion paths, 268
Microscopic elastic theory, 117–124, *see also* Lattice pair interactions
Miscibility gap, 195–197
  phase diagrams, 202
  and spinodal, 216–217, 228, 231
  ternary diagrams, 204–205
MND, *see* Mahan–Nozières–DeDominicis theory
Molecular dynamics simulation, 179
Mollenstedt analyzer, 300–301
Momentum transfer, 303–304
Monte Carlo simulation, 87, 179

N

Natural iteration method, 172–173
Newton–Raphson method, 172-173
Neutron diffuse scattering
  in anisotropic systems, 236–237
  elastic, 236
Nonequilibrium free energy, function, 147
Nonmetallic crystals, 23–31
  thermal conductivity, 1–71
Normalization condition, 237
Nucleation and growth, 210–212
  reduced concentration profiles, 212

O

Onsager's solution, 167–168, 175
Optical absorption coefficients, 348–350
Ordered domain, 113–114
Ordus lines, 223
Orthogonality catastrophy, 330
Oscillator strength density
  function of energy, 352
  K threshold, 353
Oscillator strength sum rule, 290–292

P

Pair correlation function, 235
Pair energy, 92
  in direct space, 97
  expectation value, 98
  in Fourier space, 96
  independent variables, 98–99
Pair interaction, 130–138
  angular dependence, 137
  continuum formulation, 133–138
  elastic energy, 133–134
  energy, 132, 134–136, 139
  microscopic formulation, 131–133
  and oscillating potential, 241
  parameters, 199
Pair interaction model, 190
Pair potentials, 90–99
Parseval's theorem, 112
Path probability method, 268–272
  clusters, 270–271
Path variables, 268
Pauli exclusion principle, 153
Perturbation treatment, multicomponent, 253–257
Phase diagram
  binary phase separation, 195–197
  from Bragg–Williams model, 201
  coherent, 194–210
  coherent vs. incoherent, 208, 210
  by cluster variation method, 202–203
  and configuration polyhedron, 191
  experimentally determined, 200
  free energy, 195
  grand potential, 195
  ternary, 202
Phase separation
  double spinodal, 264
  kinetics, 253
  spinodal, 229, 257–260
Phase transformations, coherent, 74
Phillips ionicity, 341
Phillips–VanVechten universal semiconductor model, 340–343
Phonon dispersion curve, 9, 23–26
  liquids, 45
Phonons, 2
  acoustic, 2–5, 9–10, 16
  longitudinal, 33–34
  optic, 9–10, 16, 21–24, 41–42
  transverse, 33–34
Plasmon
  decay, 313
  dispersion, 317–318, 320
  energy, 341
  excitation, 325, 338–339
  oscillations, 309–310
  vs. pseudopotential, 316
  ratio of intensities, 327
  screening, 309–310
  in simple metal, 309–329

spectrum, 312–315, 321–325
  width, 311–320
Point dipole approximation, 295
Polarization, defined, 244
Probability variation method, 247–249
  entropy measure, 248
Pseudopotential, 88, 90–93
  form factor, 91
PVM, *see* Probability variation method

R

Radiation damage, 280–281
Random phase approximation, 310
  plasmon dispersion, 311
Rare-gas crystals
  acoustic modes, 5
  thermal conductivity, 4–9, 38, 57–59
Relaxation energy, 95–98
  coefficient, 99
Relaxation times, inverse, 225
Replacive scattering, 82–86
  defined, 82
  geometry, 83–84
  intensity, 82–86
Rocksalt-structure crystals, 9, 17–21
  vs. adamantine crystals, 17
  thermal conductivity, 18, 56, 59–63
Root-three rule, 217
RPA, *see* Random phase approximation
Rutherford cross section, 279
Rydberg series, 351–352

S

Sample preparation, 305–309
Scattering
  of acoustic phonons, 40–41
  cross section, 285
  from isolated defects, 142–143
  multiple, 292–294
  by optic phonons, 39–44
  phonon–phonon, 2
  replacive, *see* Replacive scattering
Schrödinger equation, 88, 90
Semiconductor, 340–357
Short-range order
  by cluster variation method, 244
  intensity, 85, 140, 235, 237–243
  intensity peaks, 220
  pair correlation function, 77

Simple metals, core spectra, 329–340
Size effect term, 140
Soft-mode transitions, 163
Soft X-ray threshold theory, *see* Mahan–Nozières–DeDominicis theory
Solid solutions
  binary, 156–158, 181
  cohesive energy, 88
  configurational energy, 88
  energy of mixing, 88
  internal energy, 88–116
  k-space potential, 90
  short-range order, 77–80
  thermodynamics of, 73
Spectroscopy, comparison of types, 276–277
Spinodal
  binary, clustering, 216
  coherent, 216, 257
  concentration, 217
  decomposition, 210, 212–214
  defined, 217–218
  ordering, 260–264
  surfaces, ternary, 227
  ternary, vertical section, 228
Sputtering, 308
Square-gradient continuum approximation, 113
Stable states, 179–231
Stability
  analysis, 210–231
  families, 215
  kinetic, 253–257
  limits, 214–215, 222
  and pair interactions, 215
Stacking sequences, 245
State of order, 75–88, 164, *see also* specific types of order
Substrate
  removal of, 306
  thin carbon film, 307
  thin organic film, 306–307

T

Taylor expansion, 154–157
Ternary solution
  isothermal sections, 205–206
  subregular, 205, 209
Ternary solution models, 205

Subject Index

Ternary systems, 201–210, 224–231
  coherent, 202–203
  ordering waves, 226
  stability, 225
Thermal conductivity
  absolute value, 2–34
  acoustic, 12, 21–22, 59
  anisotropy, 70
  of complex crystals, 31–33
  expansion effects, 37–39
  experimental vs. theoretical, 15, 21, 23, 25, 28–30
  liquids, 44–46, 53, 55–57
  at melting point, 44–57
  minimum, 57–70
  effect of molar volume, 17–18
  optic mode, minimum, 59
  phonon contribution, 33
  pressure dependence, 35
  sublattice contribution, 11
  and temperature, 3, 9, 32, 37–44, 50, 54
  variation, liquid, 46
  vibratory mode, liquids, 45
  volume dependence, 34–37, 47–51
Thermal diffuse peak, 320–321
Thermal expansion coefficients, 5
Thin film evaporation, 306–307
Thinning, electrochemical, 308
Thomas–Fermi model, 280
Thomas–Fermi result, 336
Threshold, *see also* K threshold
  amorphous sample, 357
  shape, Elliott model, 51
  single crystal, 356–357
Threshold region, single crystal, 354, 356
Total cross section, 286–288
Transformation defect, 117–120
  Bain-type, 118
  Burgers-type, 118
  defined, 117–118
  replacive effects, 118

Transition
  rate, 284
  solid–liquid, 52–57
  strength, threshold region, 331
  temperature, approximation methods, 173–175

U

Uniaxial stress, 36–37
Umklapp behavior, 32
Unrelaxed lattice energy, 95–96
Unstable states, 179–231

V

Valence excitations, 340–357
Victawet 35B, 306
Virtual crystal method, 89
Void periodicity, 144

W

Wagner–Lifshitz–Slyozov theory, 137
Wannier model, 344–346
Warren–Cowley short-range order parameters, 78, 247
Wave cluster duality, 177
Wave method, 232–233, 272–273
Wentzel model, 278–280, 286–288
  atomic potential, 281
Wien filter, 301–302
  double focusing, 302

X

X-ray Raman scattering, *see* Inelastic X-ray scattering

Z

Zoom lens, schematic, 304